Diabetes Systems Biology

Quantitative methods for understanding beta-cell dynamics and function

Biophysical Society–IOP Series

About the Series

The Biophysical Society and IOP Publishing have forged a new publishing partnership in biophysics, bringing the world-leading expertise and domain knowledge of the Biophysical Society into the rapidly developing IOP ebooks program.

The program publishes textbooks, monographs, reviews, and handbooks covering all areas of biophysics research, applications, education, methods, computational tools, and techniques. Subjects of the collection will include: bioenergetics; bioengineering; biological fluorescence; biopolymers *in vivo*; cryo-electron microscopy; exocytosis and endocytosis; intrinsically disordered proteins; mechanobiology; membrane biophysics; membrane structure and assembly; molecular biophysics; motility and cytoskeleton; nanoscale biophysics; and permeation and transport.

Diabetes Systems Biology

Quantitative methods for understanding beta-cell dynamics and function

Edited by
Anmar Khadra
Department of Physiology, McGill University, Montreal, QC, Canada

IOP Publishing, Bristol, UK

ISBN 978-0-7503-3739-7 (ebook)
ISBN 978-0-7503-3737-3 (print)
ISBN 978-0-7503-3740-3 (myPrint)
ISBN 978-0-7503-3738-0 (mobi)

DOI 10.1088/978-0-7503-3739-7

Version: 20201201

IOP ebooks

British Library Cataloguing-in-Publication Data: A catalogue record for this book is available from the British Library.

Published by IOP Publishing, wholly owned by The Institute of Physics, London

IOP Publishing, Temple Circus, Temple Way, Bristol, BS1 6HG, UK

US Office: IOP Publishing, Inc., 190 North Independence Mall West, Suite 601, Philadelphia, PA 19106, USA

To Andy and Axel.

Contents

Description of the aims, scope and the audience

The aim of this book is to provide senior undergraduate students and junior scientists, interested in diabetes systems biology, with a tool to learn more about the mathematical models and methods used to understand macroscopically and microscopically beta-cell behaviour in health and disease. The book is designed to introduce the readers to the quantitative methods used to examine beta-cell dynamics, islet biology and architecture, as well as diabetes etiology and implications. The goal is to allow junior researchers in the field of mathematical biology to obtain a broad understanding of these quantitative methods, and guide them into taking the first steps into the field of diabetes systems biology. At the end of each chapter, several problem-solving exercises (that require both analytical and computational skills) are provided for the readers to help them become more proficient in this field.

Editor biography

Anmar Khadra

Anmar Khadra has been an associate professor in the Department of Physiology and associate member of the Department of Physics and Department of Mathematics and Statistics at McGill University, Montreal, QC, Canada, since 2018. He received his BSc degree in pure mathematics (honors) with distinction from Concordia University, Montreal, QC, Canada, in 1997, and his MMath and PhD degrees in applied mathematics from the University of Waterloo, Waterloo, ON, Canada, in 1999 and 2004, respectively. He obtained his postdoctoral training in computational biology from the University of British Columbia, Vancouver, BC, Canada (2005–2008) and from the Institute of Theoretical Biology, Humboldt University, Berlin, Germany (2008). Prior to joining McGill University as an assistant professor in 2012, he was a visiting fellow in the Laboratory for Biological Modeling at the National Institutes of Health, Bethesda, MD, USA (2008–2011). He is the co-director of the Centre for Applied Mathematics in Bioscience and Medicine at McGill University, and a member of the Society for Mathematical Biology and the Canadian Applied and Industrial Mathematics Society. His research interests lie at the interface of computational biology and medicine. He has a broad background in mathematical and computational biology, with proficiency in key research areas, including complex systems, nonlinear and stochastic dynamics, biophysics, numerical computations and algorithm design. He uses this expertise to study quantitatively the mechanisms underlying the behaviour of various physiological systems in immunology, neurophysiology and cell biology.

List of contributors

Giuliana Cortese
Padova University, Italy

Shouguo Gao
The National Heart, Lung, and Blood Institute (NHLBI), The National Institutes of Health (NIH), USA

Soumitra Ghosh
GlycosSmithKline, USA

Rui Hu
University of Waterloo, Canada

Junghyo Jo
Seoul National University, South Korea

Hassan Jamaleddine
McGill University, Canada

Anmar Khadra
McGill University, Canada

Anita Layton
University of Waterloo, Canada

Francesco Montefusco
University of Sassari, Italy

Chiara Dalla Man
Padova University, Italy

Morten Gram Pedersen
Padova University, Italy

Leslie S Satin
University of Michigan, USA

Arthur S Sherman
The National Institutes of Health, USA

I Johanna Stamper
University of Alabama at Birmingham, USA

Alessia Tagliavini
Menarini Ricerche S.p.A., Italy

Margaret Watts
Doane University, USA

Xujing Wang
The National Institute of Diabetes and Digestive and Kidney Diseases (NIDDK), The National Institutes of Health (NIH), USA

Nathaniel Wolanyk
Region's Bank, USA

Chapter 1

Introduction

Anmar Khadra

Insulin plays a crucial role in metabolizing carbohydrates and fat and in maintaining glucose homeostasis. It promotes glucose absorption from the blood by skeletal muscles and fat tissue to keep the glucose concentration within a narrow range, and allows fat to be stored rather than used for energy. This hormone is released by beta cells that, together with other endocrine cells including alpha, delta and epsilon cells, form the islets of Langerhans in the pancreas. Beta cells are known to be excitable, exhibiting patterns of electrical activities in their membrane voltage characterized by the presence of clusters of action potentials called bursts separated by quiescent periods. These bursts are regulated by various voltage-gated and ligand-gated ion channels expressed on their cell membranes, and are also modulated by glucose. Their frequency is thus strongly correlated with insulin release.

Although islets of Langerhans exhibit intra- and inter-variability between them in terms of islet-size and architecture, one common feature among them is that they predominantly consist of beta cells (~80% of cellular composition). Moreover, these islets lie at the heart of several regulatory mechanisms that govern not only secretion of insulin but also of other hormones (including glucagon secreted by alpha cells, somatostatin released by delta cells and ghrelin secreted by epsilon cells), all of which are essential for metabolism. The interactions of all of these cells and hormones generate a very complex network of negative and positive feedbacks that govern beta-cell activities.

An imbalance in insulin secretion that results from beta-cell dysfunction, insulin resistance and beta-cell loss due to genetic, environmental or autoimmune causes, may lead to the onset of diabetes, including type 1 and type 2 diabetes. If left untreated, these diseases can cause very serious complications and adverse effects on the rest of the body, such as neuropathy, nephropathy, ketoacidosis (accumulation of ketones in the circulation), hypertension, and even stroke. Type 1 diabetes is an autoimmune disorder, whereas type 2 diabetes is a metabolic disorder. It has been estimated that 9% of adults in the world suffer from these diseases and that

doi:10.1088/978-0-7503-3739-7ch1

1.5 million deaths can be directly attributed to them. Although both of these diseases appear in genetically predisposed individuals, the main causes and different physiological processes that underlie their progression are not very well understood. Knowledge about beta-cell mass and/or function is crucial for the early diagnosis and treatment of both diseases.

Type 1 diabetes (T1D) is an autoimmune disorder mediated by autoreactive islet-specific cytotoxic $CD4^+$ and $CD8^+$ T lymphocytes that infiltrate the pancreatic islets and cause beta-cell destruction. The disease is triggered by various environmental and genetic factors that have been touted for their etiological importance in T1D. It has been hypothesized that increased neonatal beta-cell apoptosis during the apoptotic wave and defective clearance of apoptotic bodies by macrophages are the main trigger of this disease. The subsequent activation and recruitment of T cells to the islets, along with the increased secretion of inflammatory cytokines by these immune cells, eventually drive beta-cell destruction and increase the work load on surviving beta cells. This, in turn, is suggested to elevate stress in surviving beta cells and exasperate their loss by apoptosis (or 'suicide'). The disease is also known to be associated with a series of islet-specific autoantibodies released by mature B cells that appear several years prior to the onset of T1D-related symptoms.

Type 2 diabetes (T2D) is more prevalent (comprising 90% of diabetic individuals worldwide) and appears to be positively correlated with obesity. The disease is associated with beta-cell dysfunction (and loss), insulin-release and/or insulin-signaling impairment, as well as an increase in insulin resistance in which the observed supra-normal level of insulin release remains ineffective in regulating glucose homeostasis. The disease is also characterized by having some autoimmune aspects attributed to the presence of circulating beta-cell specific autoantibodies and autoreactive T cells in elderly T2D patients that become progressively diabetic. Although treatments are available, the most recommended way to manage T2D is to maintain an active life style.

From the discussion above, we see clearly that this physiological system is very complex and difficult to analyze experimentally. Theoretical approaches and mathematical models have been systematically developed and used to study this system to tackle fundamental questions associated with beta-cell dynamics and diabetes. These (inexhaustive or detailed) systems- and data-based models have been very successful in providing insights about the cellular (including electrical) properties of these cells and in deciphering their behaviour, within the islet network, in a manner that would otherwise be difficult or impossible to perform experimentally. In fact, some of these extensive models have been used successfully as computational tools for treating diabetic patients. The aim of this book is to provide senior undergraduate students and junior scientists, interested in diabetes systems biology, with a tool to learn more about the mathematical models and methods used to understand beta-cell behaviour in health and disease on both microscopic and macroscoping levels. The book is designed to introduce the readers to the quantitative methods used to examine beta-cell dynamics, islet biology and architecture, as well as diabetes etiology and implications. The goal is to allow junior researchers in the field of mathematical biology obtain a broad understanding of these quantitative methods, and guide them into taking the first steps into the field of diabetes systems biology. At the end of each chapter, several problem-solving exercises (that require both analytical and computational skills) are provided for the readers to help them become more proficient in this field.

Part I

Beta-cell function

IOP Publishing

Diabetes Systems Biology
Quantitative methods for understanding beta-cell dynamics and function
Anmar Khadra

Chapter 2

An introduction to beta cell electrophysiology and modeling

Leslie S Satin, Margaret Watts and Arthur S Sherman

Two of the defining characteristics of beta cells are that they are electrically excitable and that their electrical activity is controlled by the metabolism of glucose, which in turn regulates their secretion of insulin. The first characteristic they share with many other cells, such as neurons and muscle cells, whereas the second makes them unique. This chapter will address the physiological bases of both of these characteristics and how they are incorporated into mathematical models. A major focus will be to understand the complex oscillations of beta-cell electrical activity called bursting (see figure 2.1). A number of other useful reviews of these topics exist in the literature (see 'References for Further Reading' at the end of this chapter), but this one will take a particular point of view of relating the models and model building to the experimental data on which the models are built. Modelers often find themselves in the position of being strongly constrained by data (at least they should aspire to this) but at the same time either lack complete data or have to rely on data that are approximate at best. They therefore need to understand the limitations of the typically available data and use general knowledge of dynamics to fill in the gaps. This knowledge is also needed to make informed choices about making approximations of the data to simplify models for analysis.

The model examples in this chapter were made using the xppaut differential equation software. Tutorials on the use of xppaut can be found in chapter 1 (McCobb and Zeeman) of [2] and [3]. Alternatively, other differential equation software, such as Matlab (Mathworks, Inc.) or Berkeley Madonna could be used by appropriately adapting the xppaut source files that we provide (to be posted at https://github.com/artielbm/artielbm.github.io/tree/master/Models/FieldsChapter).

2.1 Basic physics of membrane potential and ion conductances

Beta-cell models seek to explain time-dependent changes in membrane potential that are observed in these cells when glucose concentration is varied and the mechanisms

doi:10.1088/978-0-7503-3739-7ch2

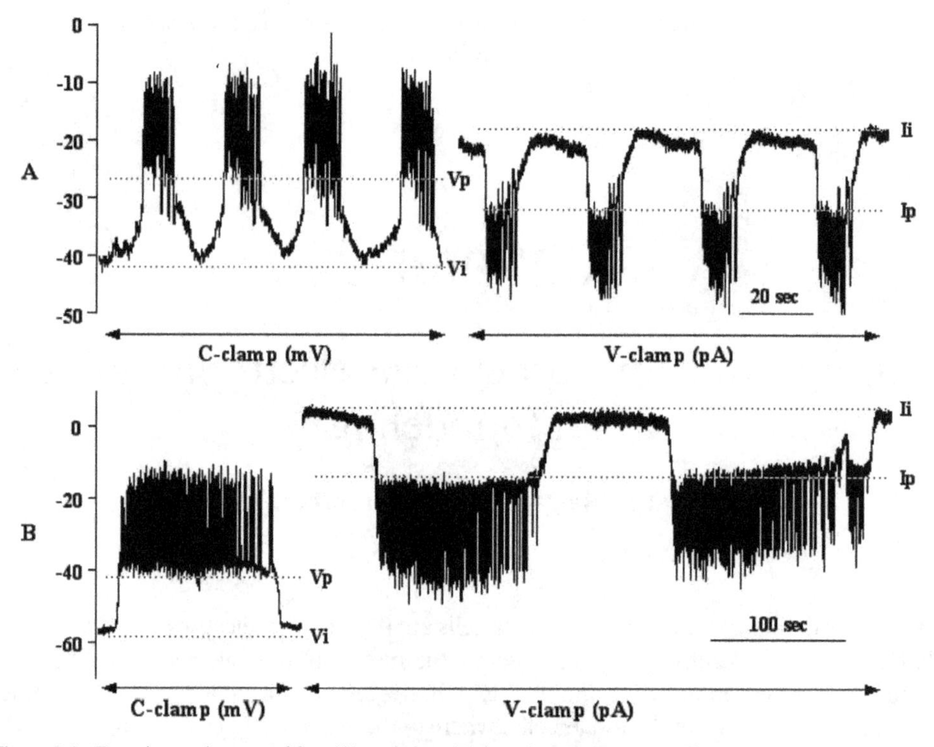

Figure 2.1. Experimental traces of fast (A) and slow (B) bursting. Left traces: V in current clamp mode; right traces: inverted burst currents in voltage-clamp mode (see [1]). V_p: plateau voltage; V_i: interburst (silent phase) voltage; I_p: plateau current; I_i: interburst current. These can be used to estimate gap junction conductance, g_c, using equation (2.28).

that in turn underlie these changes. The first task, however, is to understand how it is that cells (all cells, in fact) have a membrane potential at all. The origins of this lie deep within the history of life on Earth. Many of the organic molecules that define life, including DNA and proteins, have negative charges that must be balanced by positive charges, mainly K^+, to maintain electroneutrality [4], chapter 7. Thus, the earliest cells had a much higher K^+ concentration than the surrounding ocean because they contained these negatively charged organic molecules, an arrangement that persists today in multicellular organisms that carry around their own 'internal ocean' in the form of blood. We therefore consider an idealized scenario of two compartments containing different ionic solutions that are separated by a membrane selectively permeable to K^+ but not large organic anions; this is known as a *Donnan equilibrium* (figure 2.2(A)). We can show that this situation results in a potential difference that is due to the ionic asymmetry present.

If we assume the two compartments contain different concentrations of KCl but similar concentrations of other ions, then at $t = 0$, the chemical gradient for KCl, namely, the diffusive force, will tend to drive K^+ from the compartment containing the higher concentration of K^+ towards the compartment containing the lower concentration (figure 2.2(B)). Counterions will flow in the opposite direction

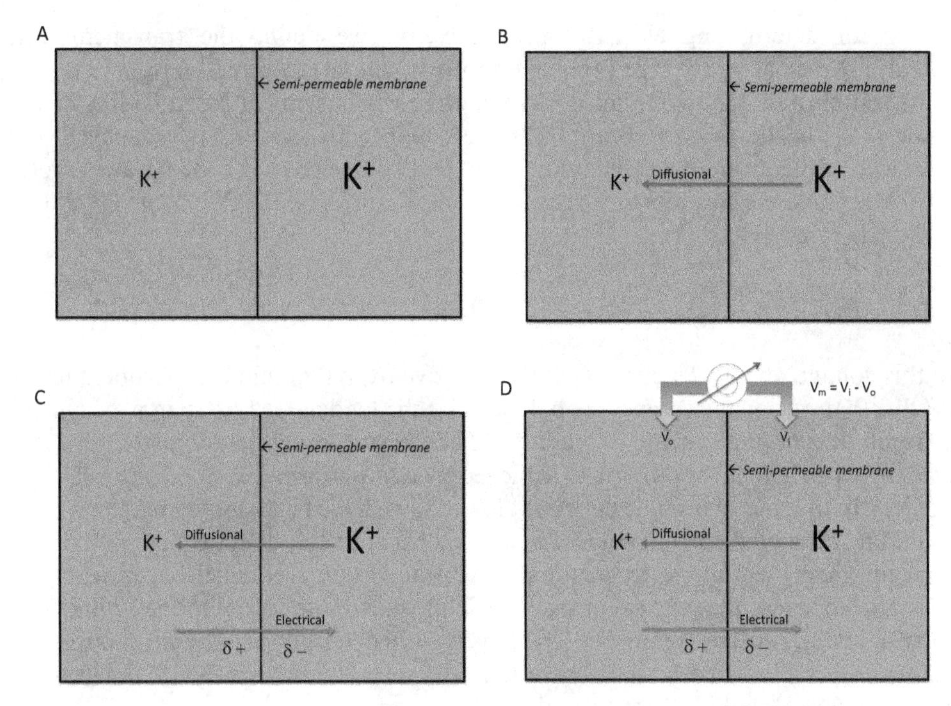

Figure 2.2. Donnan equilibrium. See text for details.

(figure 2.2(C)), and the flux of K^+ in turn will eventually, as more K^+ ions move from the side of the membrane with the higher initial concentration to the other, establish a countervailing electrical potential in the compartment from which ions are moving. Thus, the membrane will become relatively more negative on the high concentration side as more K^+ ions depart and leave the potential relatively more negative by their loss. Eventually, this growing and countervailing electronegative potential will exactly balance the chemical potential due to the diffusive term (figure 2.2(D)).

One can analyze this scenario by considering the chemical potential associated with K^+ ions on each side of the membrane. The chemical potential inside the cell (μ_i) for K^+ ions with internal concentration $[K]_i$ is

$$\mu_i = \mu^0 + RT\ \ln[\mathrm{K}]_i + zFV_i, \tag{2.1}$$

and the chemical potential outside the cell (μ_o) for K^+ ions with external concentration $[K]_o$ is

$$\mu_o = \mu^0 + RT\ \ln[\mathrm{K}]_o + zFV_o. \tag{2.2}$$

Here R is the gas constant, T is temperature in K, z is the valence ($z = 1$ for K^+ ions), F is Faraday's constant, and V_o and V_i are the electrical potentials of the outer and inner compartments, respectively. μ^o is the standard chemical potential that exists in the absence of the electrical or chemical gradients [5]. Only potential differences matter for determining electrical properties, so we define the transmembrane potential V (usually just called the membrane potential) to be $V_o - V_i$.

At equilibrium, the two terms describing the electrochemical potential for the ion inside and outside the membrane must be equal, which means the chemical and electrical gradients are balanced and the net flux of K^+ across the membrane is zero. If we rearrange the equations, we can solve for V to obtain the *Nernst potential* (for K^+ in this case):

$$V_K = \frac{RT}{zF} \ln \frac{[K]_o}{[K]_i}. \tag{2.3}$$

At this voltage, the tendency for a K^+ ion to move from the right compartment to the left due to diffusion will be exactly balanced by the tendency of K^+ to move from left to right due to the electrical gradient. The Nernst potential is also called the *equilibrium potential,* because it is achieved at equilibrium, i.e. when the net ionic flux is 0. In the case of a single permeant ion that we have been discussing, the Nernst potential is also called the *reversal potential,* because the direction of the current reverses when the voltage crosses that potential. When a channel is permeable to more than one ion, the reversal potential is in most cases not equal to the equilibrium potential of any specific permeant ion. This is similar to the situation of multiple ion channel types discussed below, where the reversal potential can be obtained using the Goldman–Hodgkin Katz formula.

For K^+ ions in a typical excitable cell, V_K is around −90 mV, as $[K]_o$ is around 5 mM, and $[K]_i$ is around 140 mM. The equilibrium potential of K^+ is a good approximation for the very negative beta-cell membrane potential observed when glucose concentration is low and the resting K^+ conductance dominates the other conductances. This yields a membrane potential not far from V_K (about −70 mV) [6]. Ca^{2+} ions on the other hand have a very positive equilibrium potential (V_{Ca}) near +120 mV, which is important for considering spiking and bursting mechanisms of beta cells.

Equation (2.3) holds at thermodynamic equilibrium, and, importantly, given the capacitance (ca. 1 μF cm^{-2}) of lipid membranes, the relationship $\Delta Q = C\Delta V$ implies that only a small number of charges must actually move to create typical membrane potentials, around 100 mV. Although even a small charge movement would seem to violate electroneutrality, the charge imbalance is localized to a layer a few nanometers thick around the plasma membrane [7]. Thus, the cell membrane can be well approximated by a capacitor that separates charge between two electroneutral compartments and which has a dielectric property between the two conductors.

While fairly simple to understand, the above scenario is highly artificial in many respects. First, real biological membranes are never permeable to only a single ionic species even in the relatively simple case of vertebrate skeletal muscle, which is also permeable to Na^+ and Cl^- ions, in addition to K^+ ions (7,8). Second, living systems are never truly at thermodynamic equilibrium while they are alive, so how can an ion truly be at its equilibrium potential? Thirdly, and related to the last point, the ion gradient we assumed to exist at the beginning of our discussion is *only possible because of membrane ion pumps* that hydrolyze ATP (and therefore must do work) to pump the ions and create the necessary gradients. This can only happen under *non-equilibrium* conditions. Thus, electrophysiology requires energy-dependent ion gradients that are maintained by continual ATP hydrolysis, and hence non-equilibrium conditions.

A further complication arises when the membrane potential departs from the steady state, such as the sharp changes that occur during action potentials, which are mediated by the transient fluxes of ions across the plasma membrane (figure 2.3(A)). During a single action potential, the effect of these fluxes is small, and the activity of the Na/K ATPase (or 'Na Pump') plays little role, as discussed by Katz in his classic text [8] (the kinetics of these pumps is analyzed in detail in chapter 10). However, after *many* action potentials, such as those that occur during beta-cell bursting (figure 2.3(B)) or continuous spiking, important changes in ion concentrations may occur. This is particularly relevant for Ca^{2+} concentration, as will be seen in the beta-cell model examples below.

Cells also have more than one ion channel; in fact, they often have *many* different types of channels. In this situation, no single channel is at equilibrium, but the cell

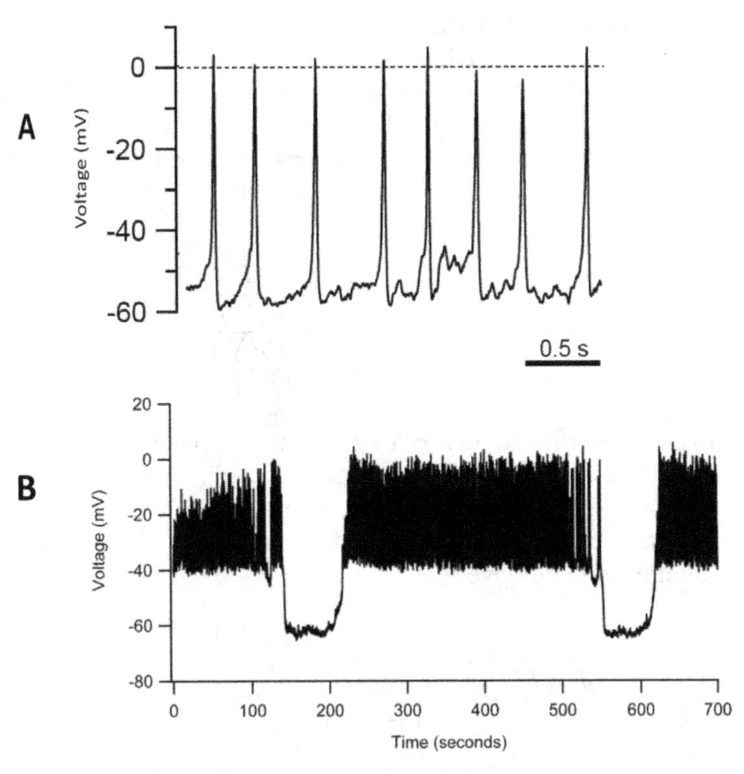

Figure 2.3. (A) A train of action potentials. (The amplitude is enhanced by the addition of the K^+ blocker paxilline.) (B) Bursts of action potentials. Each example is recorded from a peripheral cell in a mouse islet bathed in 11 mM glucose.

membrane potential can be at steady state when the sum of the fluxes through the contingent of channels is zero. The flux through any one channel can be described by the Nernst–Planck equation, which says that the flux of an ion is determined by the balance between drift due to the electrical gradient and diffusion due to the concentration gradient. Viewing the channel as a one-dimensional pore, the equation for the flux J takes the form

$$J = -D\left(\frac{dc}{dx} + \frac{zF}{RT}c\frac{d\varphi}{dx}\right), \tag{2.4}$$

where D is the diffusion coefficient, c is the ion concentration at position x, and φ is the electrical potential (see reference [7] for details and generalizations). The equation is easy to solve at steady state ($J = 0$), and the reader may check by integrating along the pore that the membrane potential $V = \varphi_i - \varphi_e$ is the Nernst potential. This is reassuring, as the previous derivation based on thermodynamics was based on the same assumption of balance between electrical and chemical gradients.

However, as stated above, we need to calculate the non-zero flux J when there is more than one channel. This is tricky because we have two dependent variables, c, and φ, that vary along the pore of the channel. A convenient approximation, pioneered by Goldman [9–11] is to assume that the electric field, and hence the gradient of φ, is constant and equal to $-V/L$, where L is the length of the pore. This is questionable, as it assumes that the electric field is unaffected by the charged residues lining the pore, but it results in a useful and reasonably accurate approximation. It reduces equation (2.4) to a linear, first order ordinary differential equation for c, which can be solved to express J in terms of c and V as follows (try to solve it or see [7]):

$$J = \frac{D}{L}\frac{zFV}{RT}\frac{c_i - c_e \exp\left(\frac{-zVF}{RT}\right)}{1 - \exp\left(\frac{-zVF}{RT}\right)}. \tag{2.5}$$

The factor D/L is the permeability P for the ionic species in question. It is more natural to work with current than flux, as current is directly measured in electrophysiological experiments. J (in moles/area × time) can be converted to current density (charge/area) by multiplying both sides of equation (2.5) by zF, the charge per ion × coulombs/mole, to give

$$I = P\frac{z^2F^2V}{RT}\frac{c_i - c_e \exp\left(\frac{-zVF}{RT}\right)}{1 - \exp\left(\frac{-zVF}{RT}\right)}, \tag{2.6}$$

or more simply in terms of conductance

$$I = g\frac{FV}{RT}\frac{c_i - c_e \exp\left(\frac{-zVF}{RT}\right)}{1 - \exp\left(\frac{-zVF}{RT}\right)}, \tag{2.7}$$

where $g = Pz^2F$.

Though equations (2.5) and (2.6) look daunting, in the simple case where $z = \pm 1$ for all monovalent ionic species, setting the sum of the fluxes or currents to 0 yields the simple and classic Goldman–Hodgkin–Katz (or 'constant field') equation

$$V = \frac{RT}{F}\ln\frac{P_{\text{Na}}[\text{Na}]_o + P_{\text{K}}[\text{K}]_o + P_{\text{Cl}}[\text{Cl}]_i}{P_{\text{Na}}[\text{Na}]_i + P_{\text{K}}[\text{K}]_i + P_{\text{Cl}}[\text{Cl}]_o}, \tag{2.8}$$

because the exponential terms drop out. For given permeabilities P_X, equation (2.8) tells us that V is a weighted combination of the ion concentrations. This accounts immediately for the well-known effect of raising $[\text{K}]_o$ to increase V. (Electrophysiologists call this *depolarization* because it reverses the resting *polarized* state at around −70 mV.) Similarly, for fixed ion concentrations, V is a weighted combination of the permeabilities. It is not as easy to read off the result in this case, but see if you can convince yourself that an increase in P_{Na}, as occurs during the upstroke of an action potential, will depolarize V, whereas an increase in P_{K}, as occurs during the downstroke of the action potential, will decrease V (that is, *repolarize* it). In the next paragraph we will introduce a simpler formulation in which these changes will become even more apparent. You may also be interested in trying the Nernst/GHK calculator at: http://nernstgoldman.physiology.arizona.edu/using/.

A second approach to deal with multiple permeant ions that can allow a steady state membrane potential to be calculated is more engineering-based, representing ion permeation by current flows through parallel membrane conductances. That is, the ion channels can be modeled as resistors in parallel, each with a voltage offset that is generated by a 'battery' to account for its respective Nernst potential. This is often referred to as the parallel conductance model (PCM; figure 2.4). In this model, conductances take the place of permeabilities, which is advantageous because

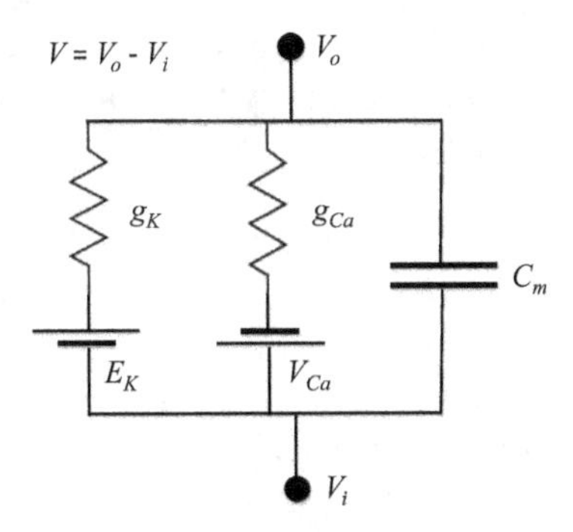

Figure 2.4. Diagram of the parallel conductance model. At equilibrium, membrane potential V is a weighted average of the K^+ and Ca^{2+} reversal potentials (equation (2.9)).

conductances can be more easily measured experimentally. At steady state, this approach yields V as an average of the Nernst potentials, weighted by the conductances. The PCM is a very useful and instructive way of thinking about electrophysiology.

For a membrane permeable to Ca^{2+} and K^+, for example, the PCM under steady-state conditions (i.e. with zero net current) yields (derivation in next section)

$$V = \frac{g_{Ca}V_{Ca} + g_K V_K}{g_{Ca} + g_K}. \tag{2.9}$$

In this case, assuming the reversal potentials are constant, one can readily understand how membrane potential varies in response to relative changes in the membrane conductances.

For example, consider again the membrane potential of an islet bathed in low glucose. Examining equation (2.9), if $g_K \gg g_{Ca}$ then the membrane potential will be near −90 mV. If a third conductance g_L representing leak with equilibrium potential of, say, −20 mV is added to the circuit, then the combination of g_L and g_K in low glucose results in a resting potential of

$$V = \frac{g_{Ca}V_{Ca} + g_K V_K + g_L V_L}{g_{Ca} + g_K + g_L}, \tag{2.10}$$

which is near the experimentally observed value of −70 mV for a suitable value of g_L. (Exercise: calculate g_L/g_K, assuming $g_{Ca} = 0$.)

The behavior of V during an action potential can also be easily understood using the PCM. The membrane starts near −70 mV because of its high relative g_K. The rising phase of the action potential corresponds to rising g_{Ca} relative to g_K, and the subsequent return to rest from the peak of the action potential is mediated by a rise in g_K until again $g_K \gg g_{Ca}$. The complex dynamics of the action potential and bursting activity are described in more detail below, along with an example in which a naïve application of the PCM can lead to wrong conclusions.

2.1.1 Voltage clamp, current clamp, and ionic currents: 'there's no place like Ohm'

Simple electrical conductors, such as metal wires, pass current when a voltage difference is placed across them. According to Ohm's law, the change in voltage is proportional to the change in current with proportionality constant, r_m, the membrane resistance. Equivalently,

$$\Delta V = \Delta I r_m. \tag{2.11}$$

As was the case with the GHK equation, this is only approximately true for cell membranes, but it is good enough for many purposes. For example, one can measure the 'input resistance', which is roughly the reciprocal of the total conductance of the open ion channels of a cell, by injecting a known current, measuring the voltage deflection, and solving equation (2.11) for r_m. This is most accurate for small excursions from rest as the induced change in V can in turn change voltage-dependent membrane conductances (and hence r_m) if it is sufficiently large. If the change is big enough, in fact, an action potential can be triggered, the hallmark of an 'excitable membrane'. If one sees small square pulses of ΔV in response to the application of small square pulses of ΔI, however, one can be reasonably confident that the conductance being monitored is linear or 'passive' and thus constant. Ignoring this guideline can result in erroneous conclusions about r_m.

We will proceed under the assumption that Ohm's law is an adequate model for ion channels and derive a dynamic equation for membrane potential. This will allow us to derive the PCM (equation (2.9)), describe how ionic conductances are actually measured experimentally using voltage clamp, and develop a simple model of more complex dynamics, such as spiking and bursting in pancreatic beta cells (figure 2.3).

In addition to Ohm's law to describe the current passing through ion channels, we need to account for the plasma membrane, which is non-conducting because it is a lipid bilayer. As already alluded to above, the membrane can be modeled as a capacitor that separates charge, which is what allows the cell interior to maintain a negative potential relative to the cell exterior (until channels open, becoming permeable to ions and eventually requiring ion pumps to restore the resting state when a sufficient number of ions move).

To model the combined resistor–capacitor (RC) circuit, similar to that in figure 2.4, requires one additional fundamental physical law, the conservation of charge, or Kirchoff's Law, which says that the total current across the circuit is the sum of the ionic and capacitive currents; i.e.

$$I_{\text{total}} = I_{\text{capacitance}} + I_{\text{ion}}, \tag{2.12}$$

where

$$I_{\text{capacitance}} = C\frac{dV}{dt} \tag{2.13}$$

and

$$I_{\text{ion}} = g_{\text{ion}}(V - V_{\text{X}}). \tag{2.14}$$

In this formulation, V_X is the Nernst potential of a permeant ion and V is the plasma membrane potential. Equation (2.14) is Ohm's law written in terms of conductance instead of resistance, and the voltage driving current flow, the so-called 'driving force', is the difference between the membrane potential and the equilibrium potential of the ion, such that when $V = V_{\text{X}}$, $I_{\text{ion}} = 0$.

To illustrate the above points, we have plotted in figure 2.5 the $I_{\text{Ca}}-V$ curves for the ohmic/Nernst and GHK representations (equations (2.14) and (2.7), respectively). We consider two different cases. First, when the internal Ca^{2+} concentration is much lower than the external concentration, the GHK formula (black, solid) is much more accurate near V_{Ca} than the ohmic formula (red). Second, when the internal and external concentrations are not very different, the ohmic curve (black, dashed) is close to the GHK curve (blue). In reality, the concentrations are very discrepant, but even then, the Nernst expression can be adjusted to closely match the GHK expression for V far from reversal, which is never approached during islet

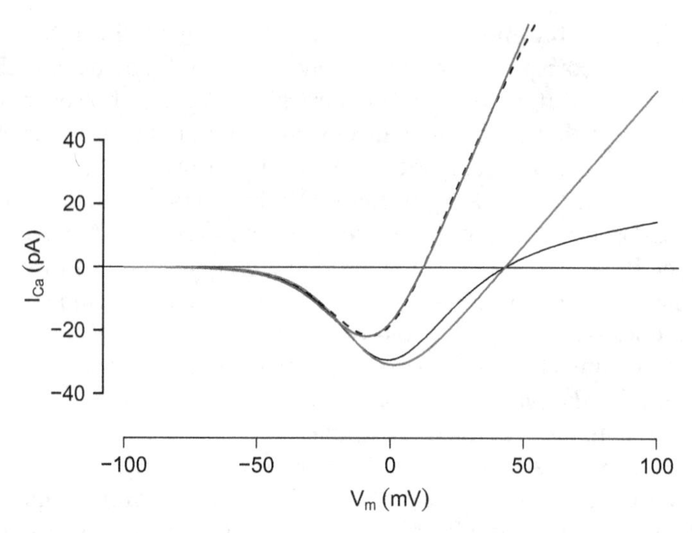

Figure 2.5. GHK versus Nernst representations of current. Plot of I_{Ca} versus V using the GHK expression (equation (2.7)) or the Nernstian expression (equation (2.14)) with $g_{\mathrm{ion}} = \bar{g}_{\mathrm{Ca}}/(1 + \exp(V_m - V/s_m))$. For GHK, let $\bar{g}_{\mathrm{Ca}} = 20$ nS, $V_m = -5$ mV, $s_m = 10$ mV. Black, solid: $c_e = 2.5$ mM, $c_i = 0.1\mu$ M; black, dashed: $c_e = 2.5$ mM, $c_i = 1$ mM. The left branches of these curves can be approximated with a Nernstian expression by setting $V_m = -15$ mV, $s_m = 10$ mV and adjusting $\bar{g}_{\mathrm{Ca}}$. For $c_e = 2.5$ mM, $c_i = 0.1\mu$M (blue), $\bar{g}_{\mathrm{Ca}} = 0.3$ nS; for $c_e = 2.5$ mM, $c_i = 1$ mM, $\bar{g}_{\mathrm{Ca}} = 1.6$ nS. When c_e is not very different from c_i, the GHK curve is almost linear near the reversal potential, and the right branches also match well.

electrical activity. We will use the Nernst representation in our examples, as in most beta-cell models.

For a cell with K^+ and Ca^{2+} channels, the current balance equation is

$$C\frac{dV}{dt} + g_{\mathrm{Ca}}(V - V_{\mathrm{Ca}}) + g_{\mathrm{K}}(V - V_{\mathrm{K}}) = I_{\mathrm{total}}. \tag{2.15}$$

If $I_{\mathrm{total}} = 0$, corresponding to no external (experimentally applied) current, then at steady state, the capacitive current is 0 and equation (2.15) can be solved to give equation (2.9).

Before we proceed to study the properties of this equation, we note that its fundamental importance lies in part in the fact that any number of currents can be added to it, both channels and electrogenic pumps, like a clothesline from which any new item can be hung. The biophysical basis of this extensibility is the HH hypothesis that the currents are independent of each other. Though this seems like second nature today, other formulations have not used it. Prominent examples are the Fitzhugh–Nagumo and Hindmarsh–Rose models, which are phenomenological models that used polynomial expressions [12, 13]. Though their simplicity makes them useful for mathematical analysis, they suffer from the limitation that the terms cannot be identified with ion currents or gating variables.

Equation (2.15) assumes that there is no spatial variation in the potential within the cell (i.e. the cytoplasm is isopotential). This is a good approximation for small round cells, such as isolated beta cells, but is inadequate for neurons with complex morphology or for beta cells when they are electrically coupled to each other, as in the pancreatic islet. We will discuss this further below.

For equation (2.15) to be useful, it is necessary to measure the conductances. This can be done by measuring the total current flowing across an excitable membrane using a feedback circuit that holds membrane potential constant by injecting just enough current to cancel out any changes in V. Suppose, for instance, that I_{Ca} is blocked in equation (2.15), so that the only ionic current remaining is I_{K}. (The block could be pharmacological or achieved by the removal of an ion.) Then, at any fixed voltage, the capacitive current is 0, which means that $I_{\mathrm{total}} = I_{\mathrm{K}}$. To obtain g_{K}, however, we would need to vary the command voltage. To avoid mixing in the capacitive current, one applies a step (actually, a very rapid) change in V, say from −70 mV to −40 mV. Then during the step rise, dV/dt is very large, but once V reaches the new stepped value, $I_{\mathrm{capacitance}} = 0$ and again the measured total current equals I_{K}. In the simple case of constant g_{K}, one can fit the currents obtained in this way to a straight line to estimate g_{K} based on equation (2.14) (we further assume that V_{K} is known and constant, as expected if the ion concentrations are constant).

However, the interesting behaviors of excitable cells, such as spiking and bursting, depend critically on conductances being not constant but voltage dependent. This makes equation (2.15) nonlinear in V and no longer so easy to understand.

To handle more realistic cases, we use a method pioneered by Hodgkin and Huxley [14–17], who expanded equation (2.15) into a system of differential equations, each of which is linear when the other variables are held fixed. In its simplest form, each conductance is decomposed into the product of a maximal conductance and the fraction of the conductance that is active

$$g_{\mathrm{Ca}} = \bar{g}_{\mathrm{Ca}} m; \quad g_{\mathrm{K}} = \bar{g}_{\mathrm{K}} n. \tag{2.16}$$

The auxiliary dimensionless quantities m and n, called *gating variables*, can be interpreted microscopically as the fraction of ion channels that are open. This is the deterministic limit of a stochastic open/close process, which can be represented for each gate, $x = m, n$, as

$$C \underset{\beta_x(V)}{\overset{\alpha_x(V)}{\rightleftharpoons}} O. \tag{2.17}$$

The average behavior when there are many channels satisfies the equation

$$\frac{dx}{dt} = \alpha_x(V)(1 - x) - \beta_x(V)x, \tag{2.18}$$

which can be rewritten as

$$\frac{dx}{dt} = \frac{x_\infty(V) - x}{\tau_x(V)}, \tag{2.19}$$

where

$$x_\infty(V) = \frac{\alpha_x(V)}{\alpha_x(V) + \beta_x(V)},$$

and

$$\tau_x(V) = \frac{1}{\alpha_x(V) + \beta_x(V)}.$$

The form of equation (2.19) highlights that, for fixed V, the solution is an exponential that approaches $x_\infty(V)$ at steady state with time constant $\tau_x(V)$:

$$x(t) = x_\infty(V) + [x(0) - x_\infty(V)]\exp\left[-\frac{t}{\tau_x(V)}\right]. \tag{2.20}$$

In the simplest form, α_x, the rate of opening of closed channels, is an exponentially increasing function of V, and β_x, the rate of closing open channels, is an exponentially decreasing function of V. In this case, x_∞ is a sigmoidally increasing function of V, as illustrated for m and n in figure 2.6.

The stochastic nature of channel opening and closing can have a strong effect on membrane potential in small patches of membrane or even at the whole cell level if the number of channels is sufficiently small. In such cases, equation (2.18) is replaced by a stochastic differential equation or Monte Carlo simulation [18]. When the number of channels is sufficiently large, however, the deterministic equation (2.18) is appropriate. We will stick to that here, but modeling has shown that some differences between the behaviors of isolated beta cells and islets may in fact result from stochastic channel fluctuations in the single cells [19, 20].

Measurements of large numbers of channels, as in whole cell clamp, and of single channels, can be related by the following equation for species X,

$$I_X = pN\gamma(V - V_X), \tag{2.21}$$

where I_X is mean current, γ is the single channel conductance, N is the number of channels in the plasma membrane, and p is the average probability that a channel is open. $N\gamma$ is the maximal conductance, i.e. $\bar{g}_X$, and $\gamma(V - V_X)$ is the single channel current, often denoted i_X. Before the advent of patch clamping for measurement of

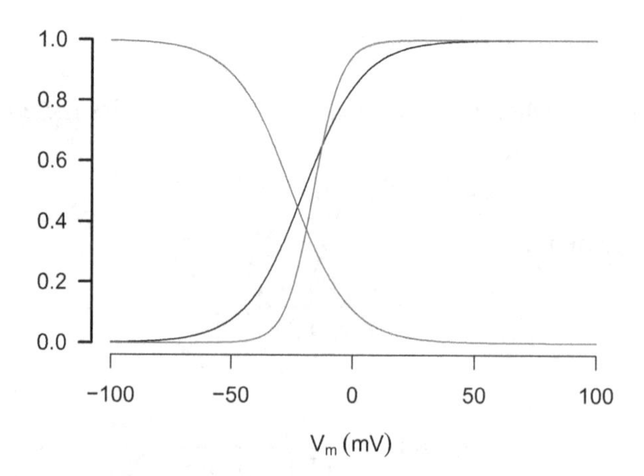

Figure 2.6. Plots of $m_\infty(V)$ (black), n_∞ (red) and $h_\infty(V)$ (blue).

single channels, noise analysis (analysis of the mean and variance of whole-cell currents) was a useful tool to get information about N and γ.

2.1.2 Hodgkin–Huxley and Morris–Lecar models

Hodgkin and Huxley, who used voltage command steps to analyze the conductances underlying the action potential of the squid giant axon [14], found that they needed somewhat more complicated equations. For g_K, they found that they had to raise n to a power to capture accurately an inflection in the rise of the current

$$g_K = \bar{g}_K n^4. \tag{2.22}$$

The exponent 4 was interpreted as concerted movement of four 'particles' in the plasma membrane. Hodgkin and Huxley did not know about protein ion channels in the membrane or their gating variables, discoveries that came much later, but in a stroke of either luck or brilliance, it turned out that both K^+ and Na^+ channels consist of four subunits acting cooperatively. Their phenomenological 'particles' could be said to have predicted the discovery of ion channels. The fact that currents flow through ion channels seems obvious today, but Hodgkin and Huxley had to overcome a prior theory proposed by Bernstein [21] that during an action potential, the membrane transiently broke down, causing the potential difference across the membrane to become zero. Hodgkin and Huxley showed that this was incorrect using more accurate measurements than those that were available to Bernstein, which revealed that the membrane potential actually became positive at the peak of a spike or was 'overshooting' [22].

The Na^+ conductance, g_{Na}, was even more complex than g_K; it first rose rapidly during the depolarizing step but then slowly decreased, which Hodgkin and Huxley interpreted as an 'inactivation' of the current. They therefore expressed g_{Na} as

$$g_{Na} = \bar{g}_{Na} m^3 h, \tag{2.23}$$

where, as before, m represents activation of the channel, but h accounts for inactivation as the fraction of channels that are not inactivated. Following the formalism of equation (2.18), we can define expressions for α_h and β_h, except that now α is a decreasing exponential and β is an increasing exponential, which makes h_∞ a decreasing sigmoidal function of V, as shown in figure 2.6.

Putting the representations for K^+ and Na^+ channels together gives the model developed by Hodgkin–Huxley to describe action potentials in the squid giant axon [14]:

$$C\frac{dV}{dt} = -\bar{g}_{Na} m^3 h(V - V_{Na}) - \bar{g}_K n^4 (V - V_K) - g_L(V - V_L) + I_{applied}, \tag{2.24a}$$

$$\frac{dx}{dt} = \frac{x_\infty(V) - x}{\tau_x(V)}, \tag{2.24b}$$

where $x = m, n, n$. The leak current, mostly carried by Cl^- ions, with conductance g_L, was added to match the observed rest potential, which was not fully accounted for by the K^+ and Na^+ currents.

During an action potential, the upstroke is mediated by an increase in m, and the downstroke by an increase in n and a decrease in h. The rise in n is slower than that of m, which allows time for V to rise [11, 23].

The Hodgkin–Huxley model, though phenomenological, was strikingly successful at accounting for both the currents observed in the squid giant axon in voltage clamp, and, more wonderfully, the all-or-none action potentials that were observed in current clamp in response to a current injection. [Under some circumstances, action potentials were seen spontaneously, but a modern effort to fix deficiencies in the HH model argues that in squid axon this was an experimental artifact [24]. Nonetheless, it occurs in many other neurons, and the HH model can explain it.] Further, the HH model can simulate the changes in conductance that mediate both standing action potentials in a patch of membrane (or axon that has been rendered isopotential by insertion of a silver wire) and propagated action potentials, whose description requires a partial differential equation version of HH

$$C_m\frac{\partial V}{\partial t} = \frac{a}{2R}\frac{\partial^2 V}{\partial x^2} - I_{\mathrm{Na}} - I_{\mathrm{K}} - I_{\mathrm{L}} + I_{\mathrm{applied}},$$

where a is the axon radius, and R is the lateral resistivity [14].

More important than the specific details of the model, Hodgkin and Huxley formulated a canonical set of equations that have become the foundation for modeling all excitable cells, including a vast array of neurons, muscle cells, cardiomyocytes, and pancreatic beta cells [22, 25]. The HH model also contributed importantly to the general theory of excitable systems, such as Ca^{2+} oscillations. One explicit and instructive example is the demonstration that the equations for the regulation of the IP3 receptor (expressed on the membrane of the endoplasmic reticulum, ER, inside the cell) can be reformulated as an analog of the HH Na^+ channel equations [26].

In 1981, Morris and Lecar developed a simplified version of the HH model (ML) to describe barnacle muscle [27]. That preparation had non-inactivating Ca^{2+} channels but no Na^+ channels, and the equations are

$$C\frac{dV}{dt} = -\bar{g}_{Ca}m(V - V_{Ca}) - \bar{g}_K n(V - V_K) - g_L(V - V_L) + I_{\mathrm{applied}}, \tag{2.25a}$$

$$\frac{dm}{dt} = \varphi_m \frac{m_\infty(V) - m}{\tau_m(V)}, \tag{2.25b}$$

$$\frac{dn}{dt} = \varphi_n \frac{n_\infty(V) - n}{\tau_n(V)}, \tag{2.25c}$$

where we have included the factors φ_m and φ_n for fine control of the dynamics, and

$$m_\infty(V) = 0.5\tanh\left[\frac{V - V_1}{V_2}\right], \tag{2.26a}$$

$$n_\infty(V) = 0.5 \tanh\left[\frac{V - V_3}{V_4}\right], \tag{2.26b}$$

$$\tau_m(V) = \frac{1}{\cosh\left[\frac{V - V_1}{2V_2}\right]}, \tag{2.26c}$$

$$\tau_n(V) = \frac{1}{\cosh\left[\frac{V - V_3}{2V_4}\right]}. \tag{2.26d}$$

We will use the ML model in the rest of this section to analyze beta-cell electrical activity.

2.1.3 A tale of tail currents

When a membrane containing a voltage-dependent ionic conductance like a Ca^{2+} channel is voltage clamped to a depolarized potential, g_{Ca} will increase because of the conductance's intrinsic voltage-dependent opening rate (due to a rise in α_m in the m version of equation (2.18)). This process is called 'activation' and reflects the voltage-dependent opening of the Ca^{2+} channels.

We can use equations (2.25a), (2.25b) with V set to prescribed values to simulate this, as shown in figure 2.7(A). Starting from a holding potential of −70 mV, where most of the Ca^{2+} channels are closed, then stepping V up to a range of more depolarized values, V_{test}, will make m rise exponentially toward $m_\infty(V_{test})$

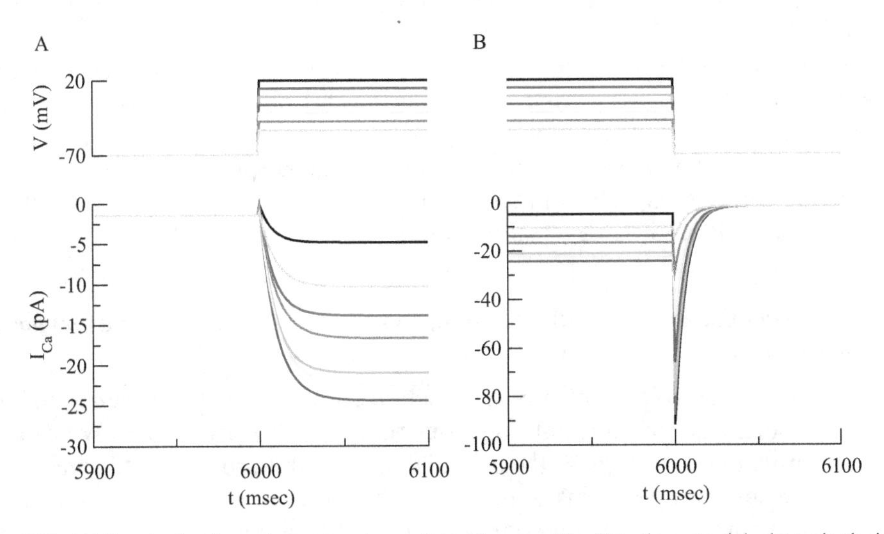

Figure 2.7. (A) Standard voltage clamp stepping up from $V_{hold} = -70$ mV to the potentials shown in the inset. (B) Tail current simulations, stepping down from the potentials shown in the inset to −70 mV. Parameters are as in table 2.1 for the Morris–Lecar model, equation (2.25), but with $\varphi_m = 0.1$ in equation (2.25b).

(see equation (2.20) and right portion of inset protocol). We can't measure m directly, but the time course of I_{Ca}, given by

$$I_{Ca}(t) = \bar{g}_{Ca} m(t)(V - V_{Ca}),$$

is also exponential for fixed V and can in principle be used to determine $m_\infty(V)$. (Note that, although $m_\infty(V)$ is monotonic in V_{test}, $I_{Ca}(V)$ is U-shaped, as already seen in figure 2.5). From this experiment alone, we cannot determine $\bar{g}_{Ca}$, but we don't need that to get m_∞ since we can normalize the currents by dividing by the largest measured current (in absolute value). On the other hand, we do not generally know V_{Ca} accurately for the cell at hand, and so the actual driving force, $V - V_{Ca}$, may be different for each trace and it can't be normalized out.

A clever solution is to flip the protocol. That is, we let the currents equilibrate using a series of V_{test} values and then step down to a common holding potential, V_{hold}; this is shown in figure 2.7(B), which simulates figure 6 in [28]. The driving force is now the same for each trace, and the normalized currents can now be used to fit $m_\infty(V)$.

The currents obtained by rapidly stepping down membrane voltage from a previous level that activated the channels are called *tail currents*. Another interesting feature of tail currents is the strong surge in inward current observed at the beginning of the downward step. This happens because the driving force changes instantaneously, while m changes more slowly. Such tail currents can be very large because stepping up to a high value of V maximizes m, and stepping down to the very negative holding potential makes the driving force large. Note that the currents in figure 2.7(B) are considerably larger than in figure 2.7(A).

Neurons exploit these large tail currents during neurotransmitter release. When an action potential enters a nerve terminal, its peak voltage is near V_{Ca}, so the rise in Ca^{2+} is limited. The bulk of the Ca^{2+} influx and concomitant transmitter release happens during the downstroke of the action potential, which approximates a tail current [29].

Tail currents can be useful in other ways. For instance, progressively increasing the size of depolarizing steps and activating the tail by stepping back to a fixed holding potential can generate a family of tails whose amplitudes describe the activation curve of a voltage-dependent channel, while keeping the depolarizing step constant and varying the value of the post-step repolarization potential can yield the 'open channel I–V' and the reversal potential of the channel under study.

2.1.4 Space clamp, and difficulties in voltage clamping structures having complex geometries

The success of the voltage clamp approach requires the voltage clamp to hold membrane potential constant, not only over time but also over space. We call this requirement *space clamp* because the potential must be uniform across the area of cell membrane being clamped. This is not the case, however, in cells having complex geometries, as current injected into one point in the cell can spread in many directions, changing the membrane potential at different points to different extents. This in turn will differentially affect the activation and deactivation of voltage-dependent ion conductances located in particular sub-regions of the cell. Currents in cells flow in loops between points that are at different potentials, just as current will flow through a conductor connected to the positive and negative poles of a dry cell battery.

Space clamp can only be achieved when the experimental approach is appropriate for the preparation being studied. For instance, a small-tipped glass patch pipette (tip diameter ≈ 1 μm) can adequately voltage clamp a small round cell that lacks membrane processes (e.g. a single beta cell of 15 μm diameter), whereas two long silver wire electrodes inserted into the cytoplasm of the axon are required to clamp a squid giant axon (length ≈ 1 mm). If there were a Maxwell's Demon that could measure the voltages all along the axon membrane, or all around the beta cell membrane, the Demon's tiny voltmeter would read the same value regardless of its location, and the space clamp requirement would be met.

Real life is not always so accommodating, however. Consider a Purkinje neuron of the cerebellum, in which the membrane area is not dominated by the cell body (or 'soma') alone but by extensive and highly branched dendritic processes, membrane extensions having a tree-like appearance that are the site for extensive synaptic contacts from neighboring pre-synaptic neurons. Similarly, the pancreatic islet contains hundreds or thousands of beta cells that form a coupled electrical network. Cell-to-cell contacts in this case are mediated not by chemical synapses, as in most neuronal networks, but by gap junctions, direct contacts between one cell's cytoplasm and another's. Like the Purkinje neuron, electrical changes that occur in a group of beta cells are unlikely to be identical because of the currents that flow between cells residing at different potentials.

These are two examples of preparations that cannot be properly space clamped due to their complex geometries or connectivity. Trying to control V will fail in each case except for a small area in space, and space clamp will not hold. Depolarizing the cell body of the large extended dendritic tree of a neuron with a voltage clamp command using strong negative feedback will cause remote currents to appear in the cell body due to unclamped portions of the neuronal processes, and the same would occur after applying a depolarizing voltage command to a beta cell that is located near the outer surface of a mouse islet. In this case, due to gap junctional coupling between beta cells, remote currents will be observed to flow into the patch clamp pipette from the unclamped part of the islet every time there is a burst or an action potential in the remote region, for the clamp is insufficient there to maintain V constant.

The contributions of inadequately clamped, electrically coupled islet cells to a patch clamp experiment designed to monitor oscillations in conductance of the ATP-sensitive K^+ (K(ATP)) channel have been analyzed [1]. In that paper, repetitive voltage ramps in voltage clamp were applied to estimate changes in beta cell K (ATP) conductance during endogenous electrical activity. As expected, remote currents flowing from unclamped neighbor cells into the voltage-clamped cell were observed because of gap junctions, which contaminated the observed voltage ramp currents.

In previous studies, the electrical network had been modeled by adding coupling terms to the V equation:

$$C\frac{dV_j}{dt} = -I_{\text{ion}}(V_j, x_i) + I_{\text{applied}} + \sum_k g_c(V_k - V_j), \tag{2.27}$$

where I_{ion} is the sum of the intrinsic ionic currents in each cell, x_i are the gating variables, j indexes the cells, and k indexes the neighboring cells directly coupled to cell j.

In [1], a simpler, approximate approach was adopted. The system was modeled as a single small beta cell, representing the voltage clamped cell, that was resistively coupled to a larger beta cell, representing the rest of the islet, which was assumed to be strongly enough coupled to remain synchronized in spite of the injected currents spreading from the clamped cell. The recorded bursting electrical activity in a cell at the islet surface (figure 2.1(A), (B), left panels) confirmed that the islet was oscillatory. Clamping the voltage in the surface cell and recording the corresponding current revealed that the remote cells continued to burst, as evidenced by oscillating inward currents that appeared in the clamped cell and looked like inverted bursts (figure 2.1(A), (B), right panels). These currents propagated from the remote cells to the clamped cell through gap junctions.

These invading currents can be used both to estimate the error introduced by the lack of space clamp and to estimate the gap junctional conductance [30]. Here we illustrate the latter with figure 2.1. The difference in current between interburst and plateau phases (I_i and I_p, respectively) is related approximately to the difference in voltage between the interburst and plateau phases (V_i and V_p, respectively) by Ohm's law (equation (2.14))

$$g_c = \frac{(I_i - I_p)}{(V_i - V_p)}. \tag{2.28}$$

For both examples in figure 2.1, g_c comes out to about 1 nS. Similar values have been reported by others using this technique [31].

The asymmetry in which the clamped cell is strongly influenced by its neighbors, but the neighbors are resistant to changes in the clamped cell, is mediated in the reduced islet model by weighting the two components of the islet based on their respective cell surface areas. This approach was previously used to model spatially complex hippocampal pyramidal cells [32], and it showed that the ratio of the surface areas of the soma and dendrites was a key parameter controlling bursting in that cell type.

2.1.5 Beta cells and islets as preparations for electrophysiological studies

There are several types of preparations that can be used in electrophysiological studies of pancreatic beta cells. The classic approach was to isolate single islets and then impale them with fine glass microelectrodes filled with KCl or K-acetate. The electrodes were themselves connected to a preamplifier to dynamically monitor V. This was the approach first used to characterize the electrical bursting of islets from the mouse, and importantly, show that glucose concentration regulated insulin secretion by modifying the bursting electrical activity, rather than by acting on a receptor, in contrast to other endocrine cells, such as pituitary cells [33, 34]. The disadvantages of the approach, which is still used but less so than in the past, are that impalement can injure small and delicate cells, and that the experimenter is limited in the types of manipulations that can be made. For instance, islets are not typically voltage clamped using this approach, in part due to the high impedance of the electrodes used for impalements [35, 36].

With the advent of patch clamping, small structures like islets or single beta cells can be patch clamped in either current- or voltage-clamp mode with low noise. This approach has been used successfully to study whole islets [37] or isolated single cells made by dispersing the islet [38]. These different techniques have been used to characterize the ion channels that functionally mediate the different currents resident in beta cells, as well as in other islet cell types, such as alpha and delta cells, along with their corresponding electrical firing properties and glucose dependencies [28, 37, 39].

2.2 Spiking, bursting and beyond with Morris–Lecar (a beta-cell DIY)

We now turn from biophysics to dynamics to building up step-by-step a model of the beta-cell burst pattern, starting from the case of a simple spiking neuron. In order to apply phase-plane methods to Morris–Lecar (ML), we use the approximation that m can be set to equilibrium because it is much faster than n: that is, we set $m = m_\infty(V)$ in equation (2.25), yielding

$$C\frac{dV}{dt} = -\bar{g}_{\mathrm{Ca}} m_\infty(V)(V - V_{\mathrm{Ca}}) - \bar{g}_{\mathrm{K}}(V - V_{\mathrm{K}}) - g_{\mathrm{L}}(V - V_{\mathrm{L}}) + I_{\mathrm{applied}}, \quad (2.29a)$$

$$\frac{dn}{dt} = \varphi_n \frac{n_\infty(V) - n}{\tau_n(V)}. \quad (2.29b)$$

Morris–Lecar now has only two dependent variables (V and n), but remains flexible enough to show all of the important behaviors of HH. Moreover, because the ML model is two-dimensional, one can analyze it in the *phase-plane* by studying the V and n nullclines (curves of steady states) defined by $\frac{dV}{dt} = 0$ and $\frac{dn}{dt} = 0$, respectively. For more background on phase-plane methods, see [40, 41] or chapter 1 in [2].

Using ML and the neuron parameters, listed in table 2.1, the time course of repetitive spiking shown in figure 2.8(A) can be readily predicted. We can gain even more understanding by plotting the V–n trajectory for this system in the phase plane, where it takes the form of a closed orbit (shown in black in figure 2.8(B)). The arrows depicted show the direction of motion of the variables in the V–n space. In the lower right quadrant, both V and n increase. When the trajectory crosses the curve specified by $dV/dt = 0$ (the V-nullcline, shown in red), V reverses direction and starts decreasing. Similarly, when the n-nullcline (the curve on which $dn/dt = 0$; shown in green) is crossed, the direction of n reverses.

The four pairs of arrows indicate the direction of change (increasing or decreasing) for V and n in each of the four regions delimited by the nullclines

Table 2.1. Parameters for the ML model.

Parameter	Figure 2.8 (spiking)	Figure 2.9 (adaptation)	Figure 2.10 (bursting)
V_K	−75 mV	−75 mV	−75 mV
V_{Ca}	25 mV	25 mV	25 mV
V_L	−70 mV	−70 mV	−20 mV
V_1	−20 mV	−20 mV	−20 mV
V_2	24 mV	24 mV	24 mV
V_3	−16 mV	−16 mV	−16 mV
V_4	11.2 mV	11.2 mV	11.2 mV
φ	0.01	0.01	0.032
$\bar{g}_K$	2700 pS	2700 pS	2700 pS
$\bar{g}_{Ca}$	1000 pS	1000 pS	1000 pS
g_L	260 pS	200 pS	15 pS
$\bar{g}_{K(Ca)}$	——	3000 pS	3000 pS
$g_{K(ATP)}$	——	——	110 pS
C_m	5300 fF	5300 fF	5300 fF
C_0	——	5 μM	5 μM
α	——	4.5e-6 μM fA^{-1} ms^{-1}	4.5e-6 μM fA^{-1} ms^{-1}
k_c	——	0.15 ms^{-1}	0.15 ms^{-1}
f	——	0.001	0.001
I_{app}	1000 fA	——	—

(magnitude of change is not represented). The directions are consistent with a counter-clockwise circulation, but a further condition is needed to guarantee a periodic orbit, that is, a sustained oscillation. We need to understand why the trajectory does not go to and stay at the point (SS) where the two nullclines cross. At that point both dV/dt and dn/dt are 0, so if the system starts at SS, it will stay there. However, SS is *unstable*: any motion away from the point is amplified rather than damped out (i.e. SS acts as a *repeller*).

The stability of point SS is determined by the eigenvalues of the Jacobian (the matrix of partial derivatives) evaluated at SS, as discussed in the appendix. If we write the ML model in a more generic form

$$\frac{dV}{dt} = f(V, n)$$

$$\frac{dn}{dt} = \varphi g(V, n),$$

then the Jacobian J is

$$J = \begin{pmatrix} f_V & f_n \\ \varphi g_V & \varphi g_n \end{pmatrix},$$

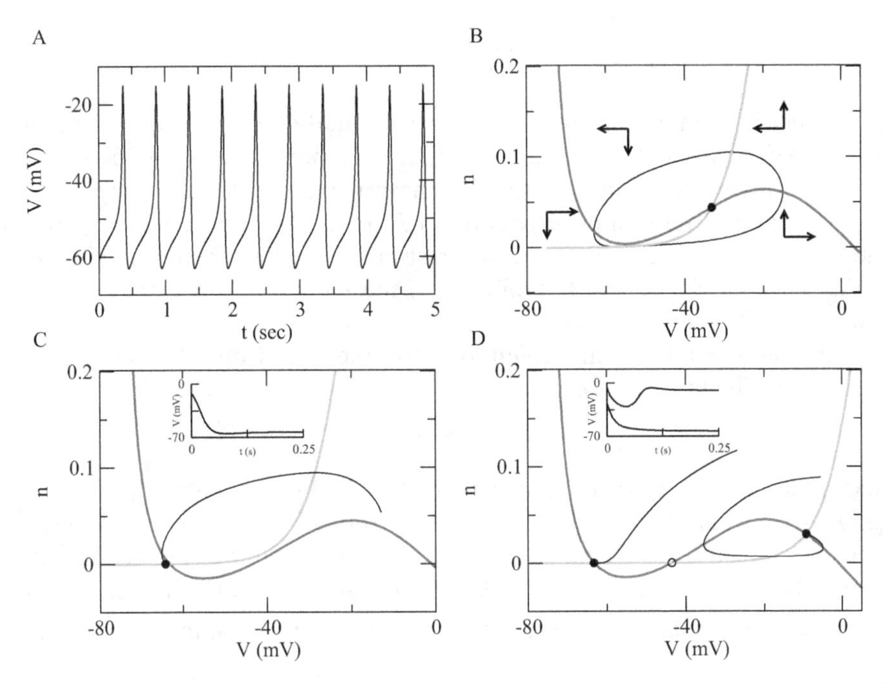

Figure 2.8. Dynamics of the Morris–Lecar model. (A) V time course (parameters are in table 2.1). (B) Phase plane corresponding to (A). Red: V-nullcline; green: n-nullcline; black: trajectory. Arrows show direction of motion only, not scaled to show speed. (C) Phase plane and trajectory with 50 pS $g_{K(Ca)}$ added; it now goes to a low-V steady state. (D) As in (C) but with V_3 increased to 10 mV to shift n-nullcline to the right. There are now three steady states, two stable nodes separated by a saddle, i.e. the system is bistable.

where the subscripts indicate partial derivatives with respect to V or n.

In two-dimensional systems, a steady state generally becomes unstable in one of two ways as a parameter is varied. One way is for one of the negative eigenvalues of the Jacobian matrix J to become positive; this gives rise to a **saddle point**, with one negative and one positive eigenvalue (such a change of stability where new behaviors are born is called a *bifurcation*; this kind is a **saddle-node bifurcation**, abbreviated **SN**, where a saddle and a node coalesce). The other way is for the real part of a complex conjugate pair to go positive (this is called a **Hopf bifurcation** or **HB**, where a periodic orbit emerges). Both pathways can define thresholds for periodic spiking. SN leads to type 1 neurons that have low frequency but large amplitude spiking near threshold, whereas HB leads to type 2 neurons that have high frequency but small amplitude spiking near threshold; see chapter 7 in [42]. These types were first identified by Hodgkin as far back as 1948 [43], but were not put into such a beautifully symmetric framework until 40 years later. We will focus here on the Hopf bifurcation (see the appendix for more on that concept).

Using the fact that the trace (sum of the diagonal elements) of a 2×2 matrix is the sum of the eigenvalues, and the determinant is the product, the condition for an HB is that the trace goes positive while the determinant stays positive. To find conditions where this happens, it is helpful to examine the sign pattern of J

$$\begin{pmatrix} \pm & - \\ + & - \end{pmatrix}. \tag{2.30}$$

The influence of n on both its own derivative and that of V is negative, the influence of V on the derivative of n is positive, and, most importantly, the influence of V on its own derivative can be either positive *or* negative. Since the derivative of V is the sum of the ionic currents, a positive entry in J can be interpreted as a negative slope of the I–V curve, and hence as a negative resistance. This is destabilizing in the same way that negative friction is destabilizing in mechanics, and it is the motive force that drives oscillations.

We can now read off the conditions for the HB from the phase plane in figure 2.8(B). The trace of J is

$$f_V + \varphi g_n.$$

As indicated in the sign pattern matrix (equation (2.30)), we know that g_n is always negative, so the only way for the trace to be positive is for f_V to be positive, which can be ascertained from a simple geometric condition without the need to analyze the complicated expression for f. On the V-nullcline, $f(V, n) = 0$, so implicit differentiation gives $f_V + f_n \frac{dn}{dV} = 0$, and the slope of the nullcline is given by

$$\frac{dn}{dV} = \frac{-f_V}{f_n}.$$

Since f_n is always negative, we must have $\frac{dn}{dV} > 0$ in order to have $f_V > 0$. Therefore, SS must lie on the middle, rising branch of the V-nullcline. This is not sufficient, however, as we also need

$$f_V > -\varphi f_n, \tag{2.31}$$

but this can be guaranteed by making φ small enough. This is natural because it means that n is slower than V, that is, the growth of the K^+ current has to be slow enough to allow V to grow before being brought back to rest.

By similar reasoning, one can show that the determinant of J is positive provided the slope of the n-nullcline is greater than the slope of the V-nullcline, which holds in figure 2.8(B). Thus, both conditions for a Hopf bifurcation are satisfied in figure 2.8(B), and this accounts for the spiking in figure 2.8(A).

To summarize, oscillations will occur when the steady state is on the middle branch of the V-nullcline, n changes slowly enough, and the n-nullcline has a greater slope than the V-nullcline. That is the most important thing (and almost the only thing) one needs to know to understand or build beta-cell models.

Another way of describing this is that oscillations are *born* via a bifurcation when SS moves onto the middle branch of the V-nullcline. For example, one can slide the V-nullcline from left to right by increasing the parameter V_3. A note of caution: the analysis above is valid near the steady state. If instead of varying V_3, a Ca^{2+}-activated K^+ (K(Ca)) current is added to the system, other steady states are introduced that can interact with the periodic orbit.

This is illustrated in figure 2.8(C), (D), and will play a big role in the next subsection. In figure 2.8(C), the added K(Ca) current has pushed down the V-nullcline and moved SS onto the left, descending branch of the V-nullcline. SS in this case is stable (exercise 9(a)), and there is no oscillation. In figure 2.8(D), a further change has been made to shift the n-nullcline to the right (by increasing V_3 from −16 to 10 mV). There are now three steady states. The left and right ones are stable nodes (two eigenvalues with negative real parts), and the middle one is a saddle point (exercise 9(b)). In the next subsection, we will exploit the third HB condition by varying the rate of change of n.

2.2.1 Negative feedback due to calcium

A two-variable model can be silent (go to a steady state) or exhibit spiking, or even be bistable as in figure 2.8(D), but it has no mechanism, and no dynamic freedom, to alternate between silent and spiking, that is, to exhibit bursting. A third degree of freedom, in the form of slow negative feedback to terminate episodes of spiking, is needed. Atwater, Rojas and colleagues proposed in the early 1980s [44] that spiking raises cytosolic Ca^{2+} concentration, c, which activates a K(Ca) channel. This was made into a mathematical model by Chay and Keizer [45], which we emulate by adding the following equation to ML

$$\frac{dc}{dt} = f(-\alpha I_{Ca}(V) - k_c c), \tag{2.32}$$

where f is the fraction of free Ca^{2+} (i.e. Ca^{2+} not bound to buffer), α converts units of current to units of flux, and k_c is the rate of pumping Ca^{2+} out of the cell. Thus, Ca^{2+} concentration is the balance between Ca^{2+} influx through voltage-gated Ca^{2+} channels and Ca^{2+} removal by pumps. For now, we ignore internal Ca^{2+} stored in the endoplasmic reticulum or taken up into the mitochondria, but this will play an important role in a more detailed model discussed below (as well as many others in the literature).

The $I_{Ca}(V)$ term links the c equation to the V equation, and, to complete the loop, we add the K(Ca) current,

$$I_{K-Ca} = \bar{g}_{K(Ca)} \frac{c}{c_0 + c}(V - V_K),$$

whose conductance increases with c, to the V equation to provide negative feedback onto membrane potential as follows

$$\begin{aligned} C\frac{dV}{dt} &= -\bar{g}_{Ca} m_\infty(V)(V - V_{Ca}) - \bar{g}_K n(V - V_K) - g_L(V - V_L) \\ &\quad + \bar{g}_{K(Ca)} \frac{c}{c_0 + c}(V - V_K) + I_{applied}. \end{aligned} \tag{2.33}$$

Figure 2.9(A) shows the result of adding the K(Ca) current to the ML model, and, as shown, it fails to result in bursting. The negative feedback is effective, as indicated by the progressive reduction in firing rate, but spiking still persists indefinitely. If we

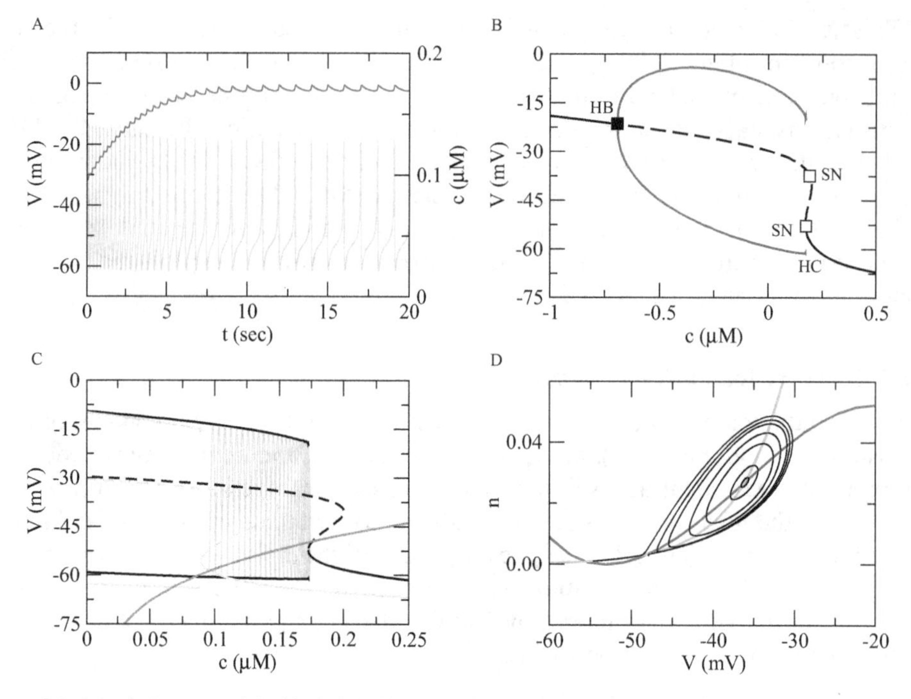

Figure 2.9. Morris–Lecar model with K(Ca) channel added fails to show bursting but does exhibit adaptation as $I_{applied}$ is stepped from 0 to 1 pA. Parameters are in table 2.1. (A) Time course of V, c; (B) Bifurcation diagram corresponding to panel A, $I_{applied} = 0$. (C) V–c trajectory showing adaptation superimposed on two bifurcation diagrams, corresponding to $I_{applied} = 0$ (grey Z-curve) and 1 pA (black Z-curve). Brown, c-nullcline. (D) Variation of periodic orbits from a Hopf bifurcation to a homoclinic orbit as φ takes on values 0.033 2, 0.033, 0.031 5, 0.028, 0.027, 0.026.

were to let c rise further by reducing the Ca^{2+} pump rate k_c, the firing would terminate (not shown) but never come back—there is no periodic alternation of firing and rest states.

Could we have anticipated this outcome? To answer this, we need to look more broadly at how solutions of the 2D V–n system vary as c changes. This is justified by the fact that c is very slow compared to V (during a single spike in figure 2.9(A), c varies only slightly). Figure 2.9(B) summarizes the behaviors of the system for a range of values of c; such a representation is called a *bifurcation diagram*. The black Z-shaped curve shows the steady states as a function of c, which generally decline as c increases, an expected outcome. Each point on this curve corresponds to a V–n phase plane like the ones in figure 2.8 (we could have shown n as a third dimension coming out of the page but have omitted it for simplicity). We can think of the bifurcation diagram as built out of a stack of phase planes like the one shown in figure 2.8, each with its own value of c and $g_{K(Ca)}$ and corresponding steady-state value or values of V. The switchback region (also known as a region of *hysteresis*) in the bifurcation diagram arises from the set of intersections of the n-nullcline with the upward sloping region of the V-nullcline in those phase planes.

Points on the solid curves are stable, and those on the dashed curve are unstable. The leftmost change of stability is mediated by a Hopf bifurcation (HB). As c increases past this point, a complex conjugate pair of eigenvalues changes from negative real part to positive real part, converting a stable spiral to an unstable spiral, as discussed in the appendix. As c increases further along the upper branch of the Z-curve, the complex conjugate eigenvalues coalesce and become real and positive, converting the unstable spiral to an unstable node. As the system passes through the upper saddle-node bifurcation, denoted SN, the unstable node becomes a saddle because one of the real positive eigenvalues becomes negative. Finally, at the lower SN, the remaining real, positive eigenvalue becomes negative, converting the saddle to a stable node. In the interval between the two SNs (shown expanded in figure 2.9(C)), three steady states co-exist, the two upper ones being unstable, and the lowest one stable.

The linear stability analysis sketched above does not by itself disclose any stable behavior for c values between HB and the lower SN. However, the nonlinear dynamics give rise to a periodic solution that emerges from the HB. Its maximum and minimum voltages are indicated by the two blue curves; its amplitude thus increases from with c. The upshot is that at every value of c, the 2D V–n subsystem has a single steady-state behavior (i.e. a single *attractor*), transitioning from a depolarized plateau, to spiking to a low-V steady state as c increases.

To make the BD in figure 2.9(B), we held c constant, but now in figure 2.9(C), we set c into motion and overlay the V–c trajectory (cyan) obtained by solving equations (2.33), (2.25c), and (2.32) on the bifurcation diagram, along with the c-nullcline (brown). For $I_{\text{applied}} = 0$, the bifurcation diagram governing the dynamics is plotted in grey. Although the diagram is not strictly speaking a phase plane, superimposing the c nullcline provides insight about the dynamics. The Z-curve functions similarly to a V-nullcline. To the left of the c-nullcline, c increases and to the right, c decreases. The intersection of the c-nullcline with the grey Z-curve, on which $\dot{V} = \dot{n} = 0$ when c is fixed, is a stable steady state. When I_{applied} is increased to 1 pA, the Z-curve is shifted to the right (black) along with its spiking branches. Thus, spiking begins and c increases, reducing spike frequency because of the increase in $g_{\text{K(Ca)}}$. Eventually the trajectory reaches a range of c values where the increase in c during the upstroke of the action potential balances the decrease during the downstroke. From that time on, the orbit oscillates with small variation around the c-nullcline. (A proper treatment of this requires a more advanced analysis called the method of averaging, which we omit here for brevity, but it is visually plausible.) If pump rate k_c were smaller, c would rise further, and the increased inhibition would terminate spiking. In this case, the c-nullcline would intersect the lower branch of the black Z-curve and result in a silent steady state (not shown).

The ML model supplemented with K(Ca) thus exhibits *spike frequency adaptation*, i.e. a reduction of spike frequency in response to a maintained stimulus. However, it does not exhibit bursting, which shows that it is not sufficient to have appropriate slow feedback. In fact, the K(Ca) mechanism can work to simulate bursting, as demonstrated by the Chay–Keizer model, from which all modern beta-cell models are descended, but the ML model needs one further modification in order to show this. As we show next, it is the fast spiking dynamics that need to be altered.

The neuron model augmented with K(Ca) also suffers from another defect—the spike amplitude produced is much too large for a beta cell (compare figure 2.9(A) and 2.1). Fixing this also fixes the spiking dynamics and is the last piece we need to get a simple bursting model.

Our road forward passes through the solution of a seemingly unrelated puzzle. If we fix c at 0.18 μM, the V–n subsystem has three steady states, with only the low-V state being stable. The phase plane for this case is shown in figure 2.9(D). The upper steady state is unstable, and even has the slope of the n-nullcline greater than the V-nullcline, which is required for obtaining complex conjugate eigenvalues, yet there is no oscillation. The reason is the interaction with the other steady states, alluded to above. Oscillations can in fact occur in this phase plane, provided we make φ smaller. Starting with a relatively large value and decreasing, we obtain a family of periodic orbits of increasing amplitude until one finally intersects the saddle point. For φ values smaller than that, there is no oscillation, just a transient spike that retains the expected counter-clockwise circulation but sweeps past the middle saddle point and then dives into the low-V stable node. The trajectory that intersects the saddle both begins and ends there, taking infinite time to go from start to finish, and is called a *homoclinic orbit*. The full theory of homoclinic bifurcations is deep [46], but intuitively it is plausible that it acts as a barrier between the family of periodic orbits and the family of transient spikes.

As φ increases beyond the values in figure 2.9, the oscillations shrink down to the upper steady state, which becomes a stable node. This is another example of a Hopf bifurcation and illustrates the principle, implied by equation (2.31), that φ must be sufficiently small for the HB to occur. Thus, as φ decreases, a periodic orbit is born in an HB and dies in a homoclinic orbit, the same sequence as shown in figure 2.9(B) for variation in c.

We varied φ, which makes it easy to depict the approach to a homoclinic orbit because the V- and n-nullclines don't change. (It also reinforces our previous assertion that oscillations cannot be predicted from the phase plane alone.) However, the same thing generally happens when other parameters, such as c, are varied. Figure 2.9(B) shows a special case, in which the homoclinic orbit occurs at the lower SN, terminating the branch of periodic solutions. Although this is special, it is actually easy to achieve and is ubiquitous in neural models, where it plays a central role in the theory of spike-frequency adaptation [47].

2.2.2 Bursting via Ca^{2+} feedback (the Chay–Keizer model)

We can now finally make a bursting model, by choosing one value of φ (any of the ones shown in figure 2.9(D) will do) that gives a small amplitude periodic orbit. The value of this configuration is that the periodic orbit coexists with a stable steady state. This bistability is the heart and soul of the type of bursting seen in beta cells.

Figure 2.10(A) shows that with smaller φ, the K(Ca) mechanism now works to produce bursting. During each active phase, c increases until sufficient K(Ca)

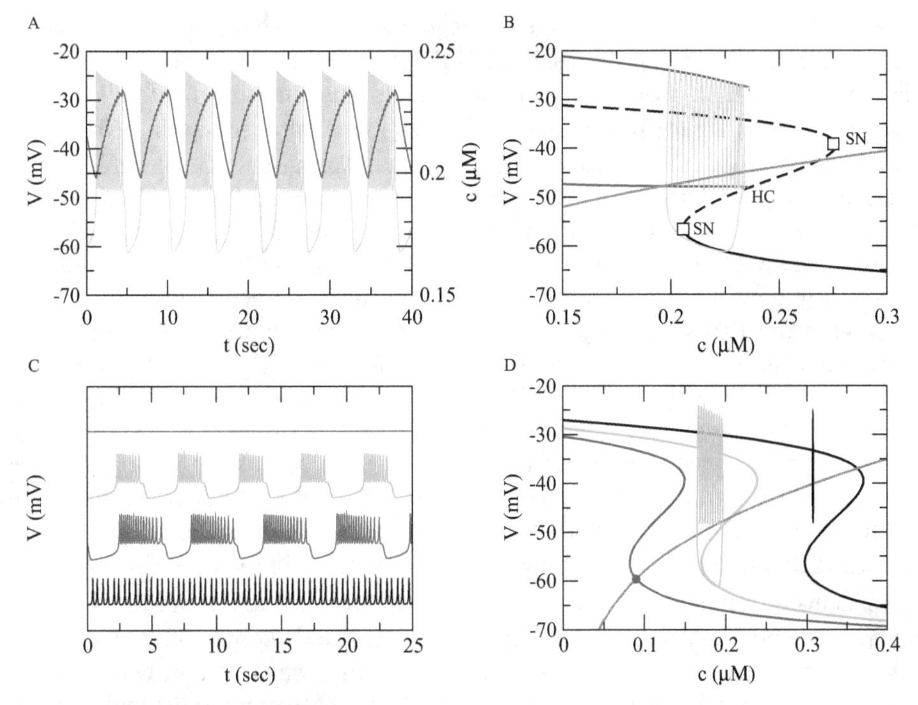

Figure 2.10. Bursting model. (A) V (cyan), c (red) time courses (parameters are in table 2.1). (B) Bifurcation diagram with trajectory corresponding to panel A. (C) V time courses for $g_{K(ATP)}$= (from top to bottom) 105, 110, 130 and 165 pS; curves are offset vertically for clarity, but would otherwise overlap. (D) V–n phase plane for reduced system with $n = n_{\infty}(V)$ for $g_{K(ATP)}$ = (from left to right) 180 (blue), 130 (green) and 60 (black) pS and corresponding trajectories of full system (blue dot, green burst cycle, and black spike cycle, respectively).

conductance is activated to terminate the burst. During the silent phase, c decreases until $g_{K(Ca)}$ is small enough to allow spiking to resume.

Figure 2.10(B) shows the bifurcation diagram with respect to c. The Z-curve of steady states of the V-n subsystem (black) and the periodic orbits (blue) are similar to figure 2.9, but now the homoclinic orbit occurs between the two SN points (the knees of the Z-curve), giving rise to bistability between the low-V steady state and the small amplitude spiking solution. The V–c trajectory (cyan) is superimposed along with the c-nullcline (brown). Because the c-nullcline now occurs between the SN points, c increases during the active phase and decreases during the silent phase.

In addition to increasing φ, we have modified equation (2.33) by dividing the leak conductance g_L of the neuron model into two components, a smaller leak with a reversal potential of −20 mV and a third K^+ current, the K(ATP) conductance alluded to earlier

$$C\frac{dV}{dt} = -\bar{g}_{Ca}m_{\infty}(V)(V - V_{Ca}) - \bar{g}_K n(V - V_K) - g_L(V - V_L) + \bar{g}_{K(Ca)}\frac{c}{c_0 + c}(V - V_K) + g_{K(ATP)}(V - V_K). \tag{2.34}$$

The K(ATP) channel is the main site where glucose acts to control the burst pattern. As glucose concentration [G] rises, so does the ATP/ADP ratio within the beta cells, which reduces $g_{K(ATP)}$. For now, we do not model the details of metabolism but use $g_{K(ATP)}$ as a surrogate for [G]. Figure 2.10(C) shows that for low [G], the beta cell is silent, then at successively greater [G], it bursts progressively more intensively (assessed as higher *plateau fraction* = (*active phase duration*)/(*burst period*)). It finally goes to a continuous, small-amplitude spiking solution at very high [G]. The signature achievement of beta-cell electrophysiology in the 1970s was to show that [G] acted by modulating the burst pattern. The signature achievement of beta-cell modeling in the 1980s was to provide a plausible dynamic mechanism for it.

It is interesting to compare the effect of reducing a K^+ conductance with the prediction of the PCM (equation (2.9)) that V should increase. Figure 2.10(C) shows that the *instantaneous* silent and active phase potentials do not vary within the bursting regime. Instead, the *mean* potentials increase because of increased plateau fraction. Geometrically, the invariance of the instantaneous potentials reflects the fact that a change in $g_{K(ATP)}$ shifts the V-nullcline (alternatively, the Z-curve of the bifurcation diagram), as $g_{K(Ca)}$ compensates by changing in the opposite direction to $g_{K(ATP)}$. This depends on the fact that the two K^+ currents are assumed to have the same reversal potentials, so a change in one can be exactly compensated by a change in the other. In contrast, if the leak current, now an inward current with a reversal potential of −20 mV, were increased, the silent phase potential would increase markedly. This is thought to be the mechanism by which acetylcholine, an important physiological modulator of electrical activity and insulin secretion, works [48] (see exercise 11).

If $g_{K(ATP)}$ is made large enough, however, the instantaneous potential does decrease, and if it is small enough, the potential does increase. These gross changes happen when the dynamics take the trajectory out of the region of bistability (hysteresis) in figure 2.10(D). Thus, the intuition provided by the steady state PCM is correct in spirit, but misses important details because it does not take into account the dynamic properties that lead to bistability.

The variation in the time course with [G] can be understood geometrically by overlaying the V–c trajectories on the bifurcation diagram. Here, we take a short cut and reduce the 3D system to a new 2D system for V and c by setting $n = n_\infty(V)$ in equation (2.34) (equivalent to making φ very large in equation (2.25c)), which eliminates the HB and spiking (figure 2.10(D)). This is only an approximation, so the critical values of $g_{K(ATP)}$ defining the various activity regimes are altered, but qualitatively the blue V-nullcline in figure 2.10(D) corresponds to the blue silent time course in figure 2.10(C), the green nullcline corresponds to the green bursting time course, and the black nullcline to the black spiking time course. The intersections of the brown c-nullcline with the V-nullclines predict the form of the solution *on the slow time scale of bursting*. The analysis is the same as that of the V–n phase plane in figure 2.8, with the exception that we have flipped the horizontal and vertical axes to conform to the convention of displaying spikes as up-and-down oscillations. Thus, intersection on the bottom branch of the V-nullcline implies a stable steady state, intersection on the middle branch implies a slow oscillation (here, bursting), and intersection on the upper branch implies a maintained depolarization (here, continuous spiking). The intersection of the nullclines shows approximately the trajectory of the full 3D system, in particular making clear that c increases as [G] increases ($g_{K(ATP)}$ decreases), which largely accounts for the increase in insulin secretion.

The addition of K(ATP) to the original Chay–Keizer model thus linked it to the known whole-body physiology of glucose-insulin homeostasis and was a theoretical tour de force. Alas, the model was soon shown to be wrong in several respects. For one thing, the first experimental measurements of cytosolic Ca^{2+} with fluorescent indicators [49], showed not the predicted sawtooth time course predicted in figure 2.7(A), but rather a square-wave-like pattern. This required including the dynamics of another Ca^{2+} compartment, the ER [48, 50, 51].

Another problem remained, however, revealed by experiments comparing the effects of glucose and the anti-diabetic drug, tolbutamide. Both reduce K(ATP) conductance, and thus, according to figure 2.10, should have equivalent effects. In reality, it is difficult or impossible to obtain bursting in tolbutamide without at least some glucose, though it could trigger continuous spike activity, as reported by Henquin [52] and Cook and Ikeuchi [53]. We address this in the next subsection.

2.2.3 Introducing metabolic oscillations

A possible explanation for the differences between glucose and tolbutamide came from the model of Keizer and Magnus [54], who proposed that the negative feedback to drive bursting came from the indirect activation of K(ATP) channels by Ca^{2+} as opposed to direct activation of KCa channels. In this model, Ca^{2+} enters the mitochondria during the active phase of bursting, depolarizes the mitochondrial membrane potential by short circuiting the proton gradient of the inner mitochondrial membrane, and thereby reducing the rate of ATP production. The resulting fall in ATP would reopen some of the K(ATP) channels closed by glucose metabolism and terminate spiking by repolarizing the cell. To describe this mathematically, following the version in [48], we rewrite the V equation with a dynamic K(ATP) conductance with open fraction a

$$\begin{aligned} C\frac{dV}{dt} = &- \bar{g}_{Ca}m_\infty(V)(V - V_{Ca}) - \bar{g}_K n(V - V_K) - g_L(V - V_L) \\ &+ \bar{g}_{K(Ca)}\frac{c}{c_0 + c}(V - V_K) + \bar{g}_{K(ATP)}a(V - V_K), \end{aligned} \tag{2.35}$$

and add a simple equation for a

$$\frac{da}{dt} = \frac{a_\infty(c) - a}{\tau_a}, \tag{2.36}$$

where

$$a_\infty(c) = \frac{1}{1 + \exp\left(\frac{r - c}{s_a}\right)},$$

r increases with [G], and s_a controls the steepness of the effect of [G].

In addition, this model has a second Ca^{2+} compartment, the endoplasmic reticulum (ER), which takes up Ca^{2+} from the cytosol via the SERCA pump with rate k_{SERCA} and releases Ca^{2+} by passive diffusion with rate p_{leak} through a leak channel, as determined by the following equation

$$\frac{dc_{ER}}{dt} = f\sigma\left[k_{SERCA}c - p_{leak}(c_{ER} - c)\right],$$

where σ is the ratio of ER to cytosolic volume, and c_{ER} is the concentration of Ca^{2+} in the ER. The equation for cytosolic Ca^{2+} has to be correspondingly modified to include the ER-cytosolic exchange fluxes, as follows

$$\frac{dc}{dt} = f\left(-\alpha I_{Ca}(V) - k_c c - k_{SERCA}c + p_{leak}(c_{ER} - c)\right).$$

The open fraction of K(ATP) channels depends mostly on the concentration of ADP, so one can think of a as roughly proportional to [ADP]. A rise in c increases a, so equation (2.36) is very similar in spirit to the K(Ca) mechanism in that a rise in c increases a K^+ conductance. In this case, however, the response to c is assumed to be slow, which, together with the above dynamics for ER Ca^{2+}, allows this model to produce much longer burst periods, which conforms to experimental observations. It also introduces a fundamentally new feature, oscillations in metabolism, which are essential for bursting oscillations with periods of tens of seconds, such as observed by Henquin [52] and Cook and Ikeuchi [53]. In the absence of ADP oscillations, Ca^{2+} acting on K(Ca) channels can only produce bursts with a period of a few seconds in this model [48]. Since metabolism would not oscillate if tolbutamide were substituted for glucose, the observation that glucose, but not tolbutamide, can support bursting is thus explained at a minimal level. However, it is instructive to examine further the effects on the dynamics of the model.

Increasing glucose, modeled as an increase in r (equation (2.36)), would decrease a and hence $g_{K(ATP)}$ at any given value of c. As shown in figure 2.11(A), increasing r leads to increased plateau fraction and finally to continuous spiking. During the active phase, a rises, and during the silent phase it falls (like c in figure 2.10(A)). Tolbutamide application is modeled not as an effect on metabolism but as a direct effect on the K (ATP) channel, namely, a decrease in $\bar{g}_{K(ATP)}$. Figure 2.11(B) shows that this drop can induce continuous spiking but not lead to bursting. This is illustrated geometrically in bifurcation diagrams with respect to the slowest variable in the model, a, for three values of r (Figure 2.11(C)) and two values of $\bar{g}_{K(ATP)}$ (figure 2.11(D)). Because a_∞ is a function of c, it is convenient to use c as the output variable on the y-axis. This is permissible because a is much slower than c, and c can therefore be considered as one

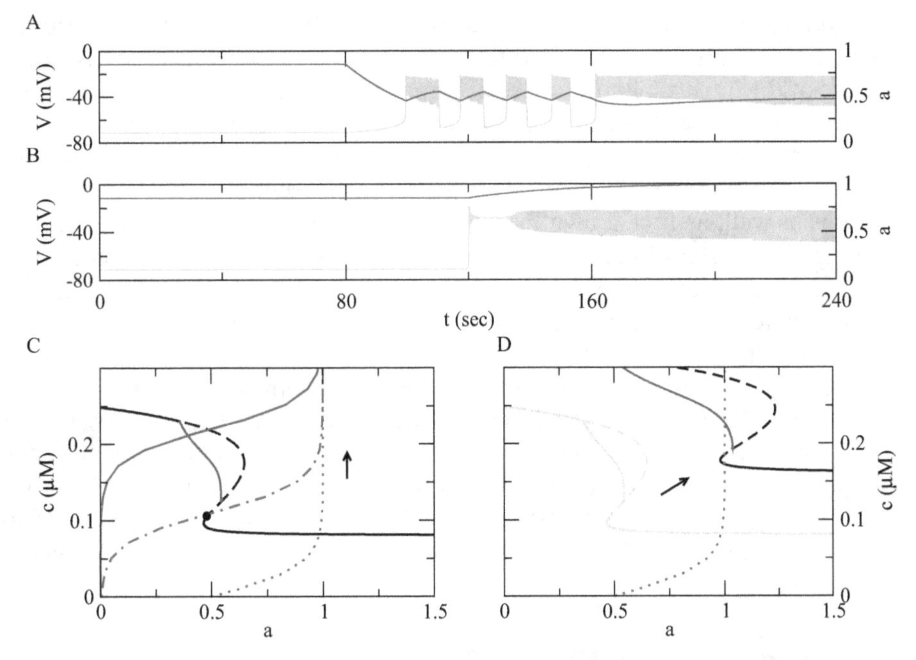

Figure 2.11. Glucose supports bursting, tolbutamide does not. The model is from [48] but with $s_a = 0.01$, $\tau_a = 30000$ ms, and r and $\bar{g}_{K(ATP)}$ as indicated below. (A) Increasing the value of the parameter r, indirectly representing an increase in glucose concentration, increases plateau fraction (r initially 0, raised to 0.11 μM at $t = 80$ s and 0.22 μM at $t = 160$ s). c_{ER} is held constant at an intermediate value, 85 μM, for the time course in panel A. (B) Decreasing $\bar{g}_{K(ATP)}$, representing application of tolbutamide while holding r (glucose) fixed at 0, triggers spiking but not bursting ($\bar{g}_{K(ATP)}$ initially 500 pS, reduced to 225 pS at $t = 120$ s). (C) Bifurcation diagram corresponding to (A). Black: steady-state values of c for each value of a (viewed as a parameter). Blue: average value of c during oscillations. Magenta: a-nullclines (dotted, $r = 0$; dot-dashed, $r = 0.1$, solid, $r = 0.22$ μM). Vertical arrow shows direction of increasing glucose (r). (D) Bifurcation diagram corresponding to (B). Grey: $\bar{g}_{K(ATP)} = 500$ pS, as in panel (C); black/blue: $\bar{g}_{K(ATP)} = 225$ pS. Magenta: a-nullcline ($r = 0$). c_{ER} held constant at 28 μM ($\bar{g}_{K(ATP)} = 500$ pS) and 180 μM ($\bar{g}_{K(ATP)} = 225$ pS). Diagonal arrow shows direction of adding tolbutamide (decreasing $\bar{g}_{K(ATP)}$).

of the fast variables. ER Ca^{2+} is also very slow and, since it is nearly constant at any given level of electrical activity, we have used its average values to calculate the diagrams. Increasing r raises the a-nullcline, which allows bursting when the rising portion of the sigmoidal a_∞ curve intersects the middle branch of the Z-curve, functioning like a c-nullcline. In contrast, reducing $\bar{g}_{K(ATP)}$ shifts the Z-curve to the right. This never results in bursting because the portion of the a-nullcline that slopes up to the right cannot intersect the middle branch of the Z-curve with the proper slope when r is low. This is the geometrical counterpart of the statement that metabolism needs to oscillate to get bursting in this model but goes further to show that the result depends on the details of how metabolism depends on Ca^{2+}.

Thus, Keizer and Magnus' model [54–56] predicted ATP/ADP oscillations. This was subsequently confirmed by several different experimental techniques. Slow oscillations in oxygen consumption due to aerobic metabolism were observed using oxygen-sensing electrodes in single islets [57–59]. Also, slow oscillations in reducing equivalents, such as NADH and FADH, were measured by taking advantage of the intrinsic fluorescence of some of these metabolites, reflecting their production by the mitochondrial citric acid cycle [60–62]. Furthermore, mitochondrial membrane potential displayed slow oscillations, as monitored using rhodamine 1,2,3 fluorescence [58, 63]. ATP levels have been shown to exhibit slow oscillations as well, as first shown by expressing the firefly luciferase gene in living insulinoma cells [64] and more recently by expressing the fluorescent ATP sensor Perceval in mouse islets [60, 65]. As expected from these findings, slow oscillations of K(ATP) conductance have been shown to occur in both single mouse beta cells [66] and individual mouse islets [1].

The existence of metabolic oscillations in beta cells is thus now well established, but uncertainty remains about the mechanism, such as whether Ca^{2+} oscillations drive metabolic oscillations through their effect on the mitochondria [54] or on ATP consumption by Ca^{2+} pumps [67–69], or alternatively Ca^{2+}-independent oscillations in glycolysis drive Ca^{2+} oscillations [59, 70, 71]. Ongoing experimental and theoretical work [60, 72, 73] indicates that the two mechanisms co-exist and interact in complex ways [74].

2.3 Concluding thoughts

Beta-cell modeling is supported by two intellectual pillars, membrane biophysics and the mathematics of dynamical systems. The reader should now be well prepared to read the electrophysiological literature and understand how even the most complex models of beta cells are constructed out of the elementary building blocks discussed here. For example, we have alluded to but not discussed glycolytic oscillations. The models we use for this follow the same design principles (fast positive feedback combined with slow negative feedback), so they are formally very similar to the spiking and bursting models described here. Readers are referred to [70, 71] for more details.

In parallel, there are two complementary methodologies for carrying out the modeling. One approach is to build models from the bottom up based on the biophysical data, as Hodgkin and Huxley did with the squid giant axon. This is in some sense the ideal, but we often lack sufficient or sufficiently precise data to build a model this way. An alternative is to fit the model to the measured output, such as the pattern of electrical activity. General principles about dynamics and bifurcations are needed to fill in the gaps in the data in this case. It has been argued [75] that we are forced to use this 'top down' approach because the inputs cannot be known precisely enough to fully constrain the model. Neither approach on its own is viable, however. 'Top-down' modeling that is divorced from data is likely to go astray, fulfilling the modeler's pre-conceived ideas of how the system *should* work, rather than how it *actually* works. 'Bottom-up' modeling divorced from the dynamical systems approach is likely to get bogged down in the details and miss the big picture. In the end, modeling is an art as well as a science, requiring extensive background knowledge and judgment. To take the next steps on the road to this noble profession, or to fill in the holes in your background, consult the references cited for further reading.

2.3.1 References for further reading

- Most elementary (assumes a year of calculus, possibly rusty): [40].
- Accessible undergraduate level survey of dynamical systems: [41].
- Advanced bifurcation theory: [46].
- A similar approach to this chapter but focused on pituitary cells, which have many similarities to beta cells: The first three chapters of [2].
- A review with another point of view about models for beta-cell electrical activity: [68].

Appendix A Linear systems

Beta-cell models are systems of nonlinear differential equations, but many of their properties can be understood by linearization.

The first building block is a single linear equation of the form

$$\dot{x} = ax,$$

with a constant. The equation is satisfied by a family of exponential solutions, of which one is selected by specifying the initial value of x

$$x(t) = ce^{at} = x(0)e^{at}.$$

The obvious fact that solutions increase if $a > 0$ and decay if $a < 0$ is a core principle that we use in the main text. To get more complex behavior, such as oscillations, requires at least a two-dimensional linear system, given by

$$\dot{x} = ax + by \tag{A.1a}$$

$$\dot{y} = cx + dy, \tag{A.1b}$$

where a, b, c, and d are constants. We use x and y because we want to think of them as the x and y coordinates moving in a plane. This can be written in matrix form as

$$\begin{pmatrix} \dot{x} \\ \dot{y} \end{pmatrix} = \begin{pmatrix} a & b \\ c & d \end{pmatrix} \begin{pmatrix} x \\ y \end{pmatrix},$$

or more compactly as

$$\dot{X} = MX.$$

The two-dimensional system for x and y can be replaced by a second-order equation in x by differentiating the x equation with respect to t and doing some algebra to eliminate y and dy/dt to yield

$$\ddot{x} - (a + d)\dot{x} + (ad - bc)x = 0. \tag{A.2}$$

Given x, y can also be calculated. Equations similar to this arise in applications of Newton's law, $F = ma$. For example, a mass on a spring with no friction satisfies the equation $\ddot{x} = -kx$. With friction, the coefficient of $\dot{x}$ is non-zero.

Like the one-dimensional differential equation, this one has exponential solutions, and an easy way to find it is to substitute a trial solution,

$$x = e^{\lambda t},$$

which gives

$$\lambda^2 e^{\lambda t} - (a + d)\lambda e^{\lambda t} + (ad - bc)e^{\lambda t} = 0.$$

We can divide by $e^{\lambda t}$, to obtain the following quadratic equation for λ:

$$\lambda^2 - (a + d)\lambda + (ad - bc) = 0,$$

or, if we factor the equation, as

$$(\lambda - \lambda_1)(\lambda - \lambda_2) = \lambda^2 - (\lambda_1 + \lambda_2)\lambda + (\lambda_1\lambda_2) = 0.$$

If $\lambda = \lambda_1$ or λ_2, then $e^{\lambda t}$ solves the differential equation. The two solutions, λ_1 and λ_2, can also be viewed as characteristic numbers called *eigenvalues* that summarize the properties of the matrix of coefficients, which can be studied as a mathematical object independent of the differential equations using the methods of linear algebra. For example, comparing the two forms of the quadratic equation shows that the sum of the eigenvalues is $a + d$, which is the sum of the diagonal elements of the matrix, known as the trace. This is true for matrices in general, but not obvious when they are bigger than 2×2. Similarly, the product of the eigenvalues is $ad - bc$, which is the determinant of the matrix, and this also holds for matrices in general. In our simple 2D case, there are two roots of the quadratic equation, so only two parameters are needed to characterize them, even though the matrix itself has four parameters. It is convenient and more revealing to parametrize the equation by the sum and product of the roots.

The theory of eigenvalues in linear algebra is deep, but we will limit our discussion to their role in summarizing the solutions of the differential equations in two dimensions. The general solution is

$$x = Ae^{\lambda_1 t} + Be^{\lambda_2 t}$$
$$y = Ce^{\lambda_1 t} + De^{\lambda_2 t},$$

where A, B, C, and D are constants determined by the initial values of x and y.

There are two main cases to consider:

1. λ_1 and λ_2 are real numbers. As for the 1D equation, the signs of λ_1 and λ_2 determine whether the solutions grow or shrink in time. If both are negative, the solution will decay to 0, but if at least one is positive, the solution will grow indefinitely. In the former case, the equation is said to be *stable*, in the latter, *unstable*.
2. λ_1 and λ_2 are complex numbers of the form $\alpha \pm i\beta$, where $i = \sqrt{-1}$.

Marvelously, exponentials of imaginary numbers can be expressed as combinations of sines and cosines following Euler's formula

$$e^{it} = \cos t + i \sin t, \tag{A.3}$$

(see exercise 15). This is the case that produces oscillating solutions, as studied in the elementary physics of masses on springs, mentioned above, and swinging pendulums.

Thus, we can rewrite a representative solution of the differential equation in case 2 as

$$x = e^{\alpha t}(\cos \beta t \pm i \sin \beta t), \tag{A.4}$$

and similarly for y. If we take the real part of x, the solutions become sinusoidal with exponentially growing or decaying amplitude, depending on whether α is positive or negative. When shown in the x–y plane, as in figure A1, they spiral in or spiral out, respectively. Maintained oscillations occur when $\alpha = 0$, which we can think as a system poised precisely between decaying and growing. There is actually an infinite family of solutions in this case, characterized by the initial amplitude, which is another way of saying that it is determined by the initial values of x and y (e.g. the black and grey curves in figure A1, middle). This corresponds to the physical observation that the amplitude of a swinging pendulum depends on how far it starts from the equilibrium state of pointing straight down. Mathematically, the solutions are neutrally stable: they stay on the initial trajectory in the absence of a perturbation, but if knocked off, they do not come back. An unending oscillation occurs in the ideal case of zero friction, which can only be realized approximately. More typically, oscillations decay because of friction, which dissipates energy. Growing oscillations require energy to be injected into the system, which can be thought of as 'negative' friction. In the main text, we consider the analogous case of

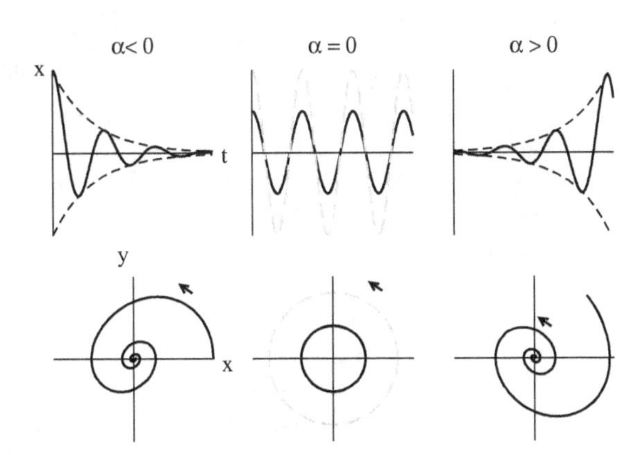

Figure A1. Solid curves: solution of linear system (equation (A.1)) with $a = 0$, $b = -1$, $c = 1$, and $d = -0.4$ (left); 0.4 (middle); 0 (right); this makes α in equation (A.4) $d/2$. Initial conditions: $x = 1$, $y = 0$. Upper: x vs t, Lower: x–y plane; motion is counter-clockwise. Left, $\alpha < 0$: decaying oscillation with envelope $\pm e^{\alpha t}$ (dashed curves). Middle, $\alpha = 0$: maintained neutrally stable oscillations. Changing the initial condition for x to 2 gives larger amplitude (grey dashed). Right, $\alpha > 0$: growing oscillation with envelope $\pm e^{\alpha t}$ (dashed curves).

negative resistance as providing the impetus for the rise of an action potential. Clocks can maintain oscillations for a while by continuously unwinding a spring or, for a grandfather clock, raising a weight that falls slowly through gravity.

A.1 Nonlinear systems

Our main goal is to understand nonlinear systems. In general, this is difficult, but near a steady state, the system can be well approximated by an appropriate linear system. Then we can apply the knowledge gained above. In some cases, especially if there is only one steady state, we can get a qualitative picture that applies globally.

A general, two-dimensional nonlinear system can be written as

$$\begin{aligned} \dot{x} &= f(x, y) \\ \dot{y} &= g(x, y), \end{aligned} \tag{A.5}$$

where f and g are nonlinear functions. At a steady state, $(\bar{x}, \bar{y})$,

$$f(\bar{x}, \bar{y}) = g(\bar{x}, \bar{y}) = 0.$$

Therefore, $\dot{x} = \dot{y} = 0$ at $(\bar{x}, \bar{y})$ and x and y remain constant. However, for a steady state to be physically realizable, it must persist if x and y are perturbed away from the steady state. That is, it must be stable. We can investigate this by setting

$$x = \bar{x} + \Delta x,\ y = \bar{y} + \Delta y.$$

If we differentiate x with respect t, we get

$$\dot{x} = \Delta\dot{x},$$

because the steady-state value $\bar{x}$ is constant, and the same for y. This allows us to derive an approximate system of differential equations for the deviations from the steady state by linearizing the equations for x, y (equivalently, by taking the first term of the Taylor series), as follows

$$\Delta\dot{x} = f(\bar{x} + \Delta x, \bar{y} + \Delta y) \approx f(\bar{x}, \bar{y}) + \Delta x\frac{\partial f}{\partial x}(\bar{x}, \bar{y}) + \Delta y\frac{\partial f}{\partial y}(\bar{x}, \bar{y}),$$

$$\Delta\dot{y} = g(\bar{x} + \Delta x, \bar{y} + \Delta y) \approx g(\bar{x}, \bar{y}) + \Delta x\frac{\partial g}{\partial x}(\bar{x}, \bar{y}) + \Delta y\frac{\partial g}{\partial y}(\bar{x}, \bar{y}),$$

or, in matrix form,

$$\begin{pmatrix} \Delta\dot{x} \\ \Delta\dot{y} \end{pmatrix} = \begin{pmatrix} f_x & f_y \\ g_x & g_y \end{pmatrix}\begin{pmatrix} \Delta x \\ \Delta y \end{pmatrix},$$

where f_x, f_y, g_x, g_y are the partial derivatives of f and g evaluated at $(\bar{x}, \bar{y})$ and hence are constant. This is thus a linear system with constant coefficients. It can be written more compactly as

$$\Delta\dot{x} = J\Delta X,$$

where $\Delta X = (\Delta x, \Delta y)$ and J is the matrix of partial derivatives, called the Jacobian.

To make this useful, we need to generalize a bit by introducing a parameter μ into equation (A.5)

$$\dot{x} = f(x, y; \mu),$$
$$\dot{y} = g(x, y; \mu),$$

so we can study how the behavior of the system changes as μ varies. For the linear system, equation (A.1), we can let $\mu = d = \alpha/2$, and figure A1 shows how the behavior of the system changes qualitatively as μ changes from negative to positive. For the nonlinear system in figure 2.9(B), the role of μ is played by c, and in figure 2.9(D), it is played by φ.

As μ varies, both the steady state $(\bar{x}, \bar{y})$ and the Jacobian J will vary, and, in particular, the eigenvalues of J will vary. In the linear case, we found that oscillations occur when the eigenvalues are complex conjugate and have real part $\alpha = 0$, or, equivalently, trace $M = 0$, det $M > 0$. The key result that drives our analysis of neurons and beta cells is that if the trace of J increases through 0 as μ passes a certain value (which we can arbitrarily take to be 0), an oscillation is born. Qualitative changes in behavior such as this are called bifurcations, and this type is a Hopf bifurcation (see [76] or [77] for a more precise statement and further discussion).

Figure A2(upper) illustrates the motion of the eigenvalues through the complex plane, and figure A2(lower) illustrates the corresponding behavior with this toy model [77]:

$$\dot{r} = \mu r - r^3, \tag{A.6a}$$

$$\dot{\theta} = 2\pi. \tag{A.6b}$$

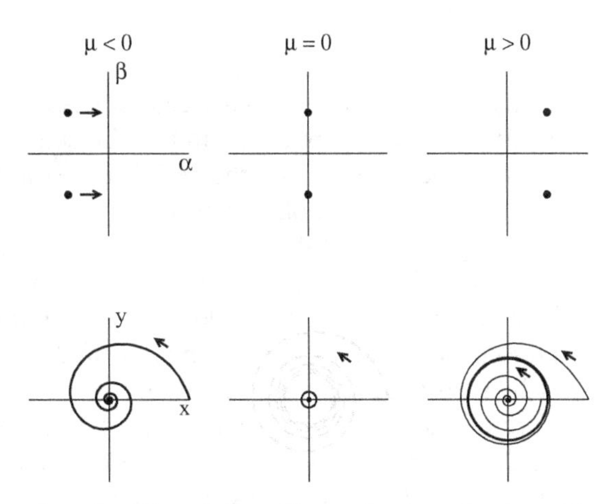

Figure A2. Hopf bifurcation of nonlinear system. Upper: the eigenvalue $\lambda = \alpha(\mu) + i\beta(\mu)$ is plotted in the complex plane. As μ increases through 0, α is assumed also to increase through 0. Lower: solution of equation (A.6). Left: for $\mu = -1$, it spirals into the origin (motion is counter-clockwise). Right: for $\mu = +1$, there is a limit cycle with radius 1. Trajectories starting inside the limit cycle spiral out to it, and those starting outside spiral into it (thin grey curves). Middle: for $\mu = 0$, there is a unique steady-state solution with zero amplitude ($r = 0$; black dot at origin), and all solutions spiral into it very slowly (see text and exercise 17). Grey: $r(0) = 1$; black: $r(0) = 0.1$.

It is convenient to use polar coordinates because the solutions near an HB are circular. We could have used the Morris–Lecar (ML) model, but it in fact behaves like equation (A.6) near an HB. When $\mu < 0$, there is a unique steady state at $r = 0$, and solutions spiral in to it, much like the linear case in figure A1. When $\mu > 0$, the solution $r = 0$ is unstable, and trajectories initially spiral out from the origin, again like the linear case. A new feature appears, however: the trajectory converges in time to a new solution that has appeared, $r = \sqrt{\mu}$. Trajectories that start outside that orbit spiral into it in the limit $t \to \infty$. The new solution is therefore called a *limit cycle*. This oscillation is born at the bifurcation as μ passes through 0 and grows as μ grows. Note that the spiking solution in figure 2.9(B) grows in a similar way near the HB. Unstable limit cycles can also occur, which can be modeled by changing the sign of the r^3 term in equation (A.6). They play important roles in neural models, but we will not have need of them.

At the bifurcation point $\mu = 0$, the behavior is subtle. The linearized version of equation (A.6) predicts a family of neutrally stable solutions, as in figure A1, but the r^3 term makes the origin stable. Trajectories starting with $r > 0$ spiral in to 0, but they do so very slowly when r is small (see exercise 17). The larger (grey) trajectory in figure A2(middle) spirals in rapidly at first, then slows down. The smaller (black) trajectory shrinks very little in 10 cycles, showing up as just a slight thickening of the curve on this zoomed out scale. However, if the integration is continued for 100 or 1000 cycles, the shrinkage becomes apparent.

In the context of excitable cells, limit cycles correspond to maintained spiking behavior. We now have enough machinery to carry out the analyses of neurons and beta cells in the main text.

Exercises

1. You are recording from a rapidly spiking cell enmeshed in a tight cellular network where the interstitial space is limited. During the recording you find that the membrane potential of the cell under study is progressively depolarizing, although spike amplitude is largely unchanged. In parallel you have studied dispersed and fully isolated cells of the same type but they never show the progressive depolarization seen in the network. Why is the membrane depolarizing and how could you potentially estimate the extracellular concentration of the most permeant ion?
2. A highly inwardly rectifying current is recorded from a cell and you need to describe the pore of the channel mediating the current measured. Which would be more desirable to use in this case, the Goldman–Hodgkin–Katz formulation or one based on the Nernst potential?
3. The pump rate of the plasma membrane Na/K ATPase is increased in a membrane having a 1 gigaohm input resistance, yielding a 10 pA net outward current. How much depolarization would be predicted for the activation of this current?

4. Using equation (2.7), show that the I–V curve becomes more linear as c_i approaches c_e (cf figure 2.5).
5. Write an expression similar to that of equation (2.8) for V if Na^+ is replaced by Ca^{2+}. Hint: the equation is no longer linear, but quadratic, because $z_{Ca} = 2$, and the result is not nearly as pretty.
6. Based on equation (2.21), what would be the whole cell conductance of the cell?
7.
   (a) Carry out the simulation in figure 2.7(B), fit an exponential to each tail current, and see how well $m_\infty(V)$ can be reconstructed.
   (b) Do the same as in (a) but with noise added to the m equation.
   (c) Which protocol, figure 2.7(A) or figure 2.7(B), is better for fitting $\tau_m(V)$ and why?
8. Explain the S-shaped rise of the spike in figure 2.8(A).
9.
   (a) In figure 2.8(C), explain why the steady state (black dot) is stable.
   (b) In figure 2.8(D), demonstrate that the middle steady state (white dot) is a saddle by showing that the determinant of the Jacobian is negative because the slope of the n nullcline is less than that of the V nullcline.
10. (Extended project) In figure 2.9(D), the amplitude of the limit cycles was made smaller by increasing the value of φ. Another way to do this is to add a large, voltage- and Ca^{2+}-dependent K(Ca) current (BK current) to the ML model, given by

$$I_{\mathrm{BK}} = \bar{g}_{\mathrm{BK}} x_\infty(V)(V - V_{\mathrm{K}}),$$

where

$$x_\infty(V) = \frac{1}{1 + \exp(-(V - V_{\mathrm{BK}})/10)}, \quad V_{\mathrm{BK}} = 0.1 - 26\ln\left(\frac{\mathrm{Ca}_d}{1.5}\right), \text{ and}$$
$$\mathrm{Ca}_d = -0.000\,15 I_{\mathrm{Ca}}(V).$$

Here, Ca_d is the Ca^{2+} concentration at the mouth of the Ca^{2+} channel, which shifts the activation of the BK channel the left when the Ca^{2+} channel is open [78]. Your task is to modify the parameters of ML from those in figure 2.9(A), guided by bifurcation diagrams, such that the addition of a suitable amount of I_{BK} converts spiking to bursting.

Historical note: Chay and Keizer [79] modeled the effect of tetraethylammonium (TEA) to convert bursting into spiking in islets by slowing the K_V channel (a decrease in φ), but Fatherazi and Cook [80] found that TEA was mainly blocking BK channels.
11. Starting with the model of figure 2.10(A), explore the effect of increasing the leak conductance, g_{L} by simulation and with bifurcation diagrams.
12.
   (a) For the model of figure 2.11, what would the effect of tolbutamide be if a_∞ were made less steep by increasing the parameter s_a to 0.1?

(b) Compare the behavior of the model as r is increased with c_{ER} held constant versus with c_{ER} removed from the model completely.

13. (Extended project) Break the bursting model of figure 2.10 by randomly changing the parameters. You could use a uniform or normal distribution that perturbs the parameters 10% on average. Compare the Ca^{2+} and K^+ currents to those with the original values; the discrepancy should be comparable to variation you might find in experimentally measured currents. Without referring to the original parameters, fix this miscalibrated model, using phase planes and bifurcation diagrams as a guide.
14. Derive equation (A.2).
15. Derive Euler's formula (equation (A.3)) by solving the complex differential equation

$$\dot{z} = iz, \, z(0) = 1,$$

and letting $z = x + iy$. (Alternatively, see Wolfram Mathworld, http://mathworld.wolfram.com/EulerFormula.html.)

16. Show that the case $\alpha = 0$ in equation (A.4) corresponds to trace $M = 0$, $\det M > 0$ and verify by direct substitution into the differential equation that the solutions are pure sinusoids.
17. Show that when $\mu = 0$ in equation (A.6), the decay is not exponential, but algebraic, given by $r = 1/\sqrt{2t}$.

References

[1] Ren J, Sherman A, Bertram R, Goforth P B, Nunemaker C S, Waters C D and Satin L S 2013 Slow oscillations of KATP conductance in mouse pancreatic islets provide support for electrical bursting driven by metabolic oscillations *Am. J. Physiol. Endocrinol. Metab.* **305** E805–17

[2] MacGregor D J and Leng G (ed) 2016 *Computational Neuroendocrinology* (New York: Wiley)

[3] Ermentrout B 2002 *Simulating, Analyzing, and Animating Dynamical Systems* (Philadelphia, PA: SIAM) http://www.math.pitt.edu/bard/xpp/xpp.html

[4] Hoppensteadt F C and Peskin C S 1992 *Mathematics in Medicine and the Life Sciences* (Texts in Applied Mathematics vol 10) (New York: Springer)

[5] Bockris J O and Reddy A K N 1970 *Modern Electrochemistry* (New York: Plenum)

[6] Ashcroft F M and Rorsman P 1989 Electrophysiology of the pancreatic beta-cell *Prog. Biophys. Mol. Biol.* **54** 87–143

[7] Keener J and Sneyd J 2009 *Cellular physiology Mathematical Physiology I* 2nd edn (New York: Springer)

[8] Katz B 1966 *Nerve, Muscle and Synapse* 1st edn (New York: McGraw-Hill)

[9] Goldman D E 1943 Potential, impedance, and rectification in membrances *J. Gen. Physiol.* **27** 37–60

[10] Hodgkin A L and Katz B 1949 The effect of sodium ions on the electrical activity of giant axon of the squid *J. Physiol.* **108** 37–77

[11] Junge D 1992 *Nerve and Muscle Excitation* (Sunderland, MA: Sinauer)

[12] Fitzhugh R 1961 Impulses and physiological states in theoretical models of nerve membrane *Biophys. J.* **1** 445–66
[13] Hindmarsh J L and Rose R M 1984 A model of neuronal bursting using three coupled first order *Proc. R. Soc. Lond. B Biol. Sci.* **221** 87–102
[14] Hodgkin A L and Huxley A F 1952 A quantitative description of membrane current and its application to conduction and excitation in nerve *J. Physiol.* **117** 500–44
[15] Hodgkin A L and Huxley A F 1952 Currents carried by sodium and potassium ions through the membrane of the giant axon of Loligo *J. Physiol.* **116** 449–72
[16] Hodgkin A L and Huxley A F 1952 The dual effect of membrane potential on sodium conductance in the giant axon of Loligo *J. Physiol.* **116** 497–506
[17] Hodgkin A L, Huxley A F and Katz B 1952 Measurement of current–voltage relations in the membrane of the giant axon of Loligo *J. Physiol.* **116** 424–48
[18] Fox R F 1997 Stochastic versions of the Hodgkin–Huxley equations *Biophys. J.* **72** 2068–74
[19] Sherman A, Rinzel J and Keizer J 1988 Emergence of organized bursting in clusters of pancreatic beta-cells by channel sharing *Biophys. J.* **54** 411–25
[20] Pedersen M G 2007 Phantom bursting is highly sensitive to noise and unlikely to account for slow bursting in beta-cells: considerations in favor of metabolically driven oscillations *J. Theor. Biol.* **248** 391–400
[21] Bernstein J 1902 Untersuchungen zur Thermodynamik der bioelektrischen Ströme: Erster Theil *Pflüger Arch. Gesammte Physiol. Menschen Thiere* **92** 521–62
[22] Rinzel J 1990 Discussion: electrical excitability of cells, theory and experiment: review of the Hodgkin-Huxley foundation and an update *Bull. Math. Biol.* **52** 3–23
[23] Hille B 2001 *Ion Channels of Excitable Membranes* 3rd edn (Sunderland, MA: Sinauer)
[24] Clay J R 1998 Excitability of the squid giant axon revisited *J. Neurophysiol.* **80** 903–13
[25] Meunier C and Segev I 2002 Playing the devil's advocate: is the Hodgkin–Huxley model useful? *Trends Neurosci.* **25** 558–63
[26] Li Y X and Rinzel J 1994 Equations for InsP3 receptor-mediated $[Ca^{2+}]_i$ oscillations derived from a detailed kinetic model: a Hodgkin–Huxley like formalism *J. Theor. Biol.* **166** 461–73
[27] Morris C and Lecar H 1981 Voltage oscillations in the barnacle giant muscle fiber *Biophys. J.* **35** 193–213
[28] Rorsman P and Trube G 1986 Calcium and delayed potassium currents in mouse pancreatic beta-cells under voltage-clamp conditions *J. Physiol.* **374** 531–50
[29] Zucker R S and Fogelson A L 1986 Relationship between transmitter release and presynaptic calcium influx when calcium enters through discrete channels *Proc. Natl Acad. Sci. USA* 83 3032–6
[30] Sherman A, Xu L and Stokes C L 1995 Estimating and eliminating junctional current in coupled cell populations by leak subtraction. A computational study *J. Membr. Biol.* **143** 79–87
[31] Göpel S, Kanno T, Barg S, Galvanovskis J and Rorsman P 1999 Voltage-gated and resting membrane currents recorded from B-cells in intact mouse pancreatic islets *J. Physiol.* **521** 717–28
[32] Pinsky P F and Rinzel J 1994 Intrinsic and network rhythmogenesis in a reduced Traub model for CA_3 neurons *J. Comput. Neurosci.* **1** 39–60
[33] Atwater I, Ribalet B and Rojas E 1978 Cyclic changes in potential and resistance of the beta-cell membrane induced by glucose in islets of Langerhans from mouse *J. Physiol.* **278** 117–39

[34] Meissner H P and Schmelz H 1974 Membrane potential of beta-cells in pancreatic islets *Pflugers Arch.* **351** 195–206
[35] Scott A M, Atwater I and Rojas E 1981 A method for the simultaneous measurement of insulin release and B cell membrane potential in single mouse islets of Langerhans *Diabetologia* **21** 470–5
[36] Cook D L 1983 Isolated islets of Langerhans have slow oscillations of electrical activity *Metabolism* **32** 681–5
[37] Rorsman P, Eliasson L, Kanno T, Zhang Q and Gopel S 2011 Electrophysiology of pancreatic β-cells in intact mouse islets of Langerhans *Prog. Biophys. Mol. Biol.* **107** 224–35
[38] Kinard T A, de Vries G, Sherman A and Satin L S 1999 Modulation of the bursting properties of single mouse pancreatic beta-cells by artificial conductances *Biophys. J.* **76** 1423–35
[39] Kanno T, Gopel S O, Rorsman P and Wakui M 2002 Cellular function in multicellular system for hormone-secretion: electrophysiological aspect of studies on alpha-, beta- and delta-cells of the pancreatic islet *Neurosci. Res.* **42** 79–90
[40] Edelstein-Keshet L 2005 *Mathematical Models in Biology* 1st edn (Philadelphia: SIAM)
[41] Strogatz S H 2015 *Nonlinear Dynamics and Chaos: With Applications to Physics, Biology, Chemistry, and Engineering* 2nd edn (Boulder, CO: Westview)
[42] Koch C and Segev I 1998 *Methods in Neuronal Modeling: From Ions to Networks. Computational Neuroscience* 2nd edn (Cambridge, MA: MIT Press)
[43] Hodgkin A L 1948 The local electric changes associated with repetitive action in a non-medullated axon *J. Physiol.* **107** 165–81
[44] Atwater I, Rosario L and Rojas E 1983 Properties of the Ca-activated K. channel in pancreatic beta-cells *Cell Calcium* **4** 451–61
[45] Chay T R and Keizer J 1983 Minimal model for membrane oscillations in the pancreatic beta-cell *Biophys. J.* **42** 181–90
[46] Kuznetsov Y A 1995 *Elements of Applied Bifurcation Theory* (Applied Mathematical Sciences vol 112) (New York: Springer)
[47] Ermentrout B 1998 Linearization of F-I curves by adaptation *Neural Comput.* **10** 1721–9
[48] Bertram R and Sherman A 2004 A calcium-based phantom bursting model for pancreatic islets *Bull. Math. Biol.* **66** 1313–44
[49] Valdeolmillos M, Nadal A, Soria B and García-Sancho J 1993 Fluorescence digital image analysis of glucose-induced $[Ca^{2+}]_i$ oscillations in mouse pancreatic islets of Langerhans *Diabetes* **42** 1210–4
[50] Chay T R 1996 Electrical bursting and luminal calcium oscillation in excitable cell models *Biol. Cybern.* **75** 419–31
[51] Chay T R 1997 Effects of extracellular calcium on electrical bursting and intracellular and luminal calcium oscillations in insulin secreting pancreatic beta-cells *Biophys. J.* **73** 1673–88
[52] Henquin J-C 1998 A minimum of fuel is necessary for tolbutamide to mimic the effects of glucose on electrical activity in pancreatic β-cells *Endocrinology* **139** 993–8
[53] Cook D L and Ikeuchi M 1989 Tolbutamide as mimic of glucose on beta-cell electrical activity. ATP-sensitive K+ channels as common pathway for both stimuli *Diabetes* **38** 416–21
[54] Keizer J and Magnus G 1989 ATP-sensitive potassium channel and bursting in the pancreatic beta cell. A theoretical study *Biophys. J.* **56** 229–42

[55] Magnus G and Keizer J 1998 Model of beta-cell mitochondrial calcium handling and electrical activity. I. Cytoplasmic variables *Am. J. Physiol.* **274** C1158–73

[56] Magnus G and Keizer J 1998 Model of beta-cell mitochondrial calcium handling and electrical activity. II. Mitochondrial variables *Am. J. Physiol.* **274** C1174–84

[57] Jung S K, Kauri L M, Qian W J and Kennedy R T 2000 Correlated oscillations in glucose consumption, oxygen consumption, and intracellular free Ca^{2+} in single islets of Langerhans *J. Biol. Chem.* **275** 6642–50

[58] Nunemaker C S, Zhang M and Satin L S 2004 Insulin feedback alters mitochondrial activity through an ATP-sensitive K+ channel-dependent pathway in mouse islets and beta-cells *Diabetes* **53** 1765–72

[59] Tornheim K 1997 Are metabolic oscillations responsible for normal oscillatory insulin secretion? *Diabetes* **46** 1375–80

[60] Merrins M J, Fendler B, Zhang M, Sherman A, Bertram R and Satin L S 2010 Metabolic oscillations in pancreatic islets depend on the intracellular Ca^{2+} level but not Ca^{2+} oscillations *Biophys. J.* **99** 76–84

[61] Luciani D S, Misler S and Polonsky K S 2006 Ca^{2+} controls slow NAD(P)H oscillations in glucose-stimulated mouse pancreatic islets *J. Physiol.* **572** 379–92

[62] Quesada I, Todorova M G and Soria B 2006 Different metabolic responses in alpha-, beta-, and delta-cells of the islet of Langerhans monitored by redox confocal microscopy *Biophys. J.* **90** 2641–50

[63] Krippeit-Drews P, Düfer M and Drews G 2000 Parallel oscillations of intracellular calcium activity and mitochondrial membrane potential in mouse pancreatic B-cells *Biochem. Biophys. Res. Commun.* **267** 179–83

[64] Ainscow E K and Rutter G A 2002 Glucose-stimulated oscillations in free cytosolic ATP concentration imaged in single islet beta-cells: evidence for a Ca^{2+}-dependent mechanism *Diabetes* **51** S162–70

[65] Li J, Shuai H Y, Gylfe E and Tengholm A 2013 Oscillations of sub-membrane ATP in glucose-stimulated beta cells depend on negative feedback from Ca^{2+} *Diabetologia* **56** 1577–86

[66] Merrins M J, Poudel C, McKenna J P, Ha J, Sherman A, Bertram R and Satin L S 2016 Phase analysis of metabolic oscillations and membrane potential in pancreatic islet β-cells *Biophys. J.* **110** 691–9

[67] Fridlyand L E, Tamarina N and Philipson L H 2003 Modeling of Ca^{2+} flux in pancreatic beta-cells: role of the plasma membrane and intracellular stores *Am. J. Physiol. Endocrinol. Metab.* **285** E138–54

[68] Fridlyand L E, Tamarina N and Philipson L H 2010 Bursting and calcium oscillations in pancreatic beta-cells: specific pacemakers for specific mechanisms *Am. J. Physiol. Endocrinol. Metab.* **299** E517–32

[69] Cha C Y, Nakamura Y, Himeno Y, Wang J, Fujimoto S, Inagaki N, Earm Y E and Noma A 2011 Ionic mechanisms and Ca^{2+} dynamics underlying the glucose response of pancreatic β cells: a simulation study *J. Gen. Physiol.* **138** 21–37

[70] Bertram R, Satin L, Zhang M, Smolen P and Sherman A 2004 Calcium and glycolysis mediate multiple bursting modes in pancreatic islets *Biophys. J.* **87** 3074–87

[71] Bertram R, Sherman A and Satin L S 2007 Metabolic and electrical oscillations: partners in controlling pulsatile insulin secretion *Am. J. Physiol. Endocrinol. Metab.* **293** E890–900

[72] Watts M, Fendler B, Merrins M J, Satin L S, Bertram R and Sherman A 2014 Calcium and metabolic oscillations in pancreatic islets: who's driving the bus? *SIAM J. Appl. Dyn. Syst.* **13** 683–703

[73] McKenna J P, Ha J, Merrins M J, Satin L S, Sherman A and Bertram R 2016 Ca^{2+} effects on ATP production and consumption have regulatory roles on oscillatory islet activity *Biophys. J.* **110** 733–42

[74] Bertram R, Satin L S and Sherman A 2018 Closing in on the mechanisms of pulsatile insulin secretion *Diabetes* **67** 351–9

[75] Gutenkunst R N, Waterfall J J, Casey F P, Brown K S, Myers C R and Sethna J P 2007 Universally sloppy parameter sensitivities in systems biology models *PLoS Comput. Biol.* **3** 1871–8

[76] Edelstein-Keshet L 2005 *Mathematical Models in Biology* 1st edn (Philadelphia: SIAM)

[77] Strogatz S H 2015 *Nonlinear Dynamics and Chaos: With Applications to Physics, Biology, Chemistry, and Engineering* 2nd edn (Boulder, CO: Westview)

[78] Tsaneva-Atanasova K, Sherman A, van Goor F and Stojilkovic S S 2007 Mechanism of spontaneous and receptor-controlled electrical activity in pituitary somatotrophs: experiments and theory *J. Neurophysiol.* **98** 131–44

[79] Chay T R and Keizer J 1983 Minimal model for membrane oscillations in the pancreatic beta-cell *Biophys. J.* **42** 181–90

[80] Fatherazi S and Cook D L 1991 Specificity of tetraethylammonium and quinine for three K channels in insulin-secreting cells *J. Membr. Biol.* **120** 105–14

IOP Publishing

Diabetes Systems Biology
Quantitative methods for understanding beta-cell dynamics and function
Anmar Khadra

Chapter 3

Recent advances in mathematical modeling and statistical analysis of exocytosis in endocrine cells

Morten Gram Pedersen, Francesco Montefusco, Alessia Tagliavini and Giuliana Cortese

This chapter will focus on modelling the interaction between calcium and the machinery triggering exocytosis of insulin-containing secretory granules. Special attention will be given to models that analyze local versus global calcium levels, since there is evidence that at least a subpopulation of the granules are colocalized with calcium channels. This has implications for the interpretation of a certain class of experiments investigating the readily releasable pool of granules with electrophysiological methods. The mathematical analysis of these types of experiments will highlight how to interpret such experiments, and point to limitations and pitfalls.

Mathematical models that describe the dynamics of several pools of granules, possibly with different calcium sensitivities, and how the pools might be controlled by different calcium compartments will be presented. This treatment will illustrate how such models can be used to combine and formalize biological insight obtained from experiments that happen to be inherently different, operating on different spatial and temporal scales. The developed models can then be used to simulate other scenarios to predict answers to what-if questions.

3.1 Cell biology of exocytosis

Sections 3.1–3.4 feature excerpts from [64].

Most endocrine cells, including pancreatic beta cells, share the fundamental cellular organization and control of hormone secretion [1]. In beta cells, the insulin molecules are contained in secretory granules that, in response to a series of cellular mechanisms culminating with an increase in the intracellular Ca^{2+} concentration (see chapter 2) fuse with the cell membrane in a process called exocytosis, which allows the insulin molecules to exit the cell and enter the blood stream. The main

doi:10.1088/978-0-7503-3739-7ch3

events underlying hormone exocytosis and release are shared with exocytosis of synaptic vesicles underlying neurotransmitter release in neurons [2, 3].

The molecular machinery involved in exocytosis is becoming increasingly more well understood, and involves isoforms of the SNARE proteins syntaxin and SNAP, which are located in the cell membrane, and VAMP (also called synaptobrevin) that is inserted into the vesicle/granule membrane [3]. The SNARE proteins can form the so-called SNARE complex, which drives fusion of the two (granular and cellular) membranes. SNARE complexes interact with many other proteins, notably Ca^{2+}-sensing proteins such as synaptotagmins, which trigger exocytosis upon Ca^{2+} binding [3]. Thus, the local Ca^{2+} concentration at the Ca^{2+} sensor of the exocytotic machinery is an important determinant of the probability (rate) of exocytosis of the secretory granule. This fact will be a recurrent theme in the present section.

Depending on their ability to undergo exocytosis, granules in endocrine cells are traditionally divided functionally into a readily releasable pool (RRP) and a number of reserve pools [4–8]. The readily releasable granules are immediately available for secretion and typically consist of 1%–5% of the total number of granules in the cell [6]. After exocytosis of these granules, the RRP is replenished by granules from the reserve pools. However, it is still unclear if the refilling process involves physical translocation of granules within the cell, chemical modification of granules already situated at the membrane, recruitment of exocytotic proteins, or a combination of these processes, which are commonly referred to as priming [9–11].

3.2 Experimental techniques for investigations of exocytosis

3.2.1 Measuring whole-cell exocytosis as capacitance increases

Monitoring the total cellular amount of exocytosis in response to various stimuli can be relatively easily performed using the patch-clamp technique. Fusion of granules with the plasma membrane effectively increases the area of the cell membrane. Since the membrane capacitance C is proportional to the area (see chapter 2) this leads to an increase in whole-cell capacitance, which can be measured using either the whole-cell or the perforated-patch variants of patch-clamping [1, 6, 12]. For instance, the fusion of a 300 nm-diameter granule with the plasma membrane yields an electrically detectable step in C of 2–3 fF [1, 6].

Two major stimulation protocols have been applied to investigate rapid exocytosis in beta cells. One option, which is typically used to investigate directly the Ca^{2+} sensitivity of the exocytotic machinery, is to load the cell via the patch pipette with 'caged' Ca^{2+}, i.e. Ca^{2+} bound to a light-sensitive buffer, which upon light stimulation is released [1, 2] (figure 3.1, upper). This so-called 'flash release' rapidly increases the Ca^{2+} concentration uniformly in the cell to levels that can be measured simultaneously with Ca^{2+} sensitive probes. By relating the Ca^{2+} levels to the increases in membrane capacitance as a measure of exocytosis, it has been revealed that endocrine exocytosis typically occurs when the Ca^{2+} concentration is raised to tens of μM [13–16]. More recently, pools with higher Ca^{2+} sensitivity (~2 μM) were found in chromaffin [17] and beta cells [18, 19].

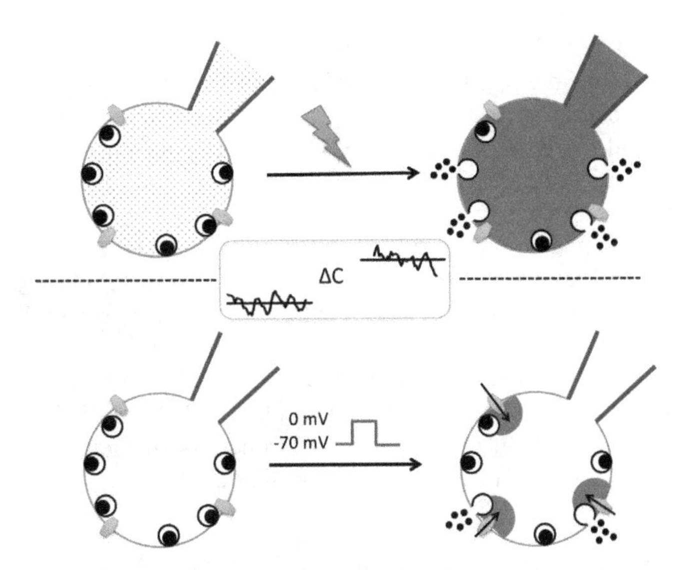

Figure 3.1. Electrophysiological methods to study exocytosis as increases in membrane capacitance (center panel). Upper: when the cell is filled with 'caged' Ca^{2+} via the patch pipette (left), a brief UV flash releases the bound Ca^{2+} and the intracellular Ca^{2+} concentration is raised uniformly, which in turn triggers release of secretory granules independently of their location. Lower: a depolarizing pulse applied via the patch pipette opens voltage-sensitive Ca^{2+} channels, causing Ca^{2+} influx into the cell. The resulting Ca^{2+} microdomains evoke exocytosis of granule located close to the Ca^{2+} channels.

Although flash-release gives important insight into the Ca^{2+}-sensitivity of the release machinery, the method is unphysiological since Ca^{2+} is raised artificially, and not because of Ca^{2+} influx via voltage-dependent Ca^{2+} channels [2] as discussed in the previous section. A more physiological protocol is to depolarize the cell, which opens Ca^{2+} channels, and subsequently triggers Ca^{2+} influx and exocytosis [1, 2] (figure 3.1, lower). During depolarization, it is possible to measure the Ca^{2+} current, whereas the cell capacitance can only be measured reliably before and after, but not during, depolarization. To investigate the kinetics of exocytosis, it is therefore necessary to apply depolarizing pulses of varying duration. A recent, detailed analysis of this so-called pulse-length protocol will be explained in section 3.4.1.

3.2.2 Recording single exocytotic events with imaging

Live-cell imaging provides an alternative experimental method for the study of exocytotic events. For example, two-photon imaging of pancreatic islets, bathed in the tracer sulforhodamine-B, allows the detection of single events of membrane fusion as the tracer enters the granule through the fusion pore, resulting in bright spots below the cell membrane [20–22]. A major advantage of this technique, due to the two-photon microscopy technique, is the possibility to monitor exocytosis in cells deep within their natural environment, i.e. within intact pancreatic islets, whereas for example capacitance recordings typically are performed on single cells or membrane patches. However, the use of sulforhodamine-B as an extracellular marker does not allow the visualization of secretory granules before they undergo exocytosis. Thus, the single-granule rate of exocytosis cannot be estimated.

In contrast, labeling of the secretory granules with one of several fluorescent markers [23, 24] allows the experimenter to follow the single granules with the use of total internal reflection fluorescence (TIRF) microscopy [10, 25–32]. TIRF imaging excites fluorescent reporters in a thin (a few hundred nanometers) layer below the cell membrane attached to the coverslip, thus allowing observation of the granules located at the membrane while minimizing the signal from granules deeper within the cells. It is therefore possible to investigate three-dimensional spatial movement of the granules as they approach the membrane, become ready for exocytosis, and eventually undergo exocytosis. Combined with other fluorophores and two-color imaging, it is possible to monitor for example Ca^{2+} levels [33–35] or protein abundance [10, 30, 35, 36] at the individual granules. With such data, it is possible to relate rates of exocytosis to signals and molecules controlling single fusion events.

3.3 Local control of exocytosis by Ca^{2+} microdomains

Mathematical modeling has played an important role for the development of current theories of the control of neurotransmitter and hormone release. Whereas it was well-established early that Ca^{2+} influx is crucial for triggering exocytosis in neurons [37] and endocrine cells [38, 39], the spatial organization of the Ca^{2+} channels and the release machinery is only beginning to be clarified (see e.g. [35, 40]). Thus, for decades, mathematical modeling was the main approach to study the relation between individual granules and single Ca^{2+} channels.

Chad and Eckert [41] and Simon and Llinas [42] presented important simulation studies of the spatiotemporal Ca^{2+} profiles below and around open Ca^{2+} channels, reaching similar conclusions. Their main results were that local Ca^{2+} *microdomains* build up in microseconds below open Ca^{2+} channels, and collapse just as rapidly when the channels close. When a Ca^{2+} channel is open, the local Ca^{2+} concentration reaches tens to hundreds of μM, which decays within tens of nm from the channel. Based on these simulation results it was proposed that the spatial localization of the synaptic vesicles compared to the Ca^{2+} channels is of great importance for determining the amount of neurotransmitter release, since vesicles near the channels will be exposed to much higher Ca^{2+} concentrations. Consequently, the relation between synaptic release and the whole-cell Ca^{2+} current, given by the product of the number of open Ca^{2+} channels and the single channel current, becomes less intuitive. Whereas the number of open channels increases when the membrane potential becomes depolarized, the Ca^{2+} driving force (and hence the single channel current) decreases. This means that there will be more Ca^{2+} microdomains in response to depolarizations, but each microdomain will have lower Ca^{2+} concentration. This will be discussed in greater detail below (section 3.4). The classical reviews of Ca^{2+} microdomains and related modelling by Neher [43, 44] also provide a good overview of this topic.

These earlier studies of microdomain dynamics explained well several aspects of neurotransmitter release, such as the speed of synaptic signalling after arrival of an action potential, and the rapid cessation of release following repolarization because of the collapse of the microdomains. It later became clear, however, that the slower release from endocrine cells might not be explained by pure microdomain release. Klingauf and Neher [45] modeled diffusion around Ca^{2+} channels in adrenaline-secreting chromaffin cells, paying particular attention to the choice of buffer parameters. By coupling the simulated Ca^{2+} profiles to a model of exocytosis developed from experiments involving flash-released Ca^{2+}, they could simulate exocytosis under various conditions, which was then compared to experimental results. They found that release from chromaffin cells is well-described by an arrangement where the Ca^{2+} sensors of most granules are located ~300 nm from Ca^{2+} channels, while the exocytotic machinery of a small (~8%) pool of granules is located at ~30 nm from the channels [45]. Since the major part of the granules are located some distance from Ca^{2+} channels, buffering parameters become important for the control of Ca^{2+} levels and exocytosis. In summary, these authors proposed that in (neuro-)endocrine cells most exocytosis occurs outside microdomains.

3.4 From microdomains to whole-cell modeling of exocytosis

Whole-cell imaging of Ca^{2+} has become a routine experiment, and many mathematical models of global Ca^{2+} levels have been published, including for beta cells (see chapter 2). However, as discussed above, local Ca^{2+} levels near Ca^{2+} channels determine the rate of exocytosis of the granules. Whole-cell Ca^{2+} levels reflect microdomain Ca^{2+} concentrations as well as the number of such microdomains in a rather simple way, as can be seen in the following analysis.

Recall that the prototype model of the bulk cytosolic Ca^{2+} concentration, c, neglecting internal stores, is given by equation (2.32), i.e.

$$\frac{dc}{dt} = f(-\alpha I_{Ca} - k_c c), \tag{3.1}$$

where f is the ratio of free (non-buffered) to total Ca^{2+}, α changes current to flux, and k is the rate parameter for Ca^{2+} extrusion. Unlike the previous section, the whole-cell Ca^{2+} current will be modeled in the following Hodgkin–Huxley fashion

$$I_{Ca} = g_{Ca} m^n h\,(V - V_{Ca}), \quad g_{Ca} = N_{Ca}\gamma, \tag{3.2}$$

with m and h activation and inactivation variables, respectively, and Ca^{2+} Nernst potential V_{Ca}. The maximal whole-cell conductance g_{Ca} is given as the product of the number of Ca^{2+} channels, N_{Ca}, and the single-channel conductance γ.

From equation (3.1), we have at steady-state,

$$c = -\alpha I_{Ca}/k_c = -\alpha/k_c[N_{Ca} m^n h][\gamma(V - V_{Ca})]. \tag{3.3}$$

Thus, the bulk Ca^{2+} level is proportional to the number of open channels, $N_{Ca}m^n h$, and to the single-channel current $i_{Ca} = \gamma(V - V_{Ca})$. Assuming a discrete spatial distribution of the Ca^{2+} channels, the number of microdomains is also equal to $N_{Ca}m^n h$. Further, the local Ca^{2+} concentration is proportional to i_{Ca} [43]. Thus, under these assumptions, the bulk cytosolic Ca^{2+} concentration c is proportional to both the number of microdomains and to the microdomain Ca^{2+} concentration c_{md}, and hence also to the average microdomain Ca^{2+} levels $\overline{c_{md}}$, i.e. the average of local Ca^{2+} levels near the Ca^{2+} channels, whether they are open or closed, i.e.

$$c \propto \overline{c_{md}} = m^n h c_{md}. \quad (3.4)$$

During electrical activity or voltage-clamp depolarizations, Ca^{2+} enters the cell via Ca^{2+} channels, and exocytosis is controlled by the non-uniform Ca^{2+} concentrations around Ca^{2+} channels and in the submembrane space. Nonetheless, a common, but erroneous (as will be discussed in details below), way of modeling exocytosis during such conditions, is to let whole-cell exocytosis be a (typically) nonlinear function of c. However, this approach is valid for the study of exocytosis in response to flash-released buffered Ca^{2+}, which elevates the Ca^{2+} concentration uniformly. Under this experimental protocol, the rate of exocytosis is generally well described by a sigmoidal function σ of the Ca^{2+} concentration, which is believed to reflect the intrinsic, biochemical, cooperativity of the exocytotic machinery. In other words, we can write the results of such flash-release experiments as

$$R_c = \sigma(c). \quad (3.5)$$

Typical half-max (K_D) parameters are of the order of tens of μM [13, 14, 16].

This relation has been (erroneously) applied (also by us [7]) to the local control of exocytosis to model whole-cell exocytosis as

$$\tilde{R}_c = \sigma(\overline{\mathrm{Ca}}_{md}) = \sigma(m^n h\ c_{md}), \quad (3.6)$$

under conditions when c is not uniform. However, the local control of exocytosis reflects the intrinsic Ca^{2+} dependence, i.e. single-microdomain exocytosis should be described by

$$R_{md} = \sigma(\mathrm{Ca}_{md})/N_{\mathrm{Ca}}, \quad (3.7)$$

and hence whole-cell release (assuming only release from microdomains with an open channel) is given by

$$R_c = N_{\mathrm{Ca}} m^n h R_{md} = m^n h\ \sigma(c_{md}). \quad (3.8)$$

Whereas equation (3.6) expresses whole-cell exocytosis as a function of average microdomain Ca^{2+}, equation (3.8) expresses whole-cell exocytosis as the sum of microdomain exocytosis. Because of the nonlinear, sigmoidal form of σ, R_c and $\tilde{R}_c$ differ. For example, when few channels are open, $\overline{\mathrm{Ca}}_{md}$ may be only modestly above the bulk Ca^{2+} concentration and well below the K_D of σ, implying $\tilde{R}_c \approx 0$ and almost no exocytosis. In contrast, the more correct interpretation would be the following. If the microdomain Ca^{2+} levels near open Ca^{2+} channels is well above the K_D of σ, then exocytosis would occur rapidly near these open channels, and whole-cell exocytosis could be non negligible. These arguments are treated in more details in exercise 2.

In summary, since some Ca^{2+} channels are open while others are closed during electrical activity or voltage-clamp depolarizations, the Ca^{2+} level near releasable granules is non-uniform. Because of the nonlinear behavior of the exocytotic machinery, whole-cell exocytosis thus can not be modeled as a function of the average microdomain Ca^{2+} concentrations or the bulk cytosolic Ca^{2+} level.

3.4.1 Distinguishing pool depletion from Ca^{2+} channel inactivation

Another widely-used electrophysiological method to study whole-cell exocytosis is to depolarize the cell membrane using voltage clamp and the patch-clamp technique [12]. The depolarization opens Ca^{2+} channels, allowing Ca^{2+} influx and exocytosis. As stated earlier, during depolarization, it is possible to measure the Ca^{2+} current, whereas the cell capacitance C can only be measured reliably before and after, but not during, depolarization. To investigate the dynamics of exocytosis, measured as the increase in capacitance ΔC, it is therefore necessary to apply depolarizing pulses of varying duration. Exocytosis typically proceeds at a higher average rate during shorter depolarizations when compared to longer ones, which has traditionally been interpreted as being the result of the depletion of a small pool of granules located near Ca^{2+} channels, the so-called immediately releasable pool (IRP) [46].

However, the Ca^{2+} currents inactivate on a timescale similar to the decay in the rate of exocytosis. Thus, it may be that exocytosis proceeds more slowly, not because of pool depletion, but rather because the exocytosis-triggering Ca^{2+} signal is reduced towards the end of longer depolarizations as a result of Ca^{2+} current inactivation. It has therefore been suggested that the amount of exocytosis (ΔC) should be related to the total amount of Ca^{2+} that entered the cell during the depolarization Q [47, 48]. For a depolarization of duration t, this quantity is described by the integral of (the absolute value of) the whole cell Ca^{2+}-current I_{Ca}, i.e. $Q(t) = \int_0^t |I_{Ca}|(s)ds$.

Recently, the interaction between pool depletion and Ca^{2+} current inactivation has been theoretically investigated in detail [48]. Assuming a single pool of granules that does not refill on the short time-scales relevant for the depolarization protocols, it can be shown that pool depletion can be revealed by relating the amount of exocytosis (ΔC) to Ca^{2+} entry (Q) as follows.

Let the pool size X be described by

$$\frac{dX}{dt} = -R_c(t)X, \qquad X(0) = X_0, \tag{3.9}$$

where $R_c(t)$ is the whole-cell rate of exocytosis, which depends on the time-varying Ca^{2+} currents, but is assumed not to depend on $X(t)$ (see equation (3.5)). Since refilling of the pool is slow [49–52], it is assumed not to take place during the short (<1 s) depolarizations studied here.

The cumulative capacitance increase during a depolarization, ΔC, reflects the amount of exocytosis and is given by

$$\Delta C(t) = X_0 - X(t), \tag{3.10}$$

since no refilling and no endocytosis are assumed. We can thus conclude, based on equations (3.9) and (3.10), that the explicit formula for the cumulative capacitance (see exercise 3) is

$$\Delta C(t) = X_0\left(1 - \exp\left(-\int_0^t R_c(u)du\right)\right). \quad (3.11)$$

One Ca^{2+}-channel type, one channel per microdomain

When granules are located in Ca^{2+}-microdomains at the inner mouth of Ca^{2+}-channels, the rate of exocytosis is controlled by the microdomain (MD) Ca^{2+}-concentration, c_{md}. Assume that all Ca^{2+}-channels are identical and spatially discrete. Then c_{md} below an open Ca^{2+}-channel is approximately proportional to the single-channel current i_{Ca} [43, 53]. During voltage-clamp depolarizations, the whole-cell Ca^{2+}-current I_{Ca} might inactivate, which is caused by the closure of single Ca^{2+}-channels, i.e. the number of open channels decrease during the depolarization but with no change to the single-channel current i_{Ca} through the remaining open channels. As discussed above, this implies that the number of MDs $N_{md}(t) = N_{Ca}m^n h$, which is proportional to $I_{Ca}(t)$, decreases, but the Ca^{2+}-concentration is unchanged in the remaining MDs. The rate of exocytosis from each MD, $R_{md} = R_{md}(c_{md})$, is therefore constant until the collapse of the MD due to the closure of the corresponding Ca^{2+}-channel, independently of the functional form of the relation between c_{md} and R_{md}. The total rate of exocytosis can then be expressed as (see equation (3.8))

$$E(t) = R_{md}N_{md}(t) = R_{md}I_{Ca}(t)/i_{Ca} = A\,|I_{Ca}|(t), \quad (3.12)$$

where $A = R_{md}/|i_{Ca}|$. From equation (3.11), we can conclude that

$$\Delta C(t) = X_0[1 - \exp(-A\,Q(t))]. \quad (3.13)$$

Note that equation (3.13) describes simple first-order pool kinetics when time is rescaled to $Q(t)$.

If $A\,Q(t)$ is small, i.e. if the cumulative exocytotic rate is sufficiently low, then the granule pool does not deplete substantially, and by Taylor expansion, equation (3.13) can be approximated by

$$\Delta C(t) = X_0 A\,Q(t). \quad (3.14)$$

Thus, in the case of no pool depletion and with exocytosis of each granule controlled by a single channel, cumulative exocytosis is linearly related to total Ca^{2+} entry [54]. However, if the granule pool is eventually depleted, then ΔC will be a concave function of Q (figure 3.2). Hence, ΔC must be analyzed as a function of Q to decide whether pool depletion occurs, since Ca^{2+}-channel inactivation might masquerade as pool depletion when considering capacitance increase as a function of depolarization length. Similar considerations and conclusions are reached also for more complex scenarios with, for example, more Ca^{2+} channels per granule, different Ca^{2+} channel types, etc [48].

Interestingly, when revisiting previously published results that concluded that beta cells possess a small, depletable pool based on capacitance-versus-duration analysis, we found that ΔC is linearly related to Q, suggesting that pool depletion is absent in single mouse beta cells [55]. In contrast, in human beta cells on the surface

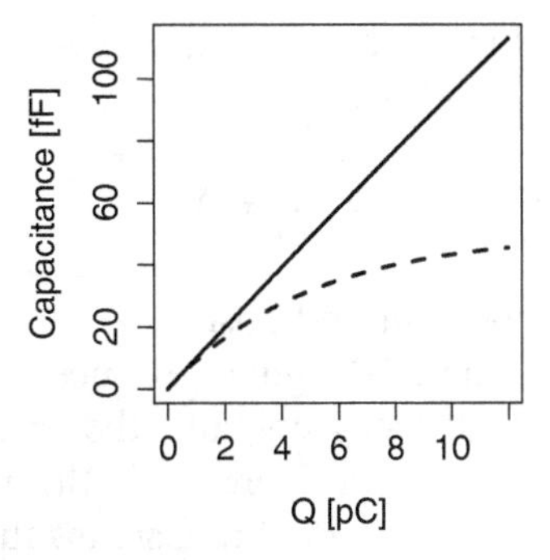

Figure 3.2. Predicted relation between Ca^{2+} influx Q and exocytosis ΔC in the presence (dashed curve) or absence (solid curve) of pool depletion.

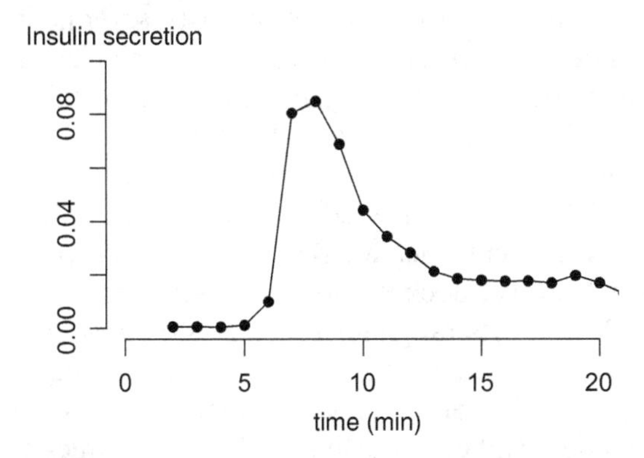

Figure 3.3. Insulin secretion measurements (% of islet content per minute) as a function of time (in minutes) following a step from 3 to 15 mM glucose at time t = 5 min. Data from [57].

of intact pancreatic islets, ΔC is a concave function of Q [56], suggesting that *in situ* human beta cells exhibit pool depletion.

3.5 Granule pool dynamics

Following insulin synthesis and packaging, secretory granules are transported to the cell membrane where they might attach to the membrane (denoted 'docking'). As a consequence of molecular and biochemical processes that are still poorly understood (denoted 'priming'), the docked granules become release-ready. The primed, release-ready granules are typically considered as a *readily releasable pool* (RRP), and it has been suggested that exocytosis of the RRP causes the first phase of insulin secretion seen in response to a step in plasma glucose levels [6] (see figure 3.3). The slower docking and priming processes then refill the RRP gradually, which causes the second phase of insulin secretion.

The simplest mathematical model corresponding to this scenario considers a single pool, the RRP, which we denote X in agreement with equation (3.9) above.

We add refilling due to docking and priming, which we describe by a variable M (representing mobilization), which yields

$$\frac{dX}{dt} = M - R_c X, \qquad X(0) = X_0, \qquad SR = R_c X, \tag{3.15}$$

where we assume that R_c is constant and positive (zero) in the presence (absence) of glucose, whereas M is a function of the glucose concentration. We have introduced the secretion rate SR, which depends on both the rate of exocytosis R_c and the number of releasable granules X. In exercise 5, the reader needs to show that equation (3.15) is sufficient for observing biphasic insulin secretion.

At a first glance, this *storage limiting* hypothesis of biphasic insulin release appears to be consistent with electrophysiological data suggesting the presence of a depletable RRP of insulin granules in single beta cells [16, 56]. However, it should be noted that the first phase of insulin secretion lasts several minutes (figure 3.3), whereas the RRP is depleted in less than a second in single cell recordings. These widely different time scales suggest that great care must be taken when interpreting the concept of pool depletion in these different experimental settings: single beta cells versus groups of islets or the entire pancreas.

Mathematical modeling has been used to link and interpret insulin release data with such different experimental scales. Bertuzzi *et al* [58] presented a model that considers four pools (reserve, docked, primed, fused) that are coupled by mechanistically meaningful processes ranging from very slow insulin synthesis and packaging, slow translocation and docking, relatively fast priming, and very fast fusion once a triggering signal in the form of calcium arrives. Chen *et al* [59] used a similar model that also included explicit modeling of Ca^{2+} dynamics to investigate where the so-called amplifying pathway [60] (which augments the amount of insulin released at a given Ca^{2+} level) should act in the pathway preparing granules for exocytosis. These models were able to reproduce several sets of experimental data ranging from capacitance recordings of exocytosis to insulin measurements occurring on a time scale of tens of minutes or hours, in various experimental settings including knock-out animals.

3.5.1 Newcomer granules and the highly Ca^{2+}-sensitive pool

We present a model [7] based on the work by Chen *et al* [59], which illustrates how modeling can be used to interpret apparently disparate experimental findings in a single, quantitative framework.

Two early studies [18, 19] have shown that substantial amounts of insulin exocytosis are triggered by elevating Ca^{2+} levels to a few μM, an order of magnitude lower than the concentration previously thought necessary for exocytosis [15, 16]. Thus, some granules appeared to have a different biophysical composition leading to a higher Ca^{2+} sensitivity of the exocytotic machinery, and this pool of granules was consequently termed the highly Ca^{2+}-sensitive pool (HCSP) [18, 19]. Since the HCSP was not depleted by depolarizations causing Ca^{2+} influx via Ca^{2+} channels, it was concluded that the HCSP is located outside microdomains controlling release [18].

Around the same time, an alternative view on exocytosis emerged from a completely different experimental technique. TIRF imaging of insulin exocytosis revealed that secretory granules do not only fuse following docking and priming, but may also undergo exocytosis immediately upon arrival at the plasma membrane [28, 61], in what has been dubbed *crash fusion*. Exocytosis of these so-called *newcomer* granules occurred mostly away from clusters of the SNARE protein syntaxin-1 [62], which is known to colocalize with Ca^{2+} channels [63]. Hence, newcomer granules undergo crash-fusion away from Ca^{2+} channels, outside Ca^{2+} microdomains. It should be noted that some experimental groups do not see newcomer granules [10, 29, 36], and that the very existence of such granules is debated.

Nonetheless, based on the spatial agreement between the HCSP and newcomer granules, it was suggested that they are in fact the same granule populations detected using different techniques [7]. Indeed, any exocytosis occurring away from Ca^{2+} channels must have high Ca^{2+} sensitivity since the Ca^{2+} concentration only few hundred nm away from the Ca^{2+} channels is at most a few μM. Also, granules reach the membrane, to engage in docking, at sites with low Ca^{2+} channel density [10], and hence crash fusion of newcomer granules would appear at sites with low Ca^{2+} levels.

We modeled this scenario by inserting a HCSP at the point when granules reach the cell membrane, into the model by Chen *et al* [59] (see figure 3.4). Exocytosis from the HCSP is controlled by bulk cytosolic Ca^{2+}, whereas IRP exocytosis is triggered

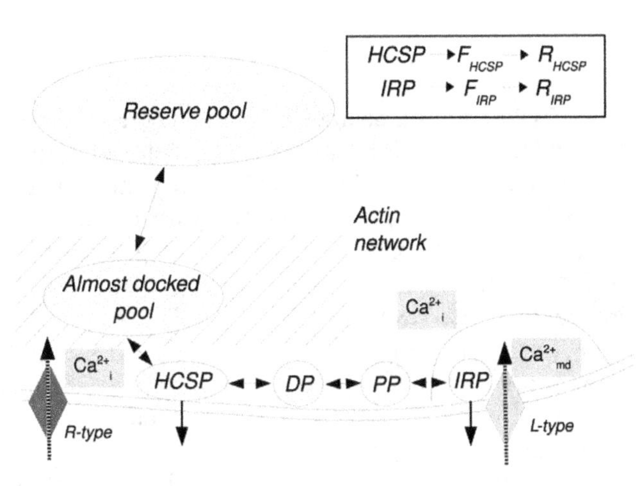

Figure 3.4. A schematic overview of the pool model including HCSP [7]. Granules from a reserve pool, via the almost docked pool, reach the cell membrane they are assumed to tether weakly to and to fuse with high affinity for bulk cytosolic Ca^{2+} (Ca^{2+}_i). Hence these granules are identified with the highly Ca^{2+} sensitive pool (*HCSP*). Tethered granules can mature further by docking (*DP* = docked pool), undergo priming (*PP* = primed pool), and attach to L-type Ca^{2+}-channels, thus entering the immediately releasable pool (*IRP*). From the IRP, granules can fuse with low affinity for microdomain Ca^{2+} (Ca^{2+}_{md}). Fusion from both HCSP and IRP are assumed to follow a Hill-function of calcium. Inset: after fusion, the granules enter a 'fused pool' (F_{HCSP} or F_{IRP}). The fusion pore can then expand, after which the granule belongs to a 'releasing pool' (R_{HCSP} or R_{IRP}). The insulin secretion rate is defined as the release flux from the two releasing pools.

by (averaged) microdomain Ca^{2+}. This model was then used to reproduce and explain a series of experimental data, including patch clamp data and insulin secretion measurements under various experimental circumstances, including knock-out animals that lack a certain protein, e.g. a certain type of Ca^{2+} channel or an exocytotic protein [7].

This model reveals typical aspects of use and construction of mathematical models of cell biology. Reproduction of available data constitutes a 'consistency check' of the hypothesis underlying the model; in the present case: is the idea of newcomer granules as a HCSP consistent with a wide range of data sets? In addition, the model was used to simulate non-observed data such a exocytosis dynamics in certain knock-out cells. Similarly, models are frequently used to simulate answers to 'what-if' questions by modifying parameter values that correspond to certain hypotheses that can or have not been experimentally tested.

Variations of the presented pool-based models that handle entire β-cell populations are given in the subsequent chapters of this book.

Exercises

1. Derive equations (3.3) and (3.4) from equations (3.1) and (3.2).
2. Assume that the sigmoidal relation σ in equation (3.5) between Ca^{2+} levels c and exocytosis is given by a Hill function

$$R_c = \sigma(c) = R_{c,\max} \frac{c^n}{K_D^n + c^n},$$

with $K_D = 17.3$ μM and Hill coefficient $n = 5$. These parameters correspond to results obtained with flash-released Ca^{2+} in beta cells [16].

   (a) Assume that the intracellular Ca^{2+} concentration is uniformly raised to a level c. Using a computer software (or draw by hand), plot normalized exocytosis $R_c/R_{c,\max}$ as a function of c for $0\ \mu M \leqslant c \leqslant 50\ \mu M$.
   (b) Assume now that exocytosis is controlled by Ca^{2+} microdomains below open Ca^{2+} channels. Furthermore, assume that all Ca^{2+} channels are open and that the microdomain Ca^{2+} concentration is $c_{md} = 30$ nμM. What is the normalized exocytosis rate? Add lines to the plot that indicate this calculation.
   (c) Suppose that there is a one-to-one relation between Ca^{2+} channels, Ca^{2+} microdomains, and releasable granules, and that 1/6 of Ca^{2+} channels are open. What is the average microdomain Ca^{2+} concentration $\overline{c_{md}}$? What are the normalized whole-cell exocytosis rates $\tilde{R}_c/R_{c,\max}$ (see equation (3.6)) and $R_c/R_{c,\max}$ (see equation (3.8))? Which one gives the correct amount of normalized whole-cell exocytosis? Explain your reasoning graphically using the plot.
3. Derive equation (3.11) from equations (3.9) and (3.10).
4. Assume that the Ca^{2+} current is described by equation (3.2).

(a) Simulate the pulse-length protocol: assume that the voltage V is stepped from a *holding potential* of −70 mV to a *test potential* of 0 mV for a *pulse duration* Δt of 10, 20, 50, 100, 200, 400 or 800 ms, and then stepped back to the holding potential. Plot $I_{Ca}(t)$ for the different pulse durations. These currents can be monitored experimentally. Calculate $Q(\Delta t)$ for the different durations.

(b) Assume that exocytosis is described by equation (3.8) with $c_{md} = 30$ μM and σ as in exercise 2, and that pool depletion is absent so that ΔC is proportional to R_c. Assume for simplicity that $\Delta C = R_c / R_{c,\max}$.

i. Plot $\Delta C(t)$ for each pulse duration. Note that only the final value $\Delta C(\Delta t)$ can be measured experimentally.

ii. Plot $\Delta C(\Delta t)$ as a function of Δt. Plot $\Delta C(\Delta t)$ as a function of $Q(\Delta t)$.

iii. What do we mean by 'Ca^{2+}-channel inactivation might masquerade as pool depletion when considering capacitance increase as a function of depolarization length'? Does the plot of $\Delta C(\Delta t)$ as a function of $Q(\Delta t)$ help?

5. Assume that the glucose concentration G is stepped from $G = 0$ mM to $G = 10$ mM at time $t = 0$ min. Suppose that $M = R_c = 0$ min^{-1} for $t < 0$ min, and $X_0 = 100$ (granules/cell). Assuming that $R_c = 0.3$ min^{-1} and $M = 15$ min^{-1} at $G = 10$ mM, solve equation (3.15) numerically. Do you see biphasic secretion? Show analytically that the secretion rate SR tends to M with time scale of $1/R_c$.

Assume now that mobilization occurs with a delay following the step in glucose concentration, as determined by the equation

$$\frac{dM}{dt} = \frac{M_\infty - M}{\tau_M},$$

with $M_\infty = 15$ min^{-1} and $\tau_M = 30$ min. Show that the second phase is now rising rather than constant. Considering the time scales, justify that for the second phase, we may assume $\frac{dX}{dt} \approx 0$, and that the secretion rate during the second phase is again approximately equal to M.

References

[1] Misler S 2009 Unifying concepts in stimulus-secretion coupling in endocrine cells and some implications for therapeutics *Adv. Physiol. Educ.* **33** 175–86

[2] Barg S 2003 Mechanisms of exocytosis in insulin-secreting B-cells and glucagon-secreting A-cells *Pharmacol. Toxicol.* **92** 3–13

[3] Burgoyne R D and Morgan A 2003 Secretory granule exocytosis *Physiol. Rev.* **83** 581–632

[4] Grodsky G M 1972 A threshold distribution hypothesis for packet storage of insulin and its mathematical modeling *J. Clin. Invest.* **51** 2047–59

[5] Voets T 2000 Dissection of three Ca^{2+}-dependent steps leading to secretion in chromaffin cells from mouse adrenal slices *Neuron* **28** 537–45

[6] Rorsman P and Renström E 2003 Insulin granule dynamics in pancreatic beta cells *Diabetologia* **46** 1029–45

[7] Pedersen M G and Sherman A 2009 Newcomer insulin secretory granules as a highly calcium-sensitive pool *Proc. Natl Acad. Sci. USA* **106** 7432–6

[8] Neher E 2015 Merits and limitations of vesicle pool models in view of heterogeneous populations of synaptic vesicles *Neuron* **87** 1131–42

[9] Barg S, Huang P, Eliasson L, Nelson D J, Obermüller S, Rorsman P, Thévenod F and Renström E 2001 Priming of insulin granules for exocytosis by granular Cl^- uptake and acidification *J. Cell Sci.* **114** 2145–54 https://jcs.biologists.org/content/114/11/2145.long

[10] Gandasi N R and Barg S 2014 Contact-induced clustering of syntaxin and munc18 docks secretory granules at the exocytosis site *Nat. Commun.* **5** 3914

[11] Neher E and Sakaba T 2008 Multiple roles of calcium ions in the regulation of neurotransmitter release *Neuron* **59** 861–72

[12] Lindau M and Neher E 1988 Patch-clamp techniques for time-resolved capacitance measurements in single cells *Pflugers Arch.* **411** 137–46

[13] Thomas P, Wong J G, Lee A K and Almers W 1993 A low affinity Ca^{2+} receptor controls the final steps in peptide secretion from pituitary melanotrophs *Neuron* **11** 93–104

[14] Heinemann C, Chow R H, Neher E and Zucker R S 1994 Kinetics of the secretory response in bovine chromaffin cells following flash photolysis of caged Ca^{2+} *Biophys. J.* **67** 2546–57

[15] Takahashi N, Kadowaki T, Yazaki Y, Miyashita Y and Kasai H 1997 Multiple exocytotic pathways in pancreatic beta cells *J. Cell Biol.* **138** 55–64

[16] Barg S *et al* 2001 Fast exocytosis with few Ca^{2+} channels in insulin-secreting mouse pancreatic B cells *Biophys. J.* **81** 3308–23

[17] Yang Y, Udayasankar S, Dunning J, Chen P and Gillis K D 2002 A highly Ca^{2+}-sensitive pool of vesicles is regulated by protein kinase c in adrenal chromaffin cells *Proc. Natl Acad. Sci. USA* **99** 17060–5

[18] Yang Y and Gillis K D 2004 A highly Ca^{2+}-sensitive pool of granules is regulated by glucose and protein kinases in insulin-secreting INS-1 cells *J. Gen. Physiol.* **124** 641–51

[19] Wan Q-F, Dong Y, Yang H, Lou X, Ding J and Xu T 2004 Protein kinase activation increases insulin secretion by sensitizing the secretory machinery to Ca^{2+} *J. Gen. Physiol.* **124** 653–62

[20] Takahashi N, Kishimoto T, Nemoto T, Kadowaki T and Kasai H 2002 Fusion pore dynamics and insulin granule exocytosis in the pancreatic islet *Science* **297** 1349–52

[21] Obermüller S *et al* 2010 Defective secretion of islet hormones in chromogranin-b deficient mice *PLoS One* **5** e8936

[22] Low J T, Mitchell J M, Do O H, Bax J, Rawlings A, Zavortink M, Morgan G, Parton R G, Gaisano H Y and Thorn P 2013 Glucose principally regulates insulin secretion in mouse islets by controlling the numbers of granule fusion events per cell *Diabetologia* **56** 2629–37

[23] Michael D J, Geng X, Cawley N X, Loh Y P, Rhodes C J, Drain P and Chow R H 2004 Fluorescent cargo proteins in pancreatic beta-cells: design determines secretion kinetics at exocytosis *Biophys. J.* **87** L03–5

[24] Gandasi N R, Vestö K, Helou M, Yin P, Saras J and Barg S 2015 Survey of red fluorescence proteins as markers for secretory granule exocytosis *PLoS One* **10** e0127801

[25] Lang T, Wacker I, Steyer J, Kaether C, Wunderlich I, Soldati T, Gerdes H H and Almers W 1997 Ca^{2+}-triggered peptide secretion in single cells imaged with green fluorescent protein and evanescent-wave microscopy *Neuron* **18** 857–63

[26] Steyer J A, Horstmann H and Almers W 1997 Transport, docking and exocytosis of single secretory granules in live chromaffin cells *Nature* **388** 474–8

[27] Oheim M, Loerke D, Stühmer W and Chow R H 1998 The last few milliseconds in the life of a secretory granule. docking, dynamics and fusion visualized by total internal reflection fluorescence microscopy (TIRFM) *Eur. Biophys. J.* **27** 83–98

[28] Ohara-Imaizumi M, Nakamichi Y, Tanaka T, Ishida H and Nagamatsu S 2002 Imaging exocytosis of single insulin secretory granules with evanescent wave microscopy: distinct behavior of granule motion in biphasic insulin release *J. Biol. Chem.* **277** 3805–8

[29] Michael D J, Xiong W, Geng X, Drain P and Chow R H 2007 Human insulin vesicle dynamics during pulsatile secretion *Diabetes* **56** 1277–88

[30] Barg S, Knowles M K, Chen X, Midorikawa M and Almers W 2010 Syntaxin clusters assemble reversibly at sites of secretory granules in live cells *Proc. Natl Acad. Sci. USA* **107** 20804–9

[31] Toomre D 2012 Generating live cell data using total internal reflection fluorescence microscopy *Cold Spring Harb. Protoc.* **2012** 439–46

[32] Midorikawa M and Sakaba T 2015 Imaging exocytosis of single synaptic vesicles at a fast CNS presynaptic terminal *Neuron* **88** 492–8

[33] Becherer U, Moser T, Stühmer W and Oheim M 2003 Calcium regulates exocytosis at the level of single vesicles *Nat. Neurosci.* **6** 846–53

[34] Hoppa M B, Collins S, Ramracheya R, Hodson L, Amisten S, Zhang Q, Johnson P, Ashcroft F M and Rorsman P 2009 Chronic palmitate exposure inhibits insulin secretion by dissociation of Ca^{2+} channels from secretory granules *Cell Metab.* **10** 455–65

[35] Gandasi N R *et al* 2017 Ca^{2+} channel clustering with insulin-containing granules is disturbed in type 2 diabetes *J. Clin. Invest.* **127** 2353–64

[36] Trexler A J, Sochacki K A and Taraska J W 2016 Imaging the recruitment and loss of proteins and lipids at single sites of calcium-triggered exocytosis *Mol. Biol. Cell* **27** 2423–34

[37] Katz B and Miledi R 1965 The effect of calcium on acetylcholine release from motor nerve terminals *Proc. R. Soc. Lond. B Biol. Sci.* **161** 496–503

[38] Douglas W W and Rubin R P 1961 The role of calcium in the secretory response of the adrenal medulla to acetylcholine *J. Physiol.* **159** 40–57

[39] Douglas W W 1968 Stimulus-secretion coupling: the concept and clues from chromaffin and other cells *Br. J. Pharmacol.* **34** 451–74

[40] Schneggenburger R, Han Y and Kochubey O 2012 Ca^{2+} channels and transmitter release at the active zone *Cell Calcium* **52** 199–207

[41] Chad J E and Eckert R 1984 Calcium domains associated with individual channels can account for anomalous voltage relations of CA-dependent responses *Biophys. J.* **45** 993–9

[42] Simon S M and Llinás R R 1985 Compartmentalization of the submembrane calcium activity during calcium influx and its significance in transmitter release *Biophys. J.* **48** 485–98

[43] Neher E 1998 Vesicle pools and Ca^{2+} microdomains: new tools for understanding their roles in neurotransmitter release *Neuron* **20** 389–99

[44] Neher E 1998 Usefulness and limitations of linear approximations to the understanding of Ca^{++} signals *Cell Calcium* **24** 345–57

[45] Klingauf J and Neher E 1997 Modeling buffered Ca^{2+} diffusion near the membrane: implications for secretion in neuroendocrine cells *Biophys. J.* **72** 674–90

[46] Horrigan F T and Bookman R J 1994 Releasable pools and the kinetics of exocytosis in adrenal chromaffin cells *Neuron* **13** 1119–29

[47] Engisch K L and Nowycky M C 1996 Calcium dependence of large dense-cored vesicle exocytosis evoked by calcium influx in bovine adrenal chromaffin cells *J. Neurosci.* **16** 1359–69

[48] Pedersen M G 2011 On depolarization-evoked exocytosis as a function of calcium entry: possibilities and pitfalls *Biophys. J.* **101** 793–802

[49] Moser T and Neher E 1997 Rapid exocytosis in single chromaffin cells recorded from mouse adrenal slices *J. Neurosci.* **17** 2314–23

[50] Gromada J, Høy M, Renström E, Bokvist K, Eliasson L, Göpel S and Rorsman P 1999 CaM kinase II-dependent mobilization of secretory granules underlies acetylcholine-induced stimulation of exocytosis in mouse pancreatic B-cells *J. Physiol.* **518** 745–59

[51] Voets T, Neher E and Moser T 1999 Mechanisms underlying phasic and sustained secretion in chromaffin cells from mouse adrenal slices *Neuron* **23** 607–15

[52] Rose T, Efendic S and Rupnik M 2007 Ca^{2+}-secretion coupling is impaired in diabetic Goto Kakizaki rats *J. Gen. Physiol.* **129** 493–508

[53] Sherman A, Keizer J and Rinzel J 1990 Domain model for Ca^{2+}-inactivation of Ca^{2+} channels at low channel density *Biophys. J.* **58** 985–95

[54] Augustine G J, Adler E M and Charlton M P 1991 The calcium signal for transmitter secretion from presynaptic nerve terminals *Ann. N. Y. Acad. Sci.* **635** 365–81

[55] Pedersen M G, Cortese G and Eliasson L 2011 Mathematical modeling and statistical analysis of calcium-regulated insulin granule exocytosis in β-cells from mice and humans *Prog. Biophys. Mol. Biol.* **107** 257–64

[56] Rorsman P and Braun M 2013 Regulation of insulin secretion in human pancreatic islets *Annu. Rev. Physiol.* **75** 155–79

[57] Pedersen M G, Tagliavini A and Henquin J-C 2019 Calcium signaling and secretory granule pool dynamics underlie biphasic insulin secretion and its amplification by glucose: experiments and modeling *Am. J. Physiol. Endocrinol. Metab.* **316** E475–86

[58] Bertuzzi A, Salinari S and Mingrone G 2007 Insulin granule trafficking in beta-cells: mathematical model of glucose-induced insulin secretion *Am. J. Physiol. Endocrinol. Metab.* **293** E396–409

[59] Chen Y, Wang S and Sherman A 2008 Identifying the targets of the amplifying pathway for insulin secretion in pancreatic beta-cells by kinetic modeling of granule exocytosis *Biophys. J.* **95** 2226–41

[60] Henquin J C 2009 Regulation of insulin secretion: a matter of phase control and amplitude modulation *Diabetologia* **52** 739–51

[61] Ohara-Imaizumi M, Nishiwaki C, Kikuta T, Nagai S, Nakamichi Y and Nagamatsu S 2004 TIRF imaging of docking and fusion of single insulin granule motion in primary rat pancreatic beta-cells: different behaviour of granule motion between normal and Goto-Kakizaki diabetic rat beta-cells *Biochem. J.* **381** 13–8

[62] Ohara-Imaizumi M *et al* 2007 Imaging analysis reveals mechanistic differences between first- and second-phase insulin exocytosis *J. Cell Biol.* **177** 695–705

[63] Yang S N, Larsson O, Bränström R, Bertorello A M, Leibiger B, Leibiger I B, Moede T, Köhler M, Meister B and Berggren P O 1999 Syntaxin 1 interacts with the L_D subtype of voltage-gated ca^{2+} channels in pancreatic beta cells *Proc. Natl Acad. Sci. USA* **96** 10164–9

[64] Pedersen M G, Tagliavini A, Cortese G, Riz M and Montefusco F 2017 Recent advances in mathematical modeling and statistical analysis of exocytosis in endocrine cells *Math. Biosci.* **283** 60–70

Part II

Modeling islet biology

Chapter 4

Islet architecture

Junghyo Jo

4.1 Islets of Langerhans

Pancreatic islets of Langerhans have special structures that appear to be important for controlling blood glucose levels. As indicated earlier, this micro-organ is composed of mainly glucagon-secreting alpha cells, insulin-secreting beta cells, and somatostatin-secreting delta cells. Since glucagon and insulin counter-regulate each other to control glucose levels, a special coordination between the two hormones is thus imperative. Their opposite response to glucose concentration guarantees their principal coordination. In addition to their global response to glucose, islet cells interact with each other through auto- and paracrine interactions. The physiological role of the cellular interactions is not yet clear, but their existence has been intensively observed [1]. The existence of such cellular interactions thus suggests the spatial distribution of islet cells should have functional implications.

Interestingly, it has been found that rodent islets have a different structure compared to human islets. While rodent islets have a beta-cell core and non-beta-cell shell structure, human islets have a mixed structure where alpha, beta, and delta cells are intermingled [2, 3]. In general, human islets have less abundant beta cells than rodent islets. Thus it is an intriguing question to determine whether the structural difference between human and rodent islets originates from the mere compositional difference or from the fundamental difference in their organization rule. How do islet cells develop to have a given organized structure within an organism? One would expect that programmed sequence of developmental processes such as cell differentiation, proliferation, and deletion are required to form the special islet structure. However, when islet cells were isolated from islets and kept in culture solution, they aggregated and spontaneously formed islet-like structures, called pseudo-islets [4, 5]. This observation as a result nullifies the contribution of the sequential development, and instead emphasizes the contribution of relative adhesion between islet cells given their motility. Indeed, the differential adhesion hypothesis (DAH) [6] can explain the spatial organization of islet cells. In this

doi:10.1088/978-0-7503-3739-7ch4

section, we introduce DAH, and show how DAH produces different organ structures. Finally, we explain how to infer the relative adhesion strengths between islet cells given islet structures.

4.2 Differential adhesion hypothesis

How do biological organs form their sophisticated structures? It has been suggested that self-organization is the key to explaining of biological organigenesis. Given relative stickiness between elements which can be determined by cell adhesion molecules on cell membrane, different populations of cells spontaneously form a well-define structure. Steinberg has proposed DAH [6], and used it to explain many organ development [7]. It has been shown that different endocrine cells in pancreatic islets express different adhesion molecules [8]. In addition, pseudo-islet formation has been also observed from dissociated islet cells. These observations suggest that islet formation may follow the DAH mechanism.

To specifically understand the DAH, let us consider a one-dimensional string with two cell species, α and β. The contact score can be formulated as

$$-E = J_{\alpha\alpha}n_{\alpha\alpha} + J_{\beta\beta}n_{\beta\beta} + J_{\alpha\beta}n_{\alpha\beta}, \tag{4.1}$$

where $n_{\sigma\sigma'}$ is the number of contacts between σ and σ' cell types for $\sigma, \sigma' \in \{\alpha, \beta, \delta\}$, and $J_{\sigma\sigma'}$ is the relative stickiness between them. Given $J_{\alpha\alpha}$, $J_{\beta\beta}$, and $J_{\alpha\beta}$, the minimal contact score (or energy) E determines the equilibrium structure of the string. In the case of a string of length 4, two extreme structures can form

$$\mathbf{s}_1: \alpha - \alpha - \beta - \beta$$
$$\mathbf{s}_2: \alpha - \beta - \alpha - \beta.$$

The first structure prefers homologous contacts ($J_{\alpha\alpha}$, $J_{\beta\beta} > J_{\alpha\beta}$), while the second structure prefers heterologous contacts ($J_{\alpha\alpha}$, $J_{\beta\beta} < J_{\alpha\beta}$). Their contact energies are

$$E(\mathbf{s}_1) = -J_{\alpha\alpha} - J_{\alpha\beta} - J_{\beta\beta}, \tag{4.2}$$

$$E(\mathbf{s}_2) = -3J_{\alpha\beta}. \tag{4.3}$$

The structural preference is determined by the balance between $E(\mathbf{s}_1)$ and $E(\mathbf{s}_2)$. Thus their equality $E(\mathbf{s}_1) = E(\mathbf{s}_2)$ gives the transition condition,

$$J_{\alpha\beta} = \frac{J_{\alpha\alpha} + J_{\beta\beta}}{2}, \tag{4.4}$$

between the two structures.

4.3 Analytical method

To understand the statistical behavior of the contact configurations $n_{\alpha\alpha}(\mathbf{s})$, $n_{\beta\beta}(\mathbf{s})$, and $n_{\alpha\beta}(\mathbf{s})$ of a string $\mathbf{s}$, we adopt approaches from statistical mechanics [9]. First, we interpret $E(\mathbf{s})$ as the *energy* of a string $\mathbf{s}$ based on the adhesion between cells. Then, we consider an equilibrium condition in which the string has an average adhesion energy $\langle E \rangle$. The motility of cells allows fluctuations in $E(\mathbf{s})$ through changing string configuration $\mathbf{s}$. However, their expectation value (or average) remains constant with $\langle E \rangle$ at equilibrium. Given this equilibrium constraint, the maximum likelihood energy distribution follows the canonical Boltzmann distribution

$$P(\mathbf{s}) = \frac{e^{-\lambda E(\mathbf{s})}}{Z}, \tag{4.5}$$

where Z is a normalization factor (for the full derivation of equation (4.5), see exercise 1b). The auxiliary parameter λ, controlling cell motility, can be determined from the constraint condition:

$$\langle E \rangle \equiv \sum_{\mathbf{s}} E(\mathbf{s})P(\mathbf{s}) = \frac{\sum_{\mathbf{s}} E(\mathbf{s})e^{-\lambda E(\mathbf{s})}}{Z}. \tag{4.6}$$

The normalization factor,

$$Z \equiv \sum_{\mathbf{s}} e^{-\lambda E(\mathbf{s})} = \sum_{\mathbf{s}} e^{j_{\alpha\alpha}n_{\alpha\alpha}(\mathbf{s})+j_{\beta\beta}n_{\beta\beta}(\mathbf{s})+j_{\alpha\beta}n_{\alpha\beta}(\mathbf{s})}, \tag{4.7}$$

is called a partition function (or generating function), which is very useful for obtaining important physical quantities. Note that, hereafter, we use dimensionless relative stickiness $j_{\sigma\sigma'} \equiv \lambda J_{\sigma\sigma'}$. Once the partition function is obtained, one can calculate the mean contact numbers and their variance by simply differentiating $\log Z$, as follows

$$\bar{n}_{\alpha\alpha} = \frac{\partial \log Z}{\partial j_{\alpha\alpha}} = \sum_{\mathbf{s}} n_{\alpha\alpha}(\mathbf{s})P(\mathbf{s}), \tag{4.8}$$

$$\delta n_{\alpha\alpha}^2 = \frac{\partial^2 \log Z}{\partial j_{\alpha\alpha}^2} = \sum_{\mathbf{s}} n_{\alpha\alpha}^2(\mathbf{s})P(\mathbf{s}) - \bar{n}_{\alpha\alpha}^2. \tag{4.9}$$

The mean and the variance of $n_{\beta\beta}$ and $n_{\alpha\beta}$ can be obtained similarly.

To further understand this statistical mechanics method, we will apply it to a simple string composed of two species, alpha and beta cells, in one dimension (figure 4.1). Given a string configuration $\mathbf{s}$, $n_{\alpha\alpha}$, $n_{\beta\beta}$, and $n_{\alpha\beta}$ are not independent. The total contact number is one less than the total cell number: $n_{\alpha\alpha} + n_{\beta\beta} + n_{\alpha\beta} = n_\alpha + n_\beta - 1$. This means that if one knows the heterologous contact number $n_{\alpha\beta}$, the homologous contact numbers $n_{\alpha\alpha}$ and $n_{\beta\beta}$ can be constrained. In particular, when $n_{\alpha\beta}$ has even numbers, $n_{\alpha\alpha}$ and $n_{\beta\beta}$ can be uniquely determined. On the other hand, when $n_{\alpha\beta}$ has odd numbers, $n_{\alpha\alpha}$ and $n_{\beta\beta}$ can have two possibilities. For example, suppose one has three alpha cells and three beta cells ($n_\alpha = n_\beta = 3$). Then the adhesion energy of a string $\mathbf{s}$ depends on the grouping and mixing of alpha and beta cells (table 4.1). In general, given n_α and n_β, one can consider every possible configuration $\mathbf{s}$, and obtain the partition function Z in equation (4.7). Since $E(\mathbf{s})$ is determined by $\mathbf{n} = (n_{\alpha\beta}, n_{\beta\beta}, n_{\alpha\beta})$, one can describe the partition function for $\mathbf{n}$ and its degeneracy $g(\mathbf{n})$ instead of considering every $\mathbf{s}$ itself, i.e.

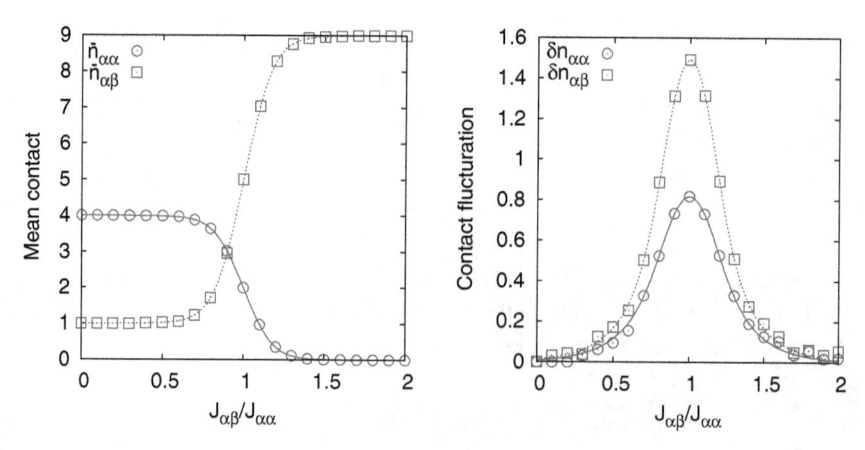

Figure 4.1. Cell-to-cell contact number and fluctuations. Five alpha cells and five beta cells form a string configuration in one dimension. For this simulation, $J_{\alpha\alpha} = J_{\beta\beta} = 1$ and $J_{\alpha\beta}$ is varied with $\lambda = 10$. Lines represent analytical calculations, while symbols represent numerical results from the Monte-Carlo sampling method with 10^5 samples.

Table 4.1. Configuration degeneracies for the one-dimensional sequence of three alpha cells and three beta cells. Every string configuration is categorized by the heterologous contact number $n_{\alpha\beta}$. Here alpha cells and beta cells are grouped by introducing partitions (bars). Then, the grouped alpha and beta cells are mixed by inserting into the predefined partitions. For each category, its degeneracy is the multiplication of possible alpha-cell partitioning and beta-cell partitioning. The degeneracy factor 2 for odd $n_{\alpha\beta}$ represents two possible orders starting from either alpha or beta cells. $\binom{n}{m}$ represents the combination number selecting m elements from a collection of n elements.

$n_{\alpha\beta}$	Category	Degeneracy	Configurations
1	$\alpha\alpha\alpha\ \beta\beta\beta$	$2\binom{2}{0}\binom{2}{0}$	$\alpha\alpha\alpha\beta\beta\beta$, $\beta\beta\beta\alpha\alpha\alpha$
2	$\alpha\|\alpha\alpha\ \beta\beta\beta$	$\binom{2}{1}\binom{2}{0}$	$\alpha\beta\beta\beta\alpha\alpha$, $\alpha\alpha\beta\beta\beta\alpha$
	$\alpha\alpha\alpha\ \beta\|\beta\beta$	$\binom{2}{0}\binom{2}{1}$	$\beta\alpha\alpha\alpha\beta\beta$, $\beta\beta\alpha\alpha\alpha\beta$
3	$\alpha\|\alpha\alpha\ \beta\|\beta\beta$	$2\binom{2}{1}\binom{2}{1}$	$\alpha\beta\alpha\alpha\beta\beta$, $\beta\alpha\beta\beta\alpha\alpha$, $\alpha\alpha\beta\alpha\beta\beta$, $\beta\beta\alpha\beta\alpha\alpha$ $\alpha\beta\beta\alpha\alpha\beta$, $\beta\alpha\alpha\beta\beta\alpha$, $\alpha\alpha\beta\beta\alpha\beta$, $\beta\beta\alpha\alpha\beta\alpha$
4	$\alpha\|\alpha\|\alpha\ \beta\|\beta\beta$	$\binom{2}{2}\binom{2}{1}$	$\alpha\beta\alpha\beta\beta\alpha$, $\alpha\beta\beta\alpha\beta\alpha$
	$\alpha\|\alpha\alpha\ \beta\|\beta\|\beta$	$\binom{2}{1}\binom{2}{2}$	$\beta\alpha\beta\alpha\alpha\beta$, $\beta\alpha\alpha\beta\alpha\beta$
5	$\alpha\|\alpha\|\alpha\ \beta\|\beta\|\beta$	$2\binom{2}{2}\binom{2}{2}$	$\alpha\beta\alpha\beta\alpha\beta$, $\beta\alpha\beta\alpha\beta\alpha$

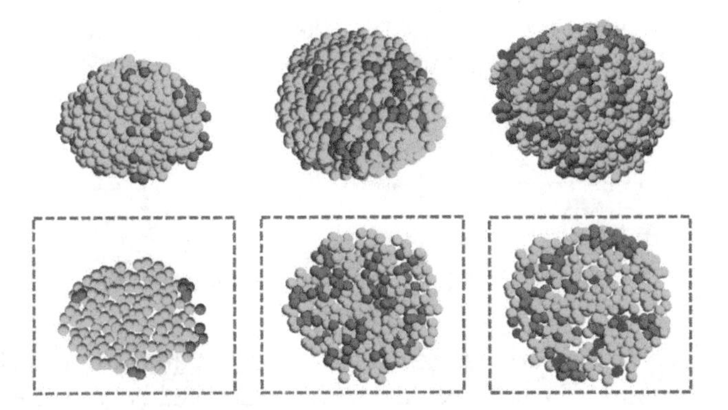

Figure 4.2. Islet structures. Three-dimensional structures and their two-dimensional cross sections (inside square) of mouse (left) and human (middle and right) islets. Islet cells are stained with insulin (green), glucagon (red), and somatostatin (blue) antibodies, and their coordinates are determined by using confocal microscopes. For the left and middle islets, only alpha and beta cells are observed without staining delta cells.

$$Z = \sum_{\mathbf{s}} e^{-\lambda E(\mathbf{s})} = \sum_{\mathbf{n}} g(\mathbf{n}) e^{-\lambda E(\mathbf{n})}. \tag{4.10}$$

As shown in table 4.1, one should separate even ($n_{\alpha\beta} = 2k$) and odd ($n_{\alpha\beta} = 2k + 1$) values for describing $Z = Z_{\text{odd}} + Z_{\text{even}}$, where

$$\begin{aligned} Z_{\text{odd}} &= 2\sum_{k=0}^{k_{\max}} \binom{n_\alpha - 1}{k}\binom{n_\beta - 1}{k} e^{j_{\alpha\alpha}(n_\alpha - 1 - k) + j_{\beta\beta}(n_\beta - 1 - k) + j_{\alpha\beta}(2k+1)}, \\ Z_{\text{even}} &= \sum_{k=1}^{k_{\max}} \binom{n_\alpha - 1}{k}\binom{n_\beta - 1}{k - 1} e^{j_{\alpha\alpha}(n_\alpha - 1 - k) + j_{\beta\beta}(n_\beta - k) + j_{\alpha\beta}(2k)} \\ &+ \sum_{k=1}^{k_{\max}} \binom{n_\alpha - 1}{k - 1}\binom{n_\beta - 1}{k} e^{j_{\alpha\alpha}(n_\alpha - k) + j_{\beta\beta}(n_\beta - 1 - k) + j_{\alpha\beta}(2k)}, \end{aligned} \tag{4.11}$$

where $k_{\max} = \min\{n_\alpha - 1, n_\beta - 1\}$. Once this partition function is obtained, one can calculate $\bar{n}_{\sigma\sigma'}$ by differentiating $\log Z$ with respect to $j_{\sigma\sigma'}$. When we increase the relative strength of heterotypic to homotypic adhesion ($j_{\alpha\beta}/j_{\alpha\alpha}$), the average heterologous contact number $\bar{n}_{\alpha\beta}$ increases as expected (figure 4.2). In the following example, we consider, for simplicity, symmetric homogeneous adhesion ($j_{\alpha\alpha} = j_{\beta\beta}$). By differentiating twice as shown in equation (4.9), contact number fluctuations can be also calculated. The fluctuation $\delta n_{\alpha\beta}$ becomes largest at $j_{\alpha\beta} = (j_{\alpha\alpha} + j_{\beta\beta})/2$ or $j_{\alpha\beta} = j_{\alpha\alpha} = j_{\beta\beta}$ in this symmetric example ($j_{\alpha\alpha} = j_{\beta\beta}$). This corresponds to the phase transition between sorting and mixing phases [10].

4.4 Monte-Carlo sampling method

In general, it is very difficult to obtain the partition function, and calculate the contact number $\mathbf{n}$. However, one can numerically solve this problem by using the Monte-Carlo sampling method. If one can faithfully sample $\mathbf{s}$ following the probability distribution $P(\mathbf{s}) = Z^{-1}e^{-\lambda E(\mathbf{s})}$, the average contact number and its variance can be directly calculated using

$$\bar{n}_{\alpha\alpha} = \frac{1}{M}\sum_{m=1}^{M} n_{\alpha\alpha}(\mathbf{s}_m), \tag{4.12}$$

$$\delta n_{\alpha\alpha}^2 = \frac{1}{M}\sum_{m=1}^{M} n_{\alpha\alpha}^2(\mathbf{s}_m) - \bar{n}_{\alpha\alpha}^2, \tag{4.13}$$

given M samples of $\mathbf{s}$. Similarly we can obtain the mean and variance of $n_{\beta\beta}$ and $n_{\alpha\beta}$.

Stationary sampling of $\mathbf{s}$ and $\mathbf{s}'$ should satisfy a detailed balance condition, given by

$$P(\mathbf{s}'|\mathbf{s})P(\mathbf{s}) = P(\mathbf{s}|\mathbf{s}')P(\mathbf{s}'), \tag{4.14}$$

where $P(\mathbf{s}'|\mathbf{s})$ represents the conditional probability for choosing $\mathbf{s}'$ given $\mathbf{s}$. Rearranging equation (4.14), we obtain

$$\frac{P(\mathbf{s}'|\mathbf{s})}{P(\mathbf{s}|\mathbf{s}')} = \frac{P(\mathbf{s}')}{P(\mathbf{s})} = e^{\lambda[E(\mathbf{s})-E(\mathbf{s}')]}. \tag{4.15}$$

A simple way to satisfy the detailed balance is to define

$$P(\mathbf{s}'|\mathbf{s}) = 1, \tag{4.16}$$

$$P(\mathbf{s}|\mathbf{s}') = e^{\lambda[E(\mathbf{s}')-E(\mathbf{s})]}, \tag{4.17}$$

for $E(\mathbf{s}') < E(\mathbf{s})$, and

$$P(\mathbf{s}'|\mathbf{s}) = e^{\lambda[E(\mathbf{s})-E(\mathbf{s}')]}, \tag{4.18}$$

$$P(\mathbf{s}|\mathbf{s}') = 1, \tag{4.19}$$

for $E(\mathbf{s}') > E(\mathbf{s})$. This is the popular Metropolis choice [11], given by

$$P(\mathbf{s}'|\mathbf{s}) = \min\left\{1,\ e^{\lambda[E(\mathbf{s})-E(\mathbf{s}')]}\right\}. \tag{4.20}$$

Once the conditional probability $P(\mathbf{s}'|\mathbf{s})$ is determined, one can sample $\mathbf{s}$ following $P(\mathbf{s})$. The detailed sampling procedure of this method is summarized in the following algorithm.

(1) Pick a random configuration $\mathbf{s}_1$ satisfying total cell number n_α and n_β (e.g. $\mathbf{s}_1$: $\alpha - \alpha - \beta - \beta$).
(2) Compute adhesion energy $E(\mathbf{s}_1) = J_{\alpha\alpha} + J_{\beta\beta} + J_{\alpha\beta}$.
(3) Generate a new configuration $\mathbf{s}'$: $\alpha - \beta - \alpha - \beta$ by switching one α cell and one β cell of the previous configuration $\mathbf{s}_1$.
(4) Compute new adhesion energy $E(\mathbf{s}') = 3J_{\alpha\beta}$.

(5) Accept $\mathbf{s}'$ with the acceptance rate, $\min\{1, e^{\lambda[E(\mathbf{s}_1)-E(\mathbf{s}')]}\}$: $\mathbf{s}_2 = \mathbf{s}'$, otherwise $\mathbf{s}_2 = \mathbf{s}_1$.
(6) Given $\mathbf{s}_2$, generate another configuration $\mathbf{s}''$, and determine $\mathbf{s}_3$ by repeating (3), (4) and (5).
(7) Iterate this until $E(\mathbf{s}_m)$ saturates.
(8) Obtain M samples after equilibration.

After obtaining M samples, the mean and fluctuations of cell-to-cell contact numbers in equation (4.13) can be computed. Based on this numerical technique, the numerical results obtained from $M = 10^5$ samples are consistent with the previous analytical results (figure 4.1).

4.5 Islet structure data

Islets have heterogeneous sizes, but their typical diameter is less than 200 μm. Thus confocal microscopes can fully scan their three-dimensional structures because they can adjust their depth of focus by about 500 μm. This imaging method can extract $\mathbf{r} = (x, y, z)$ coordinates of islets cells [12] (see figure 4.2). Given the coordinates, the number of contacts between cell types can be identified. For example, to identify the contacting (or neighboring) cells of the ith cell, one needs to measure cell-to-cell Euclidean distances, and select the jth cells satisfying

$$||\mathbf{r}_i - \mathbf{r}_j|| \equiv \sqrt{(x_i - x_j)^2 + (y_i - y_j)^2 + (z_i - z_j)^2} < d, \tag{4.21}$$

where d is a certain distance threshold. The distance ($d = a \cdot d_{\min}$) can be usually set using average minimal cell-to-cell distance,

$$d_{\min} = \frac{1}{n}\sum_{i=1}^{n} \min_j ||\mathbf{r}_i - \mathbf{r}_j||, \tag{4.22}$$

with some tolerance factor $a = \sqrt{2}$ or $\sqrt{3}$. By focusing on two dominant cell types, alpha and beta cells, one can count ($n_{\alpha\alpha}$, $n_{\beta\beta}$, $n_{\alpha\beta}$). Then it is straightforward to compute probabilities of cellular compositions and contacts based on the following equations

$$P_\alpha = \frac{n_\alpha}{n_\alpha + n_\beta} \tag{4.23}$$

$$P_\beta = \frac{n_\beta}{n_\alpha + n_\beta} \tag{4.24}$$

$$P_{\alpha\alpha} = \frac{n_{\alpha\alpha}}{n_{\alpha\alpha} + n_{\beta\beta} + n_{\alpha\beta}} \tag{4.25}$$

$$P_{\beta\beta} = \frac{n_{\beta\beta}}{n_{\alpha\alpha} + n_{\beta\beta} + n_{\alpha\beta}} \tag{4.26}$$

$$P_{\alpha\beta} = \frac{n_{\alpha\beta}}{n_{\alpha\alpha} + n_{\beta\beta} + n_{\alpha\beta}}. \tag{4.27}$$

Unlike the well-defined shell–core structure of rodent islets, the structure of human islets remains controversial, with structures suggested to be either ordered or random. If the structure is random, the contact probabilities could be directly estimated from cellular compositions using the equations

$$P_{\alpha\alpha}^{\text{rand}} = P_\alpha P_\alpha, \tag{4.28}$$

$$P_{\beta\beta}^{\text{rand}} = P_\beta P_\beta, \tag{4.29}$$

$$P_{\alpha\alpha}^{\text{rand}} = P_\alpha P_\beta + P_\beta P_\alpha = 2P_\alpha P_\beta. \tag{4.30}$$

The observed contact probabilities show that homogeneous contacts are more prevalent, while heterologous contacts are less so [12]. This means that

$$P_{\alpha\alpha} > P_{\alpha\alpha}^{\text{rand}}, \tag{4.31}$$

$$P_{\beta\beta} > P_{\beta\beta}^{\text{rand}}, \tag{4.32}$$

$$P_{\alpha\beta} < P_{\alpha\beta}^{\text{rand}}. \tag{4.33}$$

These quantitative results demonstrate that human islets are not random aggregates of alpha and beta cells. Furthermore, they imply that homologous contacts are preferred with stronger adhesion than heterologous contacts, i.e.

$$J_{\alpha\alpha}, J_{\beta\beta} > J_{\alpha\beta}. \tag{4.34}$$

This conclusion is also true for delta cells, which prefer contacts with other delta cells [12].

4.6 Inverse problem

So far we explained how to generate a string **s** or a contact configuration $\mathbf{n} = (n_{\alpha\alpha}, n_{\beta\beta}, n_{\alpha\beta})$, given the relative adhesion $\mathbf{j} = (j_{\alpha\alpha}, j_{\beta\beta}, j_{\alpha\beta})$. However, the real problem is how to infer **j**, given **n**. This is an inverse problem. It is implemented as follows. Suppose we pick a random adhesion **j**. Then, we equilibrate the islet structure as previously explained. After equilibration, we generate a sufficient number of configurations **n**, and extract their mean and variance, $\bar{\mathbf{n}}$ and $\delta\mathbf{n}$. These configurations are expected to be different from the measured data $\mathbf{n}'$. The error can be quantified as a cost function, given by

$$C(\mathbf{j}) = \frac{(\bar{n}_{\alpha\alpha} - n'_{\alpha\alpha})^2}{2\delta n_{\alpha\alpha}^2} + \frac{(\bar{n}_{\beta\beta} - n'_{\beta\beta})^2}{2\delta n_{\beta\beta}^2} + \frac{(\bar{n}_{\alpha\beta} - n'_{\alpha\beta})^2}{2\delta n_{\alpha\beta}^2}. \tag{4.35}$$

By choosing a better **j**, the cost C could be made smaller. For detailed inference, we can consider $P(\mathbf{j}|\mathbf{n}')$ which represents the probability of **j** being true given the data $\mathbf{n}'$.

Once $P(\mathbf{j}|\mathbf{n}')$ is known, one can estimate the likelihood values of $\mathbf{j}$ along with the mean and variance defined by

$$\bar{\mathbf{j}} = \sum_{\mathbf{j}} \mathbf{j}\, P(\mathbf{j}|\mathbf{n}') \tag{4.36}$$

$$\delta \mathbf{j}^2 = \sum_{\mathbf{j}} \mathbf{j}^2 P(\mathbf{j}|\mathbf{n}') - \bar{\mathbf{j}}^2. \tag{4.37}$$

To compute $P(\mathbf{j}|\mathbf{n}')$, we use Bayes' rule [13]

$$P(\mathbf{j}|\mathbf{n}') = \frac{P(\mathbf{n}'|\mathbf{j})P(\mathbf{j})}{P(\mathbf{n}')}. \tag{4.38}$$

If we lack any knowledge about $\mathbf{j}$, the prior probability $P(\mathbf{j})$ becomes uniform. In this case, the posterior probability $P(\mathbf{j}|\mathbf{n}')$ becomes equivalent to the likelihood $P(\mathbf{n}'|\mathbf{j})$. Here the likelihood is a function of the cost function, given by

$$P(\mathbf{n}'|\mathbf{j}) \propto e^{-C(\mathbf{j})}, \tag{4.39}$$

due to the maximum likelihood principle (MLP) for the condition that $\mathbf{n}'$ is matched to $\bar{\mathbf{n}}(\mathbf{j})$ given variation $\delta^2\mathbf{n}(\mathbf{j})$ [14] (the justification for MLP is kept as an exercise in exercise 1(c)). Based on this, the posterior probability is given by

$$P(\mathbf{j}|\mathbf{n}') = Ae^{-C(\mathbf{j})}, \tag{4.40}$$

where A is a normalization constant that satisfies $\int Ae^{-C(\mathbf{j})}d\mathbf{j} = 1$. Again using Monte-Carlo sampling method, we can obtain samples for $\mathbf{j}$ satisfying $P(\mathbf{j}|\mathbf{n}')$. From the detailed balance condition, we have

$$\frac{P(\mathbf{j}'|\mathbf{j})}{P(\mathbf{j}|\mathbf{j}')} = \frac{P(\mathbf{j}'|\mathbf{n}')}{P(\mathbf{j}|\mathbf{n}')} = e^{[C(\mathbf{j})-C(\mathbf{j}')]}. \tag{4.41}$$

Following the Metropolis choice introduced before, we can obtain the following conditional probability

$$P(\mathbf{j}'|\mathbf{j}) = \min\left\{1,\ e^{[C(\mathbf{j})-C(\mathbf{j}')]}\right\}. \tag{4.42}$$

The detailed sampling procedure of this method is summarized in the following algorithm:

(1) Pick a random adhesion $\mathbf{j}_1$.
(2) Obtain sufficient samples of $\mathbf{s}$ after equilibration, as described in section 4.4.
(3) Calculate cell–cell contact configurations $\mathbf{n}$ for each $\mathbf{s}$, and compute their mean and variance, $\bar{\mathbf{n}}$ and $\delta^2\mathbf{n}$.
(4) Compute the cost $C(\mathbf{j}_1)$ in equation (4.35).
(5) Generate a new $\mathbf{j}'$, and compute $C(\mathbf{j}')$ by repeating (2), (3), and (4).
(6) Accept $\mathbf{j}'$ with the rate, $\min\{1, e^{[C(\mathbf{j}_1)-C(\mathbf{j}')]}\}$: $\mathbf{j}_2 = \mathbf{j}'$, otherwise $\mathbf{j}_2 = \mathbf{j}_1$.

(7) Given $\mathbf{j}_2$, generate another $\mathbf{j}''$, and determine $\mathbf{j}_3$ by repeating (5) and (6).
(8) Iterate this until $C(\mathbf{j}_m)$ saturates.
(9) Obtain M samples after equilibration.

By applying this algorithm, we can obtain M samples of $\mathbf{j}$ that follow $P(\mathbf{j}|\mathbf{n})$. This method can be generalized to examine islet structures containing alpha, beta, and delta cells. By quantifying contact numbers, $\mathbf{n} = (n_{\alpha\alpha}, n_{\beta\beta}, n_{\alpha\beta}, n_{\delta\delta}, n_{\alpha\delta}, n_{\beta\delta})$, one can infer the relative adhesions, $\mathbf{j} = (j_{\alpha\alpha}, j_{\beta\beta}, j_{\alpha\beta}, j_{\delta\delta}, j_{\alpha\delta}, j_{\beta\delta})$.

Using this method, mouse, pig, and human islets have been examined [12]. Regardless of the species, islets have a conserved rule that homotypic attractions are slightly, but significantly, stronger than heterotypic attractions. However, depending on the compositions of islet cells, they form different structures: shell–core structures for mouse islets; and partial mixing structures for human islets. It has been known that alpha, beta, and delta cells interact and affect each other's hormone secretions. It is of particular interest to investigate how such spatial organizations of islet cells contributes to the islet functions. This question is partially addressed in chapters 5 and 6.

Exercises

1. Maximum likelihood principle.
   (a) If one has no information on a discrete variable X, its maximum likelihood distribution should be $P(X) = \text{constant}$, which is actually the safest way not to be wrong. Show this. Hint: This corresponds to maximizing the uncertainty (or entropy), $-\sum_X P(X)\log P(X)$, given the probability normalization constraint, $\sum_X P(X) = 1$.
   (b) If one knows its mean value, $\sum_X XP(X) = \bar{X}$, what is its maximum likelihood distribution?
   (c) If one knows its mean $\bar{X}$ and variation $\sum_X (X - \bar{X})^2 P(X) = \delta X^2$, what is its maximum likelihood distribution?
2. One-dimensional differential adhesion model (figure 4.1). Suppose one has five alpha cells and five beta cells. Given an inverse cell motility, $\lambda = 10$, and fixed homotypic adhesion strengths ($J_{\alpha\alpha} = J_{\beta\beta} = 1$), calculate mean and variation of cell contact numbers for different values of heterotypic adhesion strength $J_{\alpha\beta}$.
   (a) Use the analytic method with the partition function in equation (4.11).
   (b) Use the Monte-Carlo sampling method and compare the results with the analytic results.
3. The structure data of mouse and human islets are available in Supporting Information of reference [15]. Given an inverse cell motility, $\lambda = 5$, and fixed $J_{\alpha\alpha} = 1$ as a reference, estimate $J_{\beta\beta}$ and $J_{\alpha\beta}$ that can explain the mouse islet structures and human islet structures, respectively.

References

[1] Koh D S, Cho J H and Chen L 2012 *J. Mol. Neurosci.* **48** 429
[2] Orci L 1976 *Metabolism* **25** 1303
[3] Brelje T C, Scharp D W and Sorenson R L 1989 *Diabetes* **38** 808
[4] Britt L D, Stojeba P C, Scharp C R, Greider M H and Scharp D W 1981 *Diabetes* **30** 580
[5] Halban P A, Powers S L, George K L and Bonner-Weir S 1987 *Diabetes* **36** 783
[6] Steinberg M S 1963 *Science* **141** 401
[7] Steinberg M S 2007 *Curr. Opin. Genet. Dev.* **17** 281
[8] Jia D, Dajusta D and Foty R A 2007 *Dev. Dyn.* **236** 2039
[9] Lee J C 2002 *Thermal Physics: Entropy and Free Energies* (River Edge, NJ: World Scientific)
[10] Hoang D T, Song J and Jo J 2013 *Phys. Rev.* E **88** 062725
[11] Metropolis N and Ulam S 1949 *J. Am. Stat. Assoc.* **44** 335
[12] Hoang D T, Matsunari H, Nagaya M, Nagashima H, Millis J M, Witkowski P, Periwal V, Hara M and Jo J 2014 *PLoS One* **9** e110384
[13] Gregory P C 2010 *Bayesian Logical Data Analysis for the Physical Sciences: A Comparative Approach with Mathematica Support* (Cambridge: Cambridge University Press)
[14] Sivia D S and Skilling J 2006 *Data Anlaysis—A Bayesian Tutorial* (Oxford: Oxford University Press)
[15] Hoang D T, Hara M and Jo J 2016 *PLoS One* **11** e0152446

IOP Publishing

Diabetes Systems Biology
Quantitative methods for understanding beta-cell dynamics and function
Anmar Khadra

Chapter 5

Intra-islet network

Junghyo Jo and Margaret Watts

5.1 Introduction

As indicated in previous chapters, glucose homeostasis is regulated by the counter-regulatory hormones, insulin and glucagon. Pancreatic alpha cells produce glucagon in response to low blood glucose levels that stimulates the breakdown of glycogen into glucose in the liver, ultimately causing a rise in blood glucose levels. On the other hand, beta cells produce insulin in response to high blood glucose levels, which promotes synthesis of glycogen from glucose in the liver and accelerates absorption of blood glucose in peripheral tissues, causing a reduction in blood glucose levels. At first glance, these two reciprocal cell types seem to sufficiently keep blood glucose levels within a normal range. However, pancreatic islets contain a third type of endocrine cell, the delta cells, whose physiological role remains unclear. It has long been known that somatostatin, released by these cells, inhibits glucagon and insulin secretion [1–3]. It has been hypothesized that this inhibition can prevent wasteful co-secretion of glucagon and insulin at normal glucose levels [4].

In addition to the paracrine interactions of delta cells, the cross talk between alpha, beta, and delta cells has been observed (see figure 5.1 for a summary for these interactions) [5]. For example, a well-known interaction is that beta cells inhibit glucagon secretion [6–12]. This paracrine interaction has been emphasized as an important mechanism to make alpha cells silent at high glucose [9]. Other interactions have been given more attention recently. Indeed, it has been reported that alpha cells promote insulin and somatostatin secretion [13–19]. The former interaction that alpha cells can stimulate beta cells seems, however, counter intuitive. Furthermore, it is experimentally challenging to probe the changes of somatostatin secretion from delta cells, because its concentration is very low (~femto mole). Nevertheless, it has been recently confirmed that beta cells stimulate somatostatin secretion, as shown in figure 5.1 [20]. A summary of the paracrine interactions between islet cells is shown in figure 5.1.

doi:10.1088/978-0-7503-3739-7ch5

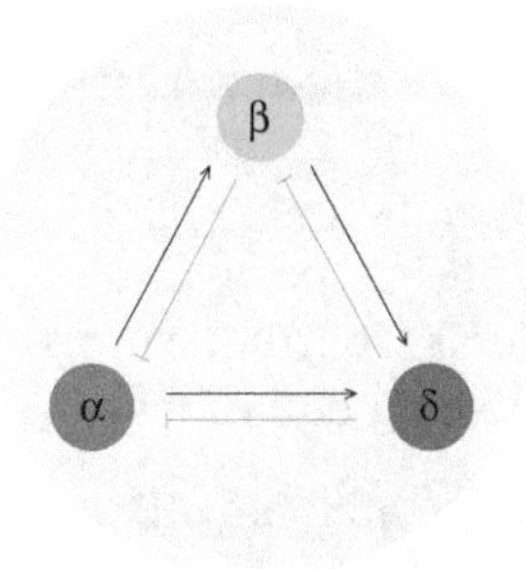

Figure 5.1. A summary of potential interactions between alpha, beta, and delta cells in the islets of Langerhans. Positive interactions are labeled with black arrows while negative interactions are labeled with red bar-headed arrows.

Why islet cells interact like this as a whole remains incompletely understood. In this section, we introduce potential roles for the paracrine interactions by adopting two different quantitative approaches: (i) electrophysiology-based cell modeling and (ii) oscillator-based cell modeling. The first approach simplifies an electrophysiological based model of pancreatic alpha cells to explore the effects of paracrine interactions on glucagon secretion. Using this model, we show how paracrine interactions are important for regulating glucagon secretion from the alpha cell and how reducing a complicated model can be useful in teasing apart complicated relationships. The second approach considers islet cells as phase oscillators that generate pulsatile hormones. By using the coupled oscillator model, we show that the interaction and organization of islet cells can generate controllable synchronization. Islet cells can produce synchronous hormone pulses under low- and high-glucose conditions, while they can produce asynchronous hormone pulses under normal glucose conditions.

5.2 Reduced islet model

Pancreatic alpha cells secrete glucagon when blood glucose is low, and stop secreting glucagon when blood glucose is elevated. However, the exact mechanism by which glucagon secretion is inhibited by glucose and stimulated by hypoglycemia is unclear. There are two main hypotheses explaining this phenomenon: (1) glucose directly modulates glucagon secretion through the closure of ATP-sensitive K^+ (K (ATP)) channels [21–24], and (2) glucose acts indirectly through paracrine mechanisms including insulin and somatostatin [9, 10, 25–27]. While each of these hypotheses have merit, they are not mutually exclusive. In fact, regulation of glucagon secretion is a complicated process that combines both intrinsic (within the cell) and paracrine regulation.

In order to model alpha cell electrical activity, the Hodgkin–Huxley formalism can be used as described in chapter 2. Pancreatic alpha cells express three different Ca^{2+} channels, three different K^+ channels and a Na^+ channel. Based on this the equation describing the dynamics of the plasma membrane voltage (V) can be expressed as

$$\frac{dV}{dt} = -(I_{\text{CaL}} + I_{\text{CaPQ}} + I_{\text{CaT}} + I_{\text{Na}} + I_{\text{K}} + I_{\text{K(ATP)}} + I_{\text{KA}} + I_{\text{L}})/C_m, \tag{5.1}$$

where C_m is the membrane capacitance, I_{CaL}, I_{CaPQ}, I_{CaT} are L-, N-, and T-type voltage-dependent Ca^{2+} currents, respectively; I_{Na} is a voltage-dependent Na^+ current; I_{K} is a delayed rectifier K^+ current, and I_{L} is the leak current. The ionic currents are defined by the following equations

$$I_{\text{CaL}} = g_{\text{CaL}} m_{\text{CaL}}^2 h_{\text{CaL}} (V - V_{\text{Ca}}), \tag{5.2}$$

$$I_{\text{CaN}} = g_{\text{CaN}} m_{\text{CaN}} h_{\text{CaN}} (V - V_{\text{Ca}}), \tag{5.3}$$

$$I_{\text{CaT}} = g_{\text{CaT}} m_{\text{CaT}}^3 h_{\text{CaT}} (V - V_{\text{Ca}}), \tag{5.4}$$

$$I_{\text{Na}} = g_{\text{Na}} m_{\text{Na}}^3 h_{\text{Na}} (V - V_{\text{Na}}), \tag{5.5}$$

$$I_{\text{K}} = g_{\text{K}} m_{\text{K}}^4 (V - V_{\text{K}}), \tag{5.6}$$

$$I_{\text{KA}} = g_{\text{KA}} m_{\text{KA}} h_{\text{KA}} (V - V_{\text{K}}), \tag{5.7}$$

$$I_{\text{L}} = g_{\text{L}} (V - V_{\text{L}}), \tag{5.8}$$

where g_x, x = CaL, CaN, CaT, Na, K, KA, L, are the maximal channel conductances, m_x are channel activation variables, h_x are channel inactivation variables, and V_x is the ionic Nernst potential associated with the channel. While the maximum conductances and Nernst potentials are constants, the activation and inactivation variables are voltage-dependent and governed by their own differential equations. These equations are given by [28]

$$\frac{m_x}{dt} = \frac{m_x^{\infty}(V) - m_x}{\tau_x^m(V)}, \tag{5.9}$$

$$\frac{h_x}{dt} = \frac{h_x^{\infty}(V) - h_x}{\tau_x^h(V)}. \tag{5.10}$$

The steady-state activation (m_x^{∞}) and inactivation (h_x^{∞}) functions are sigmoidal (sig) functions, while the time constants τ_x^m and τ_x^h are bell shaped (bell) functions with the form:

$$\text{sig}(V, V^*, S^*) = \frac{1}{1 + \exp\left(\frac{-(V - V^*)}{S^*}\right)}, \tag{5.11}$$

$$\text{bell}(V, V^*, S^*, \tau^*, \tau_0) = \frac{\tau^*}{\exp\left(\frac{-(V - V^*)}{S^*}\right) + \exp\left(\frac{(V - V^*)}{S^*}\right)} + \tau_0. \tag{5.12}$$

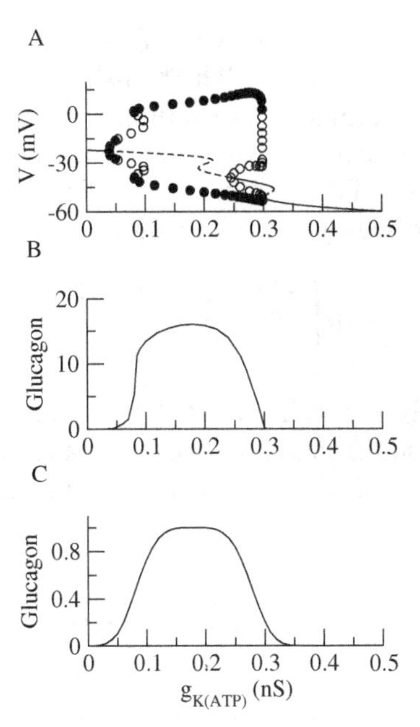

Figure 5.2. Reduction of the full alpha cell model: (A) bifurcation diagram showing the steady state dynamics of the electrophysiological alpha cell model as a function of the conductance of K(ATP) channels ($g_{K(ATP)}$). Solid and dashed lines represent stable and unstable steady state solutions, respectively, while the closed and open circles represent the envelopes of stable and unstable periodic solutions, respectively. The envelopes of periodic solutions represent the maximum and minimum voltages during spiking. (B) As $g_{K(ATP)}$ decreases, glucagon secretion first increases then decreases, producing a bell-shaped curve. (C) The reduced model in equation (5.3) can approximate the bell-shaped glucagon secretion curve from the original electrophysiological model.

Based on the first hypothesis, glucose can regulate glucagon secretion by closing K(ATP) channels. When blood glucose levels rise, glucose enters the cell through glucose transporters. After glucose enters the cell, it goes through glycolysis which increases the ATP/ADP ratio leading to the closure of K(ATP) channels. The closure of K(ATP) channels reduces the conductance of the channel ($g_{K(ATP)}$). By treating $g_{K(ATP)}$ as a parameter, a bifurcation diagram can be constructed that shows how the behavior of the system changes as $g_{K(ATP)}$ is varied. By plotting this bifurcation diagram (see figure 5.2(A)), we obtain branches of stable (solid lines) and unstable (dashed lines) steady states. We also obtain envelopes of stable (filled circles) and unstable (open circles) periodic solutions, representing the maximum and minimum of these periodic solutions. Starting on the far right, as $g_{K(ATP)}$ is decreased past a lower threshold, repetitive action potentials are produced through crossing a subcritical Hopf bifurcation. Decreasing $g_{K(ATP)}$ further, reduces the amplitude of the action potentials. Eventually, $g_{K(ATP)}$ passes an upper threshold (a supercritical Hopf bifurcation) where action potentials stop being produced and the cell is at a depolarized steady state. The electrical activity of the cell determines how much glucagon is secreted. As $g_{K(ATP)}$ is decreased, glucagon secretion first increases due to the initiation of spiking, then decreases from reduced spike amplitude and the eventual loss of spiking (figure 5.2(B)). The bell-shaped glucagon secretion curve is the fundamental feature of alpha cell secretion, so the electrophysiological model can be reduced to an algebraic model with the glucagon secretion curve modeled as a Gaussian function of K(ATP) conductance. This means that the curve in figure 5.2(C) can be described by the following equation

$$G = \exp\left(\frac{-(g_{K(ATP)} - 0.18)^4}{2(0.09)^4}\right). \tag{5.13}$$

alpha cells are not the only islet cells affected by increasing glucose concentrations. Both insulin and somatostatin increase monotonically with glucose concentration (see figure 5.3). This implies that their glucose-dependence can be described by the following equations

$$I = \frac{180}{1 + \exp(-(Glc - 7)/3)}, \tag{5.14}$$

$$S = \frac{80}{1 + \exp(-(Glc - 5)/2)}, \tag{5.15}$$

where I is insulin, S is somatostatin, and Glc is glucagon concentration.

The concentration of insulin and somatostatin in an islet has been shown to modulate glucagon secretion from alpha cells. However, insulin and somatostatin modulate glucagon secretion through different mechanisms. Specifically, it has been shown that insulin increases K(ATP) channel activity [11, 29]; in other words, the conductance of the K(ATP) channel, $g_{K(ATP)}$, is modulated by both glucose and insulin consecration and can be described by

$$g_{K(ATP)} = \frac{0.15}{1 + \exp(-(Glc - 3)/2)} + (g_{K(ATP)0} - 0.12) + E_I, \tag{5.16}$$

where $g_{K(ATP)0}$ is the value of $g_{K(ATP)}$ in 0 mM glucose and E_I is the effect of insulin on K(ATP) conductance. E_I increases with insulin and is given by the following equation:

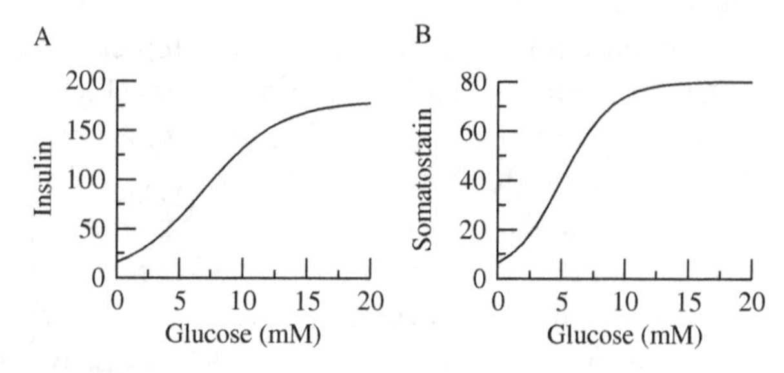

Figure 5.3. When glucose concentration rises, both (A) insulin and (B) somatostatin increase monotonically.

$$E_I = \frac{0.08}{1 + \exp(-(I - 100)/60)} - 0.015. \tag{5.17}$$

Figure 5.4 illustrates how the amount of glucagon secreted is determined by modifications to the bell-shaped glucagon secretion curve. Increasing glucose decreases $g_{K(ATP)}$ and moves the system to the left along the bell-shaped curve. Increasing insulin (due to the increase in glucose), on the other hand, increases $g_{K(ATP)}$ and moves the system to the right along the bell-shaped curve.

Instead of affecting the conductance of the K(ATP) channels, evidence suggests that somatostatin may also directly reduce the amount of glucagon available to be secreted from alpha cells [30, 31]. This effect is modeled as a decrease in the amplitude of the bell-shaped curve and is described by the following equation

$$G = E_S \exp\left(\frac{-(g_{K(ATP)} - 0.18)^4}{2(0.09)^4}\right), \tag{5.18}$$

where E_S is the effect of somatostatin on glucagon secretion and is given by:

$$E_S = \frac{1}{1 + \exp((S - 30)/30) + 0.3}. \tag{5.19}$$

As indicated, E_S increases as somatostatin increases. Figure 5.4 shows how somatostatin decreases the amplitude of the bell-shaped curve (dashed line).

Figure 5.5 demonstrates how the level of K(ATP) conductance in low glucose ($g_{K(ATP)0}$) is critical for glucagon secretion to decrease with glucose concentration. If $g_{K(ATP)}$ is chosen in such a way that secretion is maximal in 0 μM G (black circle, figure 5.4), then as glucose is increased the glucose-mediated decrease in $g_{K(ATP)}$ (figure 5.5(A), solid line) pushes the system down the bell-shaped curve (figure 5.5(C), solid line), decreasing secretion. Therefore, it is possible for pancreatic alpha cells to reduce glucagon secretion without the inhibitory effects of insulin and

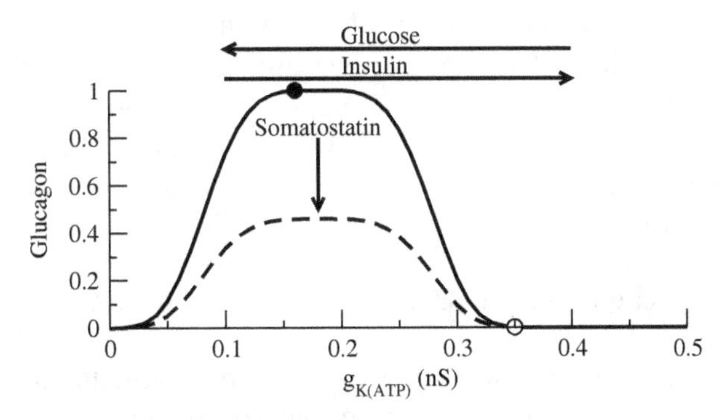

Figure 5.4. The amount of glucagon secreted is determined by modifications to the bell-shaped glucagon secretion curve. Increasing glucose decreases $g_{K(ATP)}$, while increasing insulin increases $g_{K(ATP)}$. Therefore, changes in glucose and insulin will move the phase point along the curve. Somatostatin reduces the amount of glucagon available to be secreted, reducing the amplitude of the bell-shaped curve (dashed line).

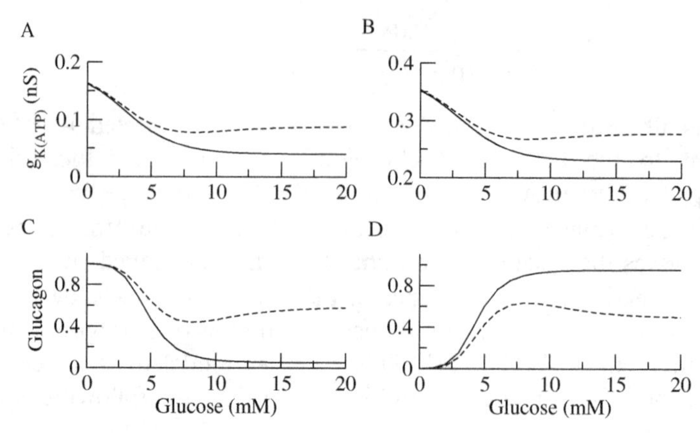

Figure 5.5. How glucose affects the maximum conductance of K(ATP) channels and glucagon secretion. If we assume that in low glucose, $g_{K(ATP)}$ is near the peak of the bell-shaped curve (black circle in figure 5.4), then increasing glucose decreases $g_{K(ATP)}$ (A, solid line), moving the phase point down the bell-shaped curve and decreasing secretion (C, solid line). When insulin concentration is high enough, $g_{K(ATP)}$ starts to increase (A, dashed line) which pushes the phase point back up the bell-shaped curve, increasing glucagon secretion (C, dashed line). If we assume that in low glucose, $g_{K(ATP)}$ is at the bottom right of the bell-shaped curve (open circle in figure 5.4), increasing glucose still decreases $g_{K(ATP)}$ (B, solid line) and the phase point moves up the bell-shaped curve increasing secretion (D, solid line). Likewise, insulin still increases $g_{K(ATP)}$ (B, dashed line). In this case, however, the phase point is pushed back down the bell-shaped curve, decreasing secretion (D, dashed line).

somatostatin. Eventually, the insulin-mediated increase in $g_{K(ATP)}$ overrides the effect of glucose (figure 5.5(A), dashed line), moving the system up the bell-shaped curve increasing secretion (figure 5.5(C), dashed line). Therefore, there is an initial decrease in secretion due to the intrinsic effects of glucose on alpha cells, followed by an increase in secretion due to the effect of insulin. This type of response has been seen experimentally.

If $g_{K(ATP)}$ is chosen to be to the far right of the bell-shaped curve (white circle, figure 5.4), $g_{K(ATP)}$ still decreases monotonically with glucose (Figure 5.5(B), solid line) and eventually increases due to the insulin-mediated increase in $g_{K(ATP)}$ (figure 5.5(B), dashed line). However, decreasing $g_{K(ATP)}$ pushes the system up the bell-shaped curve, increasing secretion (figure 5.5(D), solid line). Now, insulin's role is to keep the cell from secreting in high glucose concentrations by pushing the system back down the bell-shaped curve (figure 5.5(D), dashed line). This figure also illustrates that due to the geometry of the bell-shaped curve, the effects of insulin and glucose on secretion are ambiguous. Although glucose and insulin are normally stated to reduce glucagon secretion, they can actually increase secretion under the right conditions.

So far we have examined how glucose and insulin modulate glucagon secretion, but what role does somatostatin play? Since somatostatin acts to reduce the amount of glucagon available to be released, the effect of somatostatin on glucagon secretion is always inhibitory. Figure 5.6 shows that somatostatin reduces the amount of glucagon secreted. In figure 5.6(A), the dashed line is the same in figure 5.5(C) where

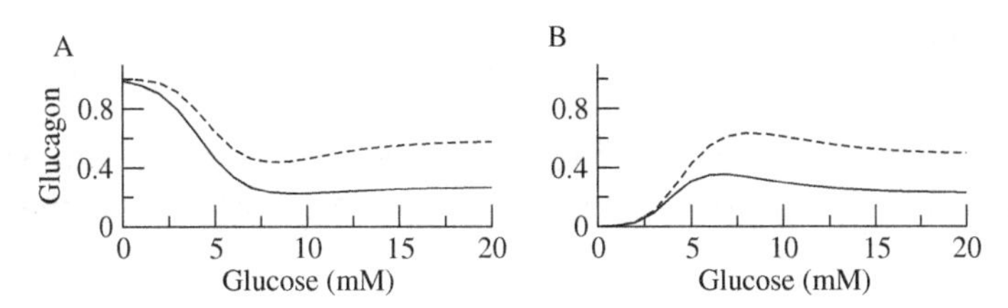

Figure 5.6. How somatostatin affects glucagon secretion. (A) First we assume that in low glucose, $g_{K(ATP)}$ is near the peak of the bell-shaped curve (black circle in figure 5.4). In the absence of somatostatin (dashed line replotted from figure 5.5(C)), glucagon secretion first decreases than increases. With somatostatin added to the model (solid line), the amount of glucagon secretion is reduced and the insulin-mediate increase in secretion at high glucose is missing. (B) Now, we assume that in low glucose, $g_{K(ATP)}$ is at the bottom right of the bell-shaped curve (open circle in figure 5.4). Now, in the absence of somatostatin (dashed line replotted from figure 5.5(D)) glucagon secretion increases. Glucagon secretion still increases when somatostatin is added to the model (solid line); however, the secretion no longer increases to the same extent.

$g_{K(ATP)}$ starts at the top of the bell-shaped curve in low glucose. The increasing effect of insulin on glucagon is negated by somatostatin (solid curve) and the dose-response curve monotonically decreases with glucose instead of rising in high glucose levels. In figure 5.6(B), the dashed curve is the same as figure 5.5(D) where $g_{K(ATP)}$ starts to the right of the bell. In this case, somatostatin adds another layer of defense against glucagon secretion at high glucose levels, where it should not be. Although, secretion rises due to the effect of glucose decreasing $g_{K(ATP)}$, the combined effects of insulin and somatostatin keep the secretion at a low level.

5.3 Oscillator-based model

Insulin, glucagon, and somatostatin secretion from pancreatic islets are pulsatile with a period of a few minutes [32, 33]. The pulse generation may be intrinsic to islet cells, since isolated cells can also generate the oscillations in Ca^{2+}, responsible for triggering hormone release [34–36]. The three pulses are not independent, but their phases are coordinated [32, 33]. The phase coordination implies communication between alpha, beta, and delta cells, secreting glucagon, insulin, and somatostatin, respectively. Paracrine interactions between those islet cells have been extensively observed [5]. Secretion factors, including hormones and neurotransmitters, act as signaling messengers for activating or suppressing neighboring cells. An active (silent) cell can push a neighboring cell to become active (silent) through positive interactions, while an active (silent) cell can pull a neighboring cell from being active (silent) through negative interactions. Here we drastically simplify the hormone secretion and interactions of islet cells by using a phenomenological coupled oscillator model [37]. The model is ideal to qualitatively explore the role of cellular interactions of hormone secretion within an islet without knowing the details of the molecular mechanisms governing it.

5.3.1 α–β model

Let us first focus on alpha and beta cells, the two major cell types within the islets, and consider the simplest case where one alpha cell is interacting with one beta cell. Their phase dynamics, θ_α and θ_β (respectively) can be described by a phase oscillator model, given by

$$\dot{\theta}_\alpha = \omega_\alpha + K_{\alpha\beta} \sin(\theta_\beta - \theta_\alpha) \tag{5.20}$$

$$\dot{\theta}_\beta = \omega_\beta + K_{\beta\alpha} \sin(\theta_\alpha - \theta_\beta), \tag{5.21}$$

where ω_α and ω_β are the intrinsic frequencies of alpha and beta cells, respectively. The sine function coupling the two oscillators governs how each oscillator pushes or pulls the phase of a neighboring oscillator. Here, the coupling strength $K_{\alpha\beta}$ represents how strongly beta cells affect a neighboring alpha cell, and vice versa for $K_{\beta\alpha}$. Since the cellular interaction is mediated by secretion factors, its strength should depend on the activity of affecter cells. In general, we assume that the strength is also inversely dependent on the activity of receiver cells. In other words,

$$K_{\alpha\beta} = K \cdot A_{\alpha\beta}\frac{r_\beta}{r_\alpha}, \tag{5.22}$$

where K is the overall scale of the interaction strength, and r_α and r_β represent the activities of alpha and beta cells, respectively. Here $A_{\alpha\beta}$ defines the interaction sign from beta to alpha cells. $A_{\alpha\beta} = -1$ because beta cells suppress alpha cells, while $A_{\beta\alpha} = 1$ because alpha cells activate beta cells. The model in equation (5.21) is a generalized version of the Kuramoto model [38], in which positive couplings are considered. If we focus on the relative phase between alpha and beta cells ($x = \theta_\alpha - \theta_\beta$), the above two equations can be reduced into one phase difference equation, given by

$$\dot{x} = (\omega_\alpha - \omega_\beta) + K\left(\frac{r_\beta}{r_\alpha} - \frac{r_\alpha}{r_\beta}\right)\sin x. \tag{5.23}$$

Assuming the intrinsic frequencies of alpha and beta cells are identical ($\omega_\alpha = \omega_\beta$), the stable solutions of equation (5.23) will depend on the sign of ($r_\beta/r_\alpha - r_\alpha/r_\beta$), i.e.

$$x = \begin{cases} 0 & \text{if } r_\alpha > r_\beta, \\ \pi & \text{if } r_\alpha < r_\beta. \end{cases}$$

This implies that hormone pulses of alpha and beta cells are in phase ($\theta_\alpha - \theta_\beta = 0$) when alpha cells are more active than beta cells, and out of phase ($\theta_\alpha - \theta_\beta = \pi$) when beta cells are more active than alpha cells. Their activities are dependent on glucose concentration. At around normal glucose conditions (when $r_\alpha \approx r_\beta$), sudden transitions between in-phase and out-of-phase activities can occur for a minute change of glucose concentration.

5.3.2 *α–β–δ* model

The previously introduced *α–β* phase oscillator model can be further generalized to the islet model of three oscillators, representing alpha, beta, and delta cells, given by

$$\dot{\theta}_\alpha = \omega_\alpha - K\frac{r_\beta}{r_\alpha}\sin(\theta_\beta - \theta_\alpha) - K\frac{r_\delta}{r_\alpha}\sin(\theta_\delta - \theta_\alpha), \tag{5.24}$$

$$\dot{\theta}_\beta = \omega_\beta + K\frac{r_\alpha}{r_\beta}\sin(\theta_\alpha - \theta_\beta) - K\frac{r_\delta}{r_\beta}\sin(\theta_\delta - \theta_\beta), \tag{5.25}$$

$$\dot{\theta}_\delta = \omega_\delta + K\frac{r_\alpha}{r_\delta}\sin(\theta_\alpha - \theta_\delta) + K\frac{r_\beta}{r_\delta}\sin(\theta_\beta - \theta_\delta), \tag{5.26}$$

where, as before, ω_i and r_i $(i = \alpha, \beta, \delta)$ represent the intrinsic frequencies and activities of alpha, beta, and delta cells. The positive/negative couplings pull/push neighboring oscillators according to the signs of intercellular interactions in pancreatic islets. Again, we are interested in the phase differences between cell types: $x = \theta_\alpha - \theta_\beta$ and $y = \theta_\alpha - \theta_\delta$. The dynamics of these phase differences are governed by

$$\dot{x} = K\left(\frac{r_\beta}{r_\alpha} - \frac{r_\alpha}{r_\beta}\right)\sin x + K\frac{r_\delta}{r_\alpha}\sin y + K\frac{r_\delta}{r_\beta}\sin(x - y), \tag{5.27}$$

$$\dot{y} = K\frac{r_\beta}{r_\alpha}\sin x + K\left(\frac{r_\delta}{r_\alpha} - \frac{r_\alpha}{r_\delta}\right)\sin y + K\frac{r_\beta}{r_\delta}\sin(x - y), \tag{5.28}$$

where identical intrinsic frequencies, $\omega_\alpha = \omega_\beta = \omega_\delta$ have been assumed. Depending on the relative activities of these three oscillators, we can obtain three possible stable solutions,

$$(x, y) = \begin{cases} (0, 0), & \text{if } r_\alpha \gg r_\beta, r_\delta, \\ (\pi, 0), (\pi/3, 2\pi/3), (5\pi/3, -2\pi/3) & \text{if } r_\alpha = r_\beta = r_\delta, \\ (\pi, \pi) & \text{if } r_\beta \gg r_\alpha, r_\delta. \end{cases}$$

These are stable fixed points in phase planes. The first case corresponds to low glucose conditions where alpha cells are dominantly active, while the third case corresponds to high glucose conditions where beta cells are dominantly active. The second condition corresponds to the normal glucose condition where alpha, beta, and delta cells are equally active. The normal glucose condition has continuous stable limit cycles around three fixed points (figure 5.7). The cycle represents the alternation between in-phase and out-of-phase coordination. Unlike the *α–β* model that shows the abrupt change between in-phase and out-of-phase coordination, the addition of delta cells allows smooth transition between the two coordination with the alternating states.

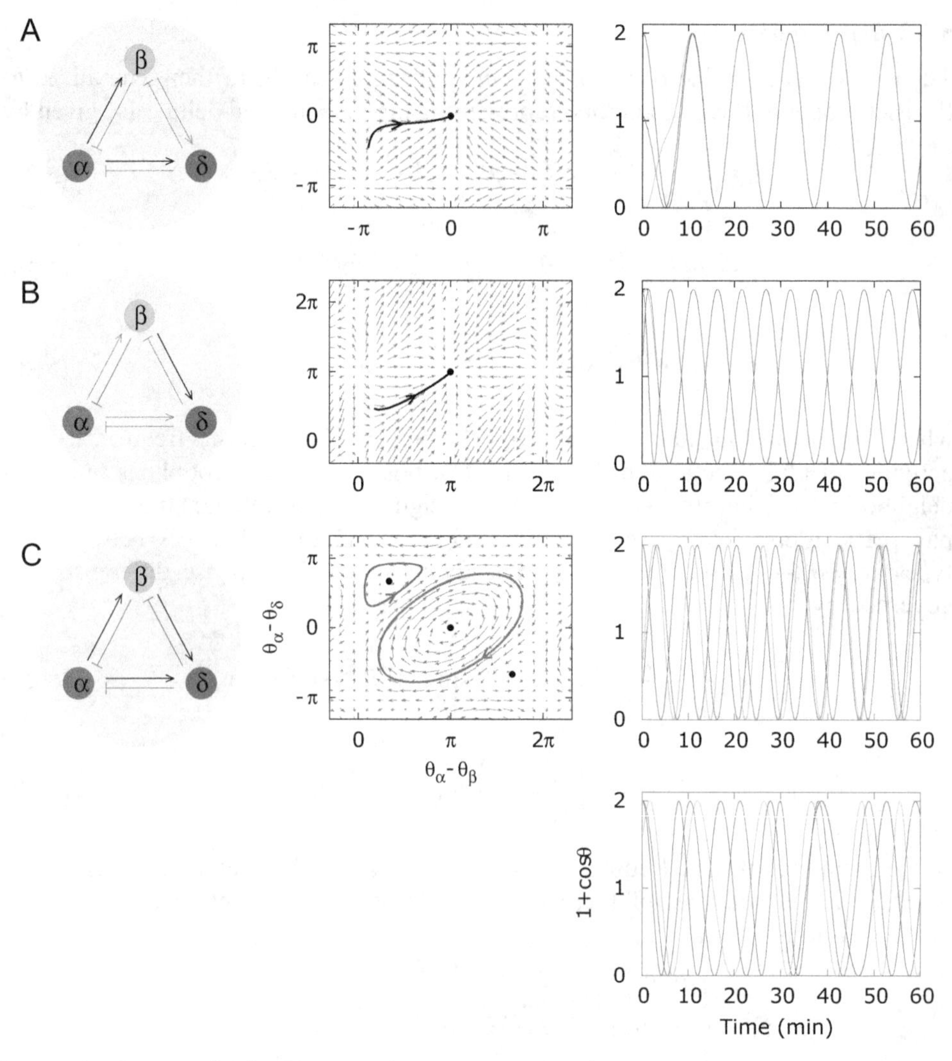

Figure 5.7. Phase coordination between alpha, beta, and delta cells depending on glucose concentrations. Coordination at (A) low ($r_\alpha = 1$, $r_\beta = r_\delta = 0.2$), (B) high ($r_\beta = 1$, $r_\alpha = r_\delta = 0.2$), and (C) normal ($r_\alpha = r_\beta = r_\delta = 1$) glucose concentrations. At low glucose in A, alpha cells are dominantly active, and their positive interactions lead beta cells and delta cells to have synchronous hormone pulses (top left). Thus their phase differences are attracted to a stable fixed point, $\theta_\alpha - \theta_\beta = \theta_\alpha - \theta_\delta = 0$ (top middle). The time trajectories of phase $\theta_\sigma(t)$ for each cell type $\sigma \in \{\alpha(\text{red}), \beta(\text{green}), \delta(\text{blue})\}$ are represented by $1 + \cos(\theta_\sigma(t))$ to consider their periodicity (top right). At high glucose in B, beta cells are dominantly active. The corresponding interactions, phase differences, and time trajectories of these three types of cells are plotted in the middle row. At normal glucose in C, every islet cell is equally active, and every interaction becomes important (bottom left). The phase differences between islet cells, in this case, have stable cycles around three fixed points (bottom middle). Their corresponding time trajectories are plotted for two cases: counter-clockwise cycle (blue, bottom right upper panel) and clockwise cycle (red, bottom right lower panel).

5.3.3 Population model

The asymmetric interactions between alpha, beta, and delta cells contribute to producing rich phase coordinations of islet cells depending on glucose concentrations. Here the cellular interactions are based on cell-to-cell contacts, highlighting the importance of identifying which cells are being coupled. Now let us consider a population model of islet cells with that takes into account their spatial organization. The general phase oscillator model can be described by

$$\dot{\theta}_k = \omega_k + K \sum_{j \in \Lambda_k} A_{\sigma_k \sigma_j} \frac{r_j}{r_k} \sin(\theta_j - \theta_k), \tag{5.29}$$

where θ_k and r_k are the phase and amplitude (or activity) of the kth cell ($k = 1, \ldots, N$, N is the size of the network). It has its intrinsic frequency ω_k, and interacts with its neighboring jth cells ($j \in \Lambda_k$). Their interaction strength is determined by three factors: the interaction scale $K(<\omega)$, the interaction sign A, and the relative activities r_j/r_k. Here $\sigma \in \{\alpha, \beta, \delta\}$ represents the three cell types, and the adjacency matrix element $A_{\sigma_k \sigma_j}$ defines the interaction sign from the jth cell to the kth cell. Given two cell types of alpha and beta cells, $A_{\alpha\alpha} = 1$, $A_{\beta\beta} = 1$, $A_{\beta\alpha} = 1$, and $A_{\alpha\beta} = -1$. Note that we used positive interactions for autocrine interactions for alpha [39–42] and beta cells [43–46].

The population dynamics depends on the spatial organization of islet cells. As a simple example, let us consider one-dimensional arrays of two alpha and two beta cells, but with different spatial organizations. Their phase dynamics follows

$$\dot{\theta}_1 = \omega + K A_{\sigma_1 \sigma_2} \frac{r_{\sigma_2}}{r_{\sigma_1}} \sin(\theta_2 - \theta_1), \tag{5.30}$$

$$\dot{\theta}_2 = \omega + K A_{\sigma_2 \sigma_1} \frac{r_{\sigma_1}}{r_{\sigma_2}} \sin(\theta_1 - \theta_2) + K A_{\sigma_2 \sigma_3} \frac{r_{\sigma_3}}{r_{\sigma_2}} \sin(\theta_3 - \theta_2), \tag{5.31}$$

$$\dot{\theta}_3 = \omega + K A_{\sigma_3 \sigma_2} \frac{r_{\sigma_2}}{r_{\sigma_3}} \sin(\theta_2 - \theta_3) + K A_{\sigma_3 \sigma_4} \frac{r_{\sigma_4}}{r_{\sigma_3}} \sin(\theta_4 - \theta_3), \tag{5.32}$$

$$\dot{\theta}_4 = \omega + K A_{\sigma_4 \sigma_3} \frac{r_{\sigma_3}}{r_{\sigma_4}} \sin(\theta_3 - \theta_4), \tag{5.33}$$

where we assumed that every cell has the same intrinsic frequency ω, and the cell activities depend only on cell types, not sites ($r_i = r_{\sigma_i}$). Again we are interested in the phase differences between cells: $x = \theta_1 - \theta_2$, $y = \theta_1 - \theta_3$, and $z = \theta_1 - \theta_4$, we obtain the following equations

$$\dot{x} = -K\left(A_{\sigma_1 \sigma_2} \frac{r_{\sigma_2}}{r_{\sigma_1}} + A_{\sigma_2 \sigma_1} \frac{r_{\sigma_1}}{r_{\sigma_2}}\right) \sin x - K A_{\sigma_2 \sigma_3} \frac{r_{\sigma_3}}{r_{\sigma_2}} \sin(x - y), \tag{5.34}$$

Table 5.1. Steady states for different cellular organization determined by the model presented in equations (5.24)–(5.26).

	(x, y, z)	
$\sigma_1 - \sigma_2 - \sigma_3 - \sigma_4$	$r_\alpha > r_\beta$	$r_\alpha < r_\beta$
$\alpha - \alpha - \beta - \beta$	$(0, 0, 0)$	$(0, \pi, \pi)$
$\alpha - \beta - \beta - \alpha$	$(0, \pi, \pi)$	$(\pi, \pi, 0)$
$\beta - \alpha - \alpha - \beta$	$(0, 0, 0)$	$(\pi, 0, 0)$
$\alpha - \beta - \alpha - \beta$	$(0, \pi, \pi)$	$(\pi, \pi, 0)$

$$\dot{y} = -KA_{\sigma_1\sigma_2}\frac{r_{\sigma_2}}{r_{\sigma_1}}\sin x + KA_{\sigma_3\sigma_2}\frac{r_{\sigma_2}}{r_{\sigma_3}}\sin(x - y) - KA_{\sigma_3\sigma_4}\frac{r_{\sigma_4}}{r_{\sigma_3}} + \sin(y - z), \tag{5.35}$$

$$\dot{z} = -KA_{\sigma_1\sigma_2}\frac{r_{\sigma_2}}{r_{\sigma_1}}\sin x + KA_{\sigma_4\sigma_3}\frac{r_{\sigma_3}}{r_{\sigma_4}} + \sin(y - z). \tag{5.36}$$

The corresponding steady states depend on the cellular organization (see table 5.1).

One can now investigate the phase coordination of islet cells given their spatial organization. Real islets have three-dimensional structures with special cellular compositions. Among the many questions that one can ask are: (i) why do mouse and human islets take shell–core and partial mixing structures rather than, sorting or complete mixing structures, (ii) why do alpha and beta cells have asymmetric interactions and what happens if they had mutually positive interactions or mutually negative interactions, (iii) why do mouse and human islets have special cellular compositions and what happens if mouse islets had more alpha cells or human islets had fewer alpha cells, and (iv) how is the phase coordination between islet cells modified under diabetic conditions in which beta cells are selectively destroyed. All these interesting questions can be explored with the population model. One key conclusion for the above questions is that islets are designed to generate asynchronous hormone secretions at normal glucose concentration, and synchronous hormone secretions at high/low glucose concentration [47]. Unlike the native design, if islets had sorting structures of alpha and beta cells or symmetric interactions between them, they will always generate synchronous hormone secretions regardless of glucose concentrations. The controllable synchronization of hormone secretions can thus have functional benefits. For example, the desynchronization at normal glucose can suppress unnecessary hormone actions, while the synchronization at high/low glucose can amplify hormone actions.

Exercises

1. α–β–δ model. Solve the coupled differential equations (5.24)–(5.25) numerically, and check their stationary solutions of $x = \theta_\alpha - \theta_\beta$ and $y = \theta_\alpha - \theta_\delta$ for (i) $r_\alpha = 1$ and $r_\beta = r_\delta = 0.2$; (ii) $r_\beta = 1$ and $r_\alpha = r_\delta = 0.2$; and (iii) $r_\alpha = r_\beta = r_\delta = 1$. Here use $\omega_\alpha = \omega_\beta = \omega_\delta = 1$ and $K = 0.1$. If alpha cells inhibit

delta cells, and delta cells activate alpha cells, how do the above results change?
2. Population model. For the one-dimensional string of two alpha cells and two beta cells, compute the stationary coordinations between cells for (i) $r_\alpha > r_\beta$ and (ii) $r_\alpha < r_\beta$, and compare the results with table 5.1.
3. The structure data of mouse and human islets are available in Supporting Information of reference [47]. Fix the alpha cell activity ($r_\alpha = 1$), and compute the phase synchronization between islet cells for varying beta cell activity ($0.1 < r_\beta < 10$).
   (a) The degree of phase synchronization can be characterized by an order parameter

$$R_\sigma e^{\Theta_\sigma} = \frac{\sum_{k=1}^{n} \delta_{\sigma,\sigma_k} e^{i\theta_k}}{\sum_{k=1}^{n} \delta_{\sigma,\sigma_k}},$$

   where R_σ measures the phase coherence (0 for perfect desynchronization and 1 for perfect synchronization) and Θ_σ represents the average phase of $\sigma \in \{\alpha, \beta, \delta\}$ cells. Given an islet structure, compute R_α, R_β, and $\Theta_\alpha - \Theta_\beta$ for varying r_β/r_α.
   (b) Repeat the above computation again after modifying the interaction between alpha and beta cells: (i) mutual activation and (ii) mutual inhibition.
   (c) Repeat the same calculation after modifying cellular compositions. One may change some beta cells to alpha cells, or *vice versa*.
   (d) Repeat the same calculation after removing some beta cells. The cell removal makes empty sites which mimic diabetic situations.

References

[1] Koerker D J, Ruch W, Chideckel E, Palmer J, Goodner C J, Ensinck J and Gale C C 1974 *Science* **184** 482

[2] Orci L and Unger R H 1975 *Lancet* **2** 1243

[3] Guillemin R and Gerich J E 1976 *Annu. Rev. Med.* **27** 379

[4] Jo J, Choi M Y and Koh D S 2009 *J. Theor. Biol.* **257** 312

[5] Koh D S, Cho J H and Chen L 2012 *J. Mol. Neurosci.* **48** 429

[6] Cherrington A D, Chiasson J L, Liljenquist J E, Jennings A S, Keller U and Lacy W W 1976 *J. Clin. Invest.* **58** 1407

[7] Samols E and Harrison J 1976 *Metabolism* **25** 1443

[8] Rorsman P, Berggren P O, Bokvist K, Ericson H, Möhler H, Ostenson C G and Smith P A 1989 *Nature* **341** 233

[9] Ishihara H, Maechler P, Gjinovci A, Herrera P L and Wollheim C B 2003 *Nat. Cell Biol.* **5** 330

[10] Ravier M A and Rutter G A 2005 *Diabetes* **54** 1789

[11] Franklin I, Gromada J, Gjinovci A, Theander S and Wollheim C B 2005 *Diabetes* **54** 1808

[12] Tudurí E, Filiputti E, Carneiro E M and Quesada I 2008 *Am. J. Physiol. Endocrinol. Metab.* **294** E952

[13] Samols E, Marri G and Marks V 1965 *Lancet* **2** 415

[14] Kawai K, Yokota C, Ohashi S, Watanabe Y and Yamashita K 1995 *Diabetologia* **38** 274
[15] Brereton H, Carvell M J, Persaud S J and Jones P M 2007 *Endocrine* **31** 61
[16] Rodriguez-Diaz R, Dando R, Jacques-Silva M C, Fachado A, Molina J, Abdulreda M H, Ricordi C, Roper S D, Berggren P O and Caicedo A 2011 *Nat. Med.* **17** 888
[17] Patton G S, Dobbs R, Orci L, Vale W and Unger R H 1976 *Metabolism* **25** 1499
[18] Weir G C, Samols E, Day J A Jr and Patel Y C 1978 *Metabolism* **27** 1223
[19] Dolais-Kitabgi J, Kitabgi P and Freychet P 1981 *Diabetologia* **21** 238
[20] van der Meulen T, Donaldson C J, Cáceres E, Hunter A E, Cowing-Zitron C, Pound L D, Adams M W, Zembrzycki A, Grove K L and Huising M O 2015 *Nat. Med.* **21** 769
[21] Göpel S O, Kanno T, Barg S, Weng X G, Gromada J and Rorsman P 2000 *J. Physiol.* **528** 509–20
[22] Gromada J, Ma X, Hoy M, Bokvist K, Salehi A, Berggren P O and Rorsman P 2004 *Diabetes* **53** S181
[23] MacDonald P E, Marinis Y Z D, Ramracheya R, Salehi A, Ma X, Johnson P R, Cox R, Eliasson L and Rorsman P 2007 *PLoS Biol.* **5** 1236
[24] Zhang Q *et al* 2013 *Cell Metab.* **18** 871
[25] Elliott A D, Ustione A and Piston D W 2014 *Am. J. Physiol. Endocrinol. Metab.* **308** E130–43
[26] Le Marchand S J and Piston D W 2010 *J. Biol. Chem.* **285** 14389
[27] Wendt A, Birnir B, Buschard K, Gromada J, Salehi A, Sewing S, Rorsman P and Braun M 2004 *Diabetes* **53** 1038
[28] Watts M and Sherman A 2014 *Biophys. J.* **106** 741
[29] Leung Y M, Ahmed I, Sheu L, Gao X, Hara M, Tsushima R G, Diamant N E and Gaisano H Y 2006 *Endocrinology* **147** 2155
[30] Gromada J, Hoy M, Buschard K, Salehi A and Rorsman P 2001 *J. Physiol.* **535** 519
[31] Kailey B, van de Bunt M, Cheley S, Johnson P R, MacDonald P E, Gloyn A L, Rorsman P and Braun P 2012 *Am. J. Physiol. Endocrinol. Metab.* **303** E1107
[32] Hellman B, Salehi A, Gylfe E, Dansk H and Grapengiesser E 2009 *Endocrinology* **150** 5334
[33] Hellman B, Salehi A, Grapengiesser E and Gylfe E 2012 *Biochem. Biophys. Res. Commun.* **417** 1219
[34] Grapengiesser E, Gylfe E and Hellman B 1991 *J. Biol. Chem.* **266** 12207 https://www.jbc.org/content/266/19/12207.long
[35] Berts A, Ball A, Gylfe E and Hellman B 1996 *Biochim. Biophys. Acta* **1310** 212
[36] Berts A, Ball A, Dryselius G, Gylfe E and Hellman B 1996 *Endocrinology* **137** 693
[37] Hong H, Jo J and Sin S J 2013 *Phys. Rev. E Stat. Nonlinear Soft Matter Phys.* **88** 032711
[38] Acebrón J A, Bonilla L L, Vicente C J P, Ritort F and Spigler R 2005 *Rev. Mod. Phys.* **77** 137
[39] Cabrera O *et al* 2008 *Cell Metab.* **7** 545
[40] Cho J H, Chen L, Kim M H, Chow R H, Hille B and Koh D S 2010 *Endocrinology* **151** 1541
[41] Leibiger B, Moede T, Muhandiramlage T P, Kaiser D, Vaca Sanchez P, Leibiger I B and Berggren P O 2012 *Proc. Natl Acad. Sci. USA* **109** 20925
[42] Gilon P, Cheng-Xue R, Lai B K, Chae H Y and Gómez-Ruiz A 2015 *Islets of Langerhans* (Berlin: Springer) pp 175–247
[43] Aspinwall C A, Lakey J R and Kennedy R T 1999 *J. Biol. Chem.* **274** 6360
[44] Leibiger I B, Leibiger B and Berggren P O 2002 *FEBS Lett.* **532** 1
[45] Braun M, Ramracheya R, Bengtsson M, Clark A, Walker J N, Johnson P R and Rorsman P 2010 *Diabetes* **59** 1694
[46] Jacques-Silva M C *et al* 2010 *Proc. Natl Acad. Sci. USA* **107** 6465
[47] Hoang D T, Hara M and Jo J 2016 *PLoS One* **11** e0152446

IOP Publishing

Diabetes Systems Biology
Quantitative methods for understanding beta-cell dynamics and function
Anmar Khadra

Chapter 6

The role of islet cell network in insulin secretion

I Johanna Stamper, Nathaniel Wolanyk and Xujing Wang

6.1 Rhythmicity in insulin and glucose and the basis for it

In this subsection we present recent progress made in investigating one basic topic in glucose homeostasis and diabetes: specifically, the regulation of insulin secretion and how that depends on islet architecture in health and disease. This investigation is centered on a fundamental concept in complex systems: that structure defines function, and that the function is often manifested in the form of emergent spatiotemporal orders. Studying glucose homeostasis from this perspective shows how nonlinearities arise, elucidates unsolved mysteries and controversies in the study of glucose tolerance and related pathological conditions such as diabetes, and sheds light on new directions for research and clinical intervention protocol development.

In complex systems, regulatory signals are often transmitted through emergent temporal rhythms. Utilizing emergent temporal orders may be the most efficient way to coordinate signals and regulations within a complex system. In living systems oscillations are observed across a wide range of temporal and spatial scales; these oscillations are believed to play an important role in maintaining homeostasis and in delivering encoded information [1–3]. As highlighted in previous chapters, glucose homeostasis is a basic physiological process that provides energy to all cells in our body. It is subject to regulation by a number of endogenous and exogenous variables. Both the blood concentration of glucose and that of its primary regulating hormone insulin oscillate at multiple time scales [4]. The most noteworthy oscillations include the ~6 h rhythm that is in part dictated by mealtimes, the high-frequency rhythms in the 10–15 min range, the ultradian rhythms in the 120–150 min range, and the low-amplitude circadian rhythm [4].

Extensive evidence suggests that the dynamic rhythmicity is essential for glucose control [5–7]. Both the high-frequency (10–15 min) and the ultradian (120–150 min) insulin oscillations have been shown to enhance glucose utilization in peripheral tissues [8]. The enhanced effect of the high-frequency pulses was found likely to be due to an inhibition of hepatic glucose production [9]. Type 2 Diabetes (T2D) patients and subjects with impaired glucose tolerance (see chapter 7) both show alteration in the ultradian rhythmicity of their insulin secretion [10–12]. In studies of clinically healthy subjects, where the endogenous insulin was suppressed by somatostatin, and glucose was infused at a constant rate intravenously, ultradian oscillatory insulin infusion was observed to be more efficient at reducing and regulating the blood glucose concentration compared to a situation where a steady, continuous insulin administration was utilized [13]. In clinical intervention studies, insulin delivered in a pulsatile fashion has been shown to exert a greater hypoglycemic effect than continuous delivery [7, 14].

The origin of the high-frequency pulsations in insulin is synchronized beta cell oscillations upon glucose stimulation, which give rise to synchronized oscillatory insulin secretion. The high frequency and ultradian oscillations, when coupled with other oscillatory systems, including the glycolysis metabolic oscillation, meal cycles, day/night cycles, and circadian rhythms in gene expression, etc, generate complex multi-scale oscillations in insulin and glucose [15–18].

In this subsection, we first present studies of beta cell oscillation and show how it is affected by the beta cell network structure inside an intact islet. We then present a model of glucose-stimulated intracellular insulin granule trafficking (similar to that highlighted in chapter 3) and show how it explains some of the rate limiting factors determining the insulin secretion rate. An integration of the insulin granule trafficking model and the islet cell network model is then presented to investigate directly how islet cell network structure affects insulin secretion rate and dynamic rhythmicity. A discussion of these studies along with potential clinical applications and significance is provided towards the end.

6.2 Islet cell network and beta cell oscillation

6.2.1 High and ultradian insulin rhythms originate at the level of pancreatic islet beta cells

As described in the previous chapters, the insulin-releasing beta cells reside in the pancreas (figure 6.1), a tadpole-shaped endocrine organ, located in the abdominal cavity behind the stomach. The beta cells comprise about 1% of the total pancreatic mass [19]. Rather than being distributed evenly throughout the pancreas, or forming a single big cluster like cardiac or liver cells, beta cells are organized in a hierarchical way as depicted in figure 6.1. They reside in the highly organized islets of Langerhans, with specific 3D morphostructure, and tight intercellular couplings. Inside a human pancreas, immersed in the exocrine tissue sea, there are several millions of islets (figure 6.1, top left), constituting a few % of the total organ volume. Each islet on average consists of 10^2–10^3 cells, with the majority (~50%–70%) being beta cells, surrounded by the glucagon-secreting alpha and somatostatin-secreting delta cells (figure 6.1, bottom left).

It is well known that, in complex systems, structure defines function. Thus a natural question one would ask is whether the hierarchical organization of beta-cell mass serves a purpose. Circumstantial knowledge of islet size conservation across different mammal species suggests that it does. Across mammals of different sizes,

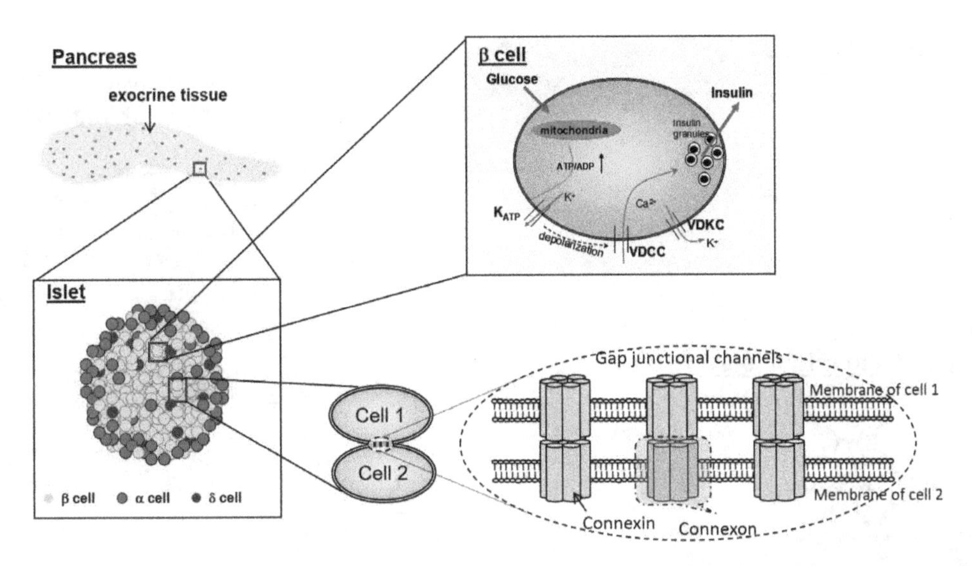

Figure 6.1. Hierarchical organization of beta-cell mass within islets of Langerhans. Islets are scattered throughout the pancreas; each islet is a tight cluster of cells, with the majority being beta cells. Inside an intact islet, beta cells are electrically coupled to each other through gap junctional channels. The exocytosis of insulin granules is driven by a glucose-initiated membrane depolarization process (see chapter 3 and the previous subsection). Figure is adapted from figure 1 of [37] and figure 1 of [157].

islet size is relatively conserved at 10^2–10^3 cells. Larger mammals compensate for the higher demand of insulin by having more islets, rather than larger islets. For instance, a rodent has ~10^3 islets; a human body, which has a body weight about 1000 fold greater, has ~10^6 islets [20, 21].

Examination of the relationship between beta-cell mass organization and oscillatory insulin release further suggests the functional significance of beta cells' being organized in islets. We have seen before that inside intact islets, beta cells are electrically coupled to each other through gap-junctional channels [22–27] (figure 6.1, bottom right), forming a beta cell network. Individual beta cells are heterogeneous. Both laboratory observations and numerical simulations have demonstrated that due to their heterogeneity, isolated individual beta cells exhibit a wide range of responses upon stimulation, which can be classified in three categories (figure 6.2, top panel): bursting, spiking, or silent (see chapter 2 for more details). Bursters in particular are thought to be preferable in that they are capable of generating a dose-dependent response to stimulus (glucose). A cluster of isolated beta cells responds poorly to glucose, which is characterized by higher basal secretion, and lower secretion under stimulatory glucose [18, 28–31] (figure 6.2 top panel). Laboratory studies have also found that islet function is compromised if beta cells lose their ability to couple with neighboring beta cells, even if they are otherwise normal [24, 26]. Gap-junctional coupling enables beta cells to entrain each other to reach synchronized bursting upon glucose stimulation [18, 31–36] (figure 6.2, bottom panel), to respond appropriately to a varying glucose dose, and to release oscillatory insulin [18, 25, 29, 35, 37, 38]. Pulsatile insulin release from islets leads to oscillations in blood-borne insulin [5, 39],

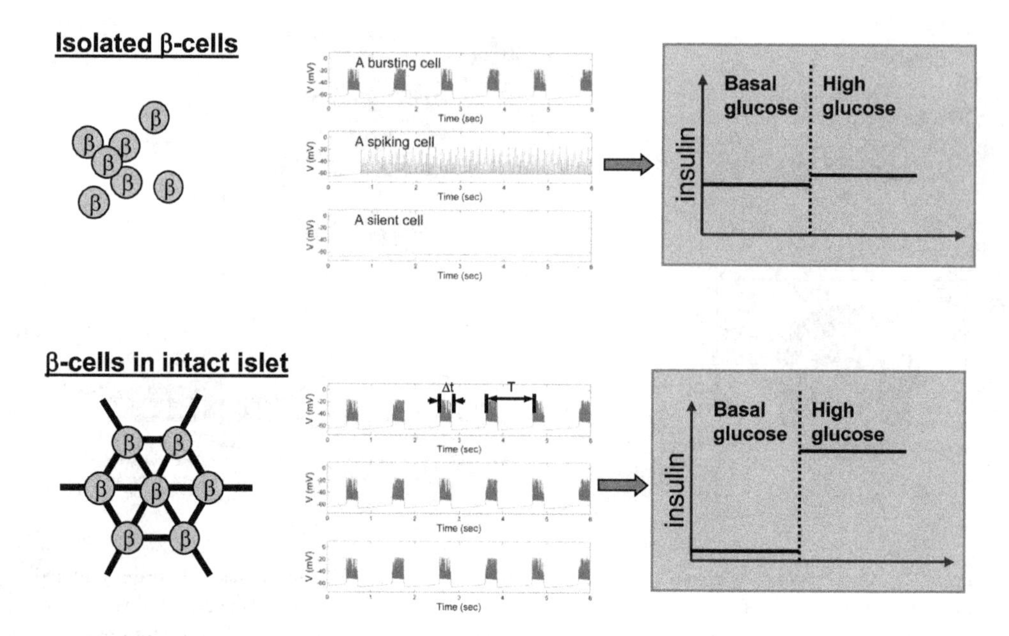

Figure 6.2. The functional importance of inter-beta cell coupling. Top: the glucose response from a cluster of isolated beta cells is heterogeneous, with some cells bursting, some spiking and some silent. For such a cluster of beta cells, the overall sensitivity to glucose dose changes is low. Bottom: in an intact islet, coupled beta cells burst synchronously in response to elevated glucose, with high sensitivity to glucose dose changes. The amount of insulin secretion is believed to be proportional to $\frac{\Delta t}{T}$, the 'plateau fraction'.

which is believed to be critically important for insulin action [9, 40, 41]. Loss of oscillation results in insulin resistance [5–7], and has been observed in obese and diabetic individuals [42], and individuals at risk of diabetes [5, 43–45].

These observations suggest that the islet may be not only an anatomical unit but also a functional one. Further supporting this notion, recent studies found that the morphostructural integrity of the islets, namely the right 3D architecture and interactions among cell populations, were predictive of the *in vivo* function and clinical outcome of islet transplantation [46]. It is intriguing to hypothesize that 10^2–10^3 cells as observed in mammals may be the optimal size for the beta cell network to generate functionally desirable spatio-temporal rhythms. Therefore, a comprehensive understanding of the generation of dynamic rhythms in blood insulin and glucose needs to start from an examination of the emergent beta cell synchrony in pancreatic islets and of the factors which determine the characteristics of their oscillations.

6.2.2 Beta cell oscillation and its mathematical model

As highlighted in chapter 2, pancreatic islet beta cells are electrically excitable cells that can be described electrophysiologically using the Hodgkin–Huxley formalism [48]. Figure 6.1, top right panel, depicts the major membrane ion channels involved in beta cell excitation. In brief, glucose is the main regulator of electrical activity. When the glucose concentration is substimulatory, the beta cell is generally inactive due to hyperpolarization of the membrane potential. The membrane potential stays at its resting value of roughly −70 mV due to an outward K^+ current through open ATP-dependent K_{ATP} channels [49–51]. As glucose reaches a stimulatory concentration, these channels close in response to an increased ATP:ADP ratio as a result of increased mitochondrial metabolism of glucose and a net generation of ATP. Channel closure reduces the efflux of K^+ and depolarizes the cell. This depolarization causes voltage-dependent Ca^{2+} channels to open and Ca^{2+} to enter the cell, which drives the membrane potential to depolarize further, thus initiating action potential spiking. Due to voltage-dependent K^+ channels, which will open in response to membrane depolarization, an outwardly rectifying K^+ current arises that serves to repolarize the cell membrane, eventually making the cell electrically silent [52, 53]. In this way, bursting, consisting of active phases (characterized by a depolarized membrane potential and spiking), followed by inactive phases (characterized by a stable, hyperpolarized membrane potential), can be sustained. As we will see in subsection 2.2, the rate of insulin release is roughly proportional to the 'plateau fraction,' defined as the fraction of time spent in the active phase [54] (figure 6.2), which increases with increasing glucose concentration.

For an isolated beta cell, the rate of change in membrane potential is given by what is essentially Ohm's law for currents:

$$C_m(dV/dt) = -\sum I_{\text{ion}}, \tag{6.1}$$

where C_m is cell membrane capacitance, and $\sum I_{\text{ion}}$ are all the currents through the membrane ion channels of the beta cell. For a minimal description, usually four current I_{ion} terms are included [18, 32]: (1) the glucose-sensitive K_{ATP} channel, I_{KATP} [55]; (2) the voltage-dependent L-type Ca^{2+}-channels, I_{Ca} [56, 57]; (3) delayed rectifier K^+ current, I_{Kn} [55, 58]; and (4) a slow inhibitory K+ current, I_{Ks} [55, 58]. Their expressions are given by

$$\begin{aligned} I_{K_{ATP}} &= g_{K_{ATP}} O_{K_{ATP}}(V - V_K) \\ I_{Ca} &= g_{Ca} \cdot m_\infty(V - V_{Ca}) \\ I_K &= g_K \cdot n(V - V_K) \\ I_S &= g_S \cdot s(V - V_K), \end{aligned} \tag{6.2}$$

where $g_{K_{ATP}}$, g_{Ca}, g_K, g_S are the maximum channel conductances, and V_K and V_{Ca} are K^+ and Ca^{2+} Nernst potentials, respectively, as previously defined. The constant $O_{K_{ATP}}$ is the fraction of open K_{ATP}-channels. A lower value of $O_{K_{ATP}}$ signifies a higher glucose concentration. We remark that a constant value of $O_{K_{ATP}}$ means that glucose is assumed to be a constant as well. In section 6.4 we extend our modeling to allow for a varying glucose concentration. n and s in equation (6.2) are activation variables whose rates of change are given by

$$\begin{aligned} \frac{dn}{dt} &= \frac{1}{\tau_n}(n_\infty - n) \\ \frac{ds}{dt} &= \frac{1}{\tau_s}(s_\infty - s), \end{aligned} \tag{6.3}$$

where the expressions m_∞, n_∞, s_∞ describe the fraction of open channels at steady state for Ca^{2+}, fast K^+ and slow K^+ currents, respectively, given by

$$m_\infty = \frac{1}{1 + \exp((V_m - V)/\theta_m)}$$
$$n_\infty = \frac{1}{1 + \exp((V_n - V)/\theta_n)} \tag{6.4}$$
$$s_\infty = \frac{1}{1 + \exp((V_s - V)/\theta_s)}.$$

The constants V_m, V_n, V_s, and θ_m, θ_n, θ_s describe how steady-state channel activations depend on membrane voltage, V. The intracellular Ca^{2+} concentration is modeled by

$$\frac{d[\mathrm{Ca}^{2+}]}{dt} = f(\alpha I_{\mathrm{Ca}} - k_{\mathrm{Ca}}[\mathrm{Ca}^{2+}]), \tag{6.5}$$

where f is the fraction of free Ca^{2+}, α is a factor converting from chemical to electrical gradient and k_{Ca} is the removal rate of Ca^{2+} in the intracellular space. Note that in chapter 2, Ca^{2+} concentration was denoted using 'c' for simplicity. Details on all parameters and their values are listed in table 6.1.

6.2.3 The role of islet cell network structure in beta-cell oscillation

When coupled to each other, for the ith beta cell in an islet, the rate of change of its membrane potential is given by

$$C_m(dV_i/dt) = -\sum I_{\mathrm{ion},i} + \sum_j g_c(V_i - V_j), \tag{6.6}$$

where $\sum_j g_c(V_i - V_j)$ describes the intercellular coupling, summed over all the nearest neighbors of i. The gap-junction channel conductance g_c determines the coupling strength [59]; under normal physiological conditions $g_c \sim$ 150–250 pS [25, 60, 61].

A number of mathematical and biophysical studies of islet oscillation have been carried out in the past several decades [25, 29, 33, 35, 39, 55, 58, 62–67]. Most efforts have focused on the nonlinearities brought about by properties of individual beta cells, including the kinetics of ion channels affecting membrane voltage and metabolic oscillations inside beta cells (see table 6.1 in [68], for instance). The role of network structure has largely been underappreciated until recently [18, 32]. On the other hand, islet biologists have long speculated about the importance of islet cytoarchitectural integrity [46], In [18, 32], it was proposed that islet cell network integrity can be captured by three key islet network parameters, including:

1. number of beta cells, n,
2. number of couplings each beta cell has, n_c, and
3. coupling strength, g_c.

Table 6.1. Parameters used in equations (6.1)–(6.5) and their values. (Adapted from table 1A of [18].)

Quantity	Description	Value (or mean value)	Unit
C_m	Membrane capacitance of beta cell	5.3	pF
r	Radius of beta cell	6.5	μm
g_K	Maximal conductance of K^+ channel	2500	pS
g_{KCa}	Maximal conductance of Ca^{2+}-activated K^+ channel	30 000	pS
g_{Ca}	Maximal conductance of Ca^{2+} channel	1400	pS
g_{KATP}	Maximal conductance of ATP-activated K^+ channel	150	pS
g_s	Maximal conductance of slow inhibiting K^+ channel	200	pS
V_K	Reversal potential for K^+ channel	−75	mV
V_{Ca}	Reversal potential for Ca^{2+} channel	25	mV
α	Conversion factor that converts electrical gradient to chemical gradient	4.5e-6	$\mu M\, f A^{-1}\, ms^{-1}$
k_{Ca}	Rate constant for removal of intracellular Ca^{2+}	0.04	ms
f	Fraction of free Ca^{2+} in the intracellular space	0.001	
$O_{K_{ATP}}$	Fraction of free open channels	0.5	
m	Steady-state m		
n	Steady-state n		
h	Steady-state h		
V_m	Half-maximal potential for the m curve	−20	mV
V_n	Half-maximal potential for the n curve	−17	mV
V_s	Half-maximal potential for the s curve	−22	mV
S_m	Slope of the m curve at $V = V_m$	12	mV
S_n	Slope of the n curve at $V = V_n$	5.6	mV
S_s	Slope of the h curve at $V = V_s$	8	mV
τ_c	Mean closed time of KCa channel	1000	ms
τ_o	Mean open times of KCa channel		ms
τ_m	Time constant for the K^+ current	0.84	ms
g_c	Gap junctional conductance	[25–1000]	pS

Inside an islet, there is a significant proportion of non-beta cells (mainly alpha and delta cells) that do not couple to beta cells or among themselves [69]; their presence modifies n_c. The proportion of non-beta cells, as well as n and g_c, is readily altered under physiological conditions, such as pregnancy, or pathological conditions, such as obesity and diabetes [70, 71]. Therefore, studying how changes in these measures of network integrity impact network function will not only shed light on the regulation of insulin secretion, but also on islet functional changes under (patho-)physiological conditions.

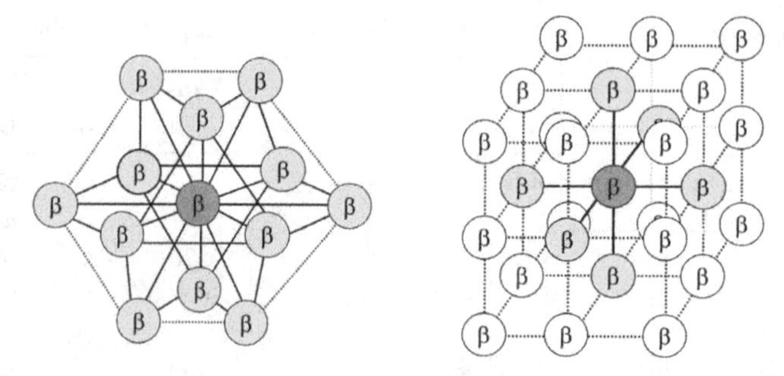

Figure 6.3. Intercellular connectivity (represented by lines between cells) is significantly higher in Hexagonal Closest Packing (HCP) clusters (left) than in simple cubic packing (SCP) clusters (right). In an HCP cluster, a cell (like the one highlighted in orange) is connected to its 12 nearest neighbors (highlighted in yellow), while in an SCP cluster, this number is only 6.

Modeling work on islet oscillation has typically assumed a simple cubic packing (SCP) of beta cells [25, 72], with each beta cell coupled to its n_c nearest neighbors ($n_c = 4$ in 2D and $n_c = 6$ in 3D), as shown in figure 6.3, right panel. When compared to laboratory measurements [38, 73], we found that this assumption regarding SCP significantly underestimates the real degree of inter-beta cell coupling [18], limiting the study of the functional consequences of the diversity and complexity in islet network topology [18, 46, 74, 75]. Physically, a biological cell is more reminiscent of a sphere than a cube, and cells can be packed more efficiently using the hexagonal closest packing lattice (HCP). Recently a new HCP beta cell cluster model was introduced which has $n_c = 6$ in 2D and $n_c = 12$ in 3D (figure 6.3). When compared to experimental measurements [38, 73, 76], HCP captures accurately the degree of intercellular coupling, and approximates much more closely the cytoarchitectural organization of islet cells [18, 32]. Such a model is the first of its kind that examines the H–H model of beta-cell electrophysiology in a lattice geometry different from cubic packing.

We simulated in total over 1000 HCP and SCP beta cell clusters with n ranging from 1–587, g_c from 0–1000 pS, and n_c from 0-12 (by randomly removing beta cells from the network). We introduced islet cell heterogeneity by randomly assigning a subset of the parameters values from normal distributions with standard deviations equal to five percent of the means. Five islet functional measures are then examined [18, 32]:

1. fraction of burster cells f_b,
2. synchronization index λ,
3. bursting period T_b,
4. plateau fraction p_f, and
5. Ca^{2+} oscillation amplitude [Ca].

The results clearly demonstrate that islet function depends nonlinearly both on its size, n, and on its cytoarchitectural organization and network integrity (n_c, g_c). Overall, if the values of (n, n_c, g_c) are small, then higher values of (n, n_c, g_c) make beta

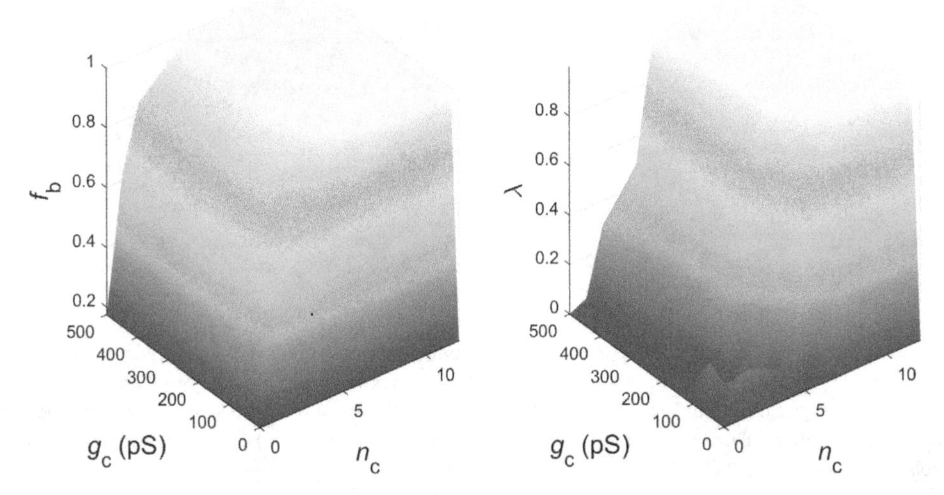

Figure 6.4. The fraction of burster beta cells f_b and their synchronization index λ are plotted against coupling strength g_c and network degree n_c.

cell clusters functionally more desirable and robust [18]. Figure 6.4 depicts how f_b and λ depend on (n_c, g_c). However, such improvements in function plateau around $n > \sim 10^2$, $n_c > \sim 6$, $g_c > \sim 100$ pS. Interestingly the normal physiological values are $n \sim 10^2$–10^3, $n_c \sim 6$–9, $g_c \sim 100$–300 pS across different mammal species [61, 69]. Presumably, normal islets arrive at and function in these ranges through evolution. Normal islets are robust against loss of beta cells, up to roughly 70% loss [18]. The threshold value of beta cell loss corresponding to impaired or lost function depends on coupling strength. When coupling is weak, the islet is less tolerant to beta-cell loss, as evident in figure 6.4.

6.3 Intracellular insulin granule trafficking during glucose-stimulated insulin secretion

The previous subsection described how islet network structure affects beta-cell oscillation, and potentially how network structure, through its regulation of the functional characteristics (such as the five functional measures in section 6.2), affects glucose dose response and the temporal orders in insulin released into blood. In order to study directly how islet cell network structure determines the insulin secretion rate and its temporal characteristics, we need to first examine how insulin granules inside beta cells are mobilized and secreted.

Normal beta cells contain a large amount, ~10 000, of insulin granules, organized into functionally different pools, most notably the reserve pool, the (membrane) docked pool, and the readily releasable pool (RRP). The majority of the granules (> 90%) reside in the reserve pool. At any given time, it is estimated that only a small fraction (~50–100 granules) are in the RRP and in a readily releasable mode [77, 78]. The discrete pool organization of intracellular insulin granules is linked to biphasic insulin secretion and other secretion patterns, such as the staircase experiment.

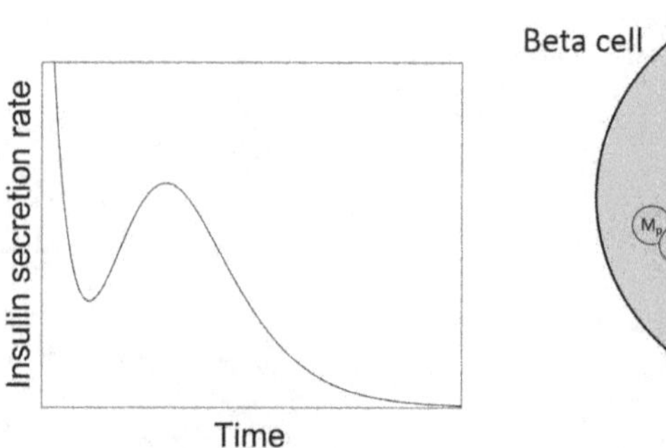

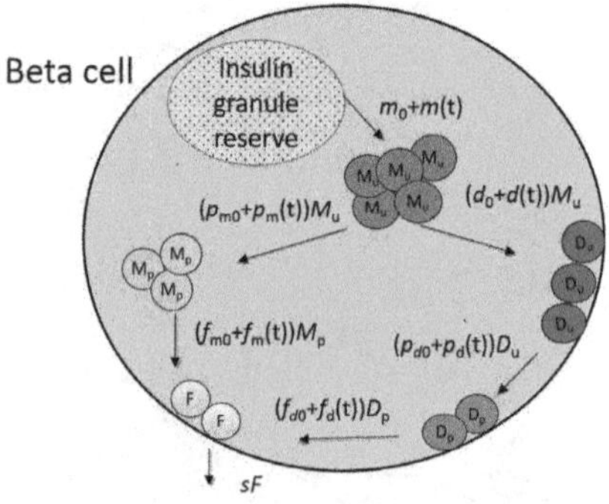

Figure 6.5. Schematic diagram of a biphasic insulin secretion profile (left) and the new model developed in [86] (right). Insulin secretion occurs via granule fusion and exocytosis following a sequence of events involving mobilization of granules from the reserve pool and priming, the latter of which may or may not include docking as an intermediate step. The rates at which granules are transferred between the different granule pools are shown next to the arrows.

Under glucose stimulation, insulin secretion from islets exhibits a biphasic pattern [77–79] (figure 6.5, left), consisting of a rapidly initiated and transient first phase of ~10–15 min, followed by a sustained second phase lasting ~60–120 min, during which secretion continues at a lower rate [77] (further analysis of this secretion profile will be provided in chapter 7). The biphasic insulin response was first reported by Grodsky's group almost five decades ago [80], and first quantitatively analyzed by the same group several years later [81]. It is now generally believed that the granules secreted during the 1st phase mainly come from the RRP [78, 80, 82, 83]; thus the size and the refilling rate of the RRP limit the rate of first phase insulin release. The insulin release slowly rises during the second phase, and this is generally attributed to glucose-dependent replenishment of the RRP [77, 78, 84, 85].

6.3.1 A new mathematical model of insulin granule trafficking and insulin secretion with glucose-dependent granule mobilization, priming and fusion

A new mathematical model that describes intracellular insulin granule trafficking and insulin secretion has been proposed [86]; its general scheme is presented in figure 6.5, right panel. Compared to other existing efforts it is different in the number of functional pools included, and in the assumption of a homogeneous, graded glucose response from all beta cells in an intact islet.

In Grodsky's first mathematical model of insulin secretion [81], two pools of insulin were proposed: one small pool ready for release and a larger pool responsible for refilling the first pool. In the new model [86], five major functional compartments of insulin granules are defined depending on their functional states and location: mobilized unprimed granules $M_u(t)$, mobilized primed granules $M_p(t)$, docked unprimed granules $D_u(t)$, docked primed granules $D_p(t)$, and fused granules $F(t)$. Secretion comes from the fused pool, while the RRP consists of primed pools $M_p(t) + D_p(t)$. Note that, in this model, it is assumed that the fused pool is refilled both from the docked primed pool and the mobilized primed pool. This is different from some existing models. Not long ago it was believed that granules had to be pre-docked at the plasma membrane to be ready for fusion [78, 84]. However, recent experimental observations have revealed that stable docking may not be necessary for granules to fuse to the plasma membrane [85, 87–95]. It has been observed that mobilized granules from the cell interior, sometimes called 'newcomers', can rapidly fuse and exocytose [85, 91, 93, 95]. The exact mechanisms underlying these signaling pathways are still under investigation and remain incompletely understood [96–99]. The likely prerequisites for this process, however, include the remodeling of filamentous actin (F-actin) [84], and ATP-dependent granular acidification or other chemical reactions (a process called *priming*), which convert the granules into the release-ready state [84, 100].

It is assumed in this new model [86] that mobilization of granules from the core of a beta cell, docking of granules to the cell membrane, priming of unprimed granules, and fusion of primed granules are all potentiated in a glucose-dose-dependent fashion. More specifically, the dynamics of each pool can be described by [86]:

$$
\begin{aligned}
\dot{M}_u &= \underbrace{m_0 + \tilde{m}(t)}_{\text{mobilization}} - \underbrace{(p_{m0} + \hat{p}_m(t))M_u}_{\text{priming}} - \underbrace{(d_0 + \hat{d}(t))M_u}_{\text{docking}} \\
\dot{M}_p &= \underbrace{(p_{m0} + \hat{p}_m(t))M_u}_{\text{priming}} - \underbrace{(f_{m0} + \hat{f}_m(t))M_p}_{\text{fusion}} \\
\dot{D}_u &= \underbrace{(d_0 + \hat{d}(t))M_u}_{\text{docking}} - \underbrace{(p_{d0} + \hat{p}_d(t))D_u}_{\text{priming}} \\
\dot{D}_p &= \underbrace{(p_{d0} + \hat{p}_d(t))D_u}_{\text{priming}} - \underbrace{(f_{d0} + \hat{f}_d(t))D_p}_{\text{fusion}} \\
\dot{F} &= \underbrace{(f_{m0} + \hat{f}_m(t))M_p}_{\text{fusion}} + \underbrace{(f_{d0} + \hat{f}_d(t))D_p}_{\text{fusion}} - \underbrace{sF}_{\text{secretion}},
\end{aligned}
\tag{6.7}
$$

where the parameters are depicted in figure 6.5, right panel. According to equations (6.7), the rates of all the processes are in proportion to size of the originating granule pools, with the exception of mobilization. Size of the reserve pool is large compared to the other pools ($\sim 10^4$ versus $\sim 10^2$), and thus considered a constant and absorbed by the rates m_0 and $\tilde{m}(t)$. Additionally, glucose-induced mobilization, $\tilde{m}(t)$, is assumed to take place only when the number of mobilized granules is below a saturation level, $M_{u,\max}$, as defined by:

$$
\tilde{m}(t) = \hat{m}(t)\left(1 - \frac{M_u}{M_{u,\max}}\right). \tag{6.8}
$$

In equations (6.7), the parameters containing a zero in their subscript describe the basal rate constants in the absence of glucose. Parameters in the form of $\hat{z}(t)$ (i.e. $\hat{m}$, $\hat{d}$, $\hat{p}_m$, $\hat{f}_m$, $\hat{p}_d$, $\hat{f}_d$), denote the glucose-dependent components, and are given by equation (6.9),

$$\hat{z} = \frac{z[G(t-\delta_x) - G_x]^n}{[G_{x50} - G_x]^n + [G(t-\delta_x) - G_x]^n} H(G(t-\delta_x) - G_x), \tag{6.9}$$

where $H(x)$ is the Heaviside step function, with $H(x) = 0$ if $x < 0$, and $H(x) = 1$ if $x \geqslant 0$. According to equation (6.9), the glucose-dependent processes are assumed to occur when the level of extracellular glucose, $G(t)$ exceeds a certain threshold, G_x, and with a time delay of δ_x (representing the time it takes glucose to enter the cells and become metabolized [84]). The rate of each glucose-dependent process is modeled as a saturating function of extracellular glucose, taking the form of a Hill equation (frequently utilized to model biological processes) [101, 102]. All model parameters, what they describe, their units and normal value ranges, are summarized in table 6.2 (adapted from table A2 of [86]).

Equations (6.7) describe the insulin granule trafficking inside each individual beta cell. For description of insulin granule release from a population of beta cells in an islet, a heterogeneous recruitment model is often used. The concept was first proposed by Grodsky to reproduce the staircase experiment (where the glucose concentration increased multiple times in steps) [81]. The idea was that the RRP consisted of individual 'packets' that have different sensitivity to glucose and are released at different glucose concentrations. In a later model proposed by Pedersen *et al* (see chapter 3), it was assumed that different beta cells respond to glucose at different activation or recruitment thresholds. In addition, the fraction of the beta cells that is activated increases with glucose dose in a sigmoidal fashion, and all the active cells release insulin with the same rate constant [103]. The overall glucose dose response in this framework relied on an increasing fraction of beta cells' being recruited at higher glucose levels. The model was able to reproduce the staircase experiment [103].

Later, a homogeneous recruitment model was then proposed in which all individual beta cells were assumed to have similar recruitment thresholds of G_*, and the glucose dose response relies on every beta cell's response upsurge with increasing glucose dose [97, 104]. While experiments have shown that individual beta cells indeed exhibit heterogeneity in glucose sensitivity [105], it has also been observed that for beta cells in an intact islet, or for cells interconnected in clusters, the recruitment occurs within a narrower range of glucose concentrations [105]. In addition, even for glucose concentrations over the value corresponding to maximal recruitment, so that all beta cells are presumed to be activated, insulin secretion still increases in a dose-dependent way [105]. These observations together with the knowledge of the importance of beta cell synchrony to function (discussed in the last subsection) prompted the investigation of the homogeneous recruitment hypothesis, by incorporating it in the new mathematical model described in equations (6.7) [86].

The number of granules secreted by a single β-cell per unit time is given by sF_i for cell i. Denoting the insulin content of one granule by I_g and the number of beta cells in an islet by n_β, it follows that the total insulin secretion rate, $I(t)$, from an islet is given by

$$I(t) = \sum_{i=1}^{n_\beta} I_g s F_i(t), \tag{6.10}$$

Table 6.2. Parameters used in equations (6.7) and their values. (Adapted from table A2 of [86].)

Parameter	Unit	Values	Parameter description
C_m	pF	5300 (mean)	Membrane capacitance
g_c	pS	200	Gap-junctional conductance
f	1	0.01	Fraction of free Ca^{2+} in the intracellular space
α_i	μM [fA]$^{-1}$ [ms]$^{-1}$	4.5×10^{-6} (mean)	Conversion factor for electrical into chemical gradient
$k_{\mathrm{Ca},i}$	[ms]$^{-1}$	0.2–0.6 (mean)	Removal rate of Ca^{2+} from the intracellular space
g_{Ca}	pS	1000 (mean)	Maximal conductance of the Ca^{2+}-channel
$g_{\mathrm{K_{ATP}}}$	pS	150 (mean)	Maximal conductance of the ATP-activated K^+-channel
g_{K}	pS	2700 (mean)	Maximal conductance of the K^+-channel
g_{S}	pS	200 (mean)	Maximal conductance of the slow-inhibiting K^+-channel
V_{Ca}	mV	25	Reversal potential for the Ca^{2+}-channel
V_{K}	mV	−75	Reversal potential for the K^+-channel
V_m	mV	−20	Half-maximal potential for the m_∞ curve
θ_m	mV	12	Voltage constant
V_n	mV	−16	Half-maximal potential for the n_∞ curve
θ_n	mV	5.6	Voltage constant
V_s	mV	−52	Half-maximal potential for the s_∞ curve
θ_s	mV	5–10	Voltage constant
τ_n	ms	20	Time constant
τ_s	ms	2×10^4	Time constant
$G_{\mathrm{K_{ATP}}50}$	mmol l^{-1}	7–10	Glucose concentration at which fraction of free K_{ATP}-channels is half-maximal
γ	1	6	Hill coefficient
m_0	[ms]$^{-1}$	$1/6 \times 10^{-4}$	Basal mobilization rate of unprimed granules
p_{m0}	[ms]$^{-1}$	$4/6 \times 10^{-7}$	Basal rate of priming of unprimed mobilized granules
f_{m0}	[ms]$^{-1}$	$8/15 \times 10^{-6}$	Basal rate of fusion of primed mobilized granules
s	[ms]$^{-1}$	$1/2 \times 10^{-3}$	Rate of insulin secretion
$M_{u\max}$	1	7×10^3	Threshold value of M_u when glucose-induced mobilization of granules toward the plasma membrane ceases
m	[ms]$^{-1}$	1×10^{-3}	Maximum glucose-induced mobilization rate of unprimed granules
G_m	mmol l^{-1}	2.8	Glucose concentration at which glucose-induced mobilization of unprimed granules is activated

(*Continued*)

Table 6.2. (*Continued*)

Parameter	Unit	Values	Parameter description
κ_m	1	1	Hill coefficient
κ_p	1	1	Hill coefficient
κ_f	1	1–12	Hill coefficient
G_{m50}	mmol l^{-1}	14	Glucose concentration at which mobilization of unprimed granules is half-maximal
p_m	$[\text{ms}]^{-1}$	$1/6 \times 10^{-7}$	Maximum glucose-induced priming rate of unprimed mobilized granules
G_p	mmol l^{-1}	2.8	Glucose concentration at which glucose-induced priming of unprimed granules is activated
G_{p50}	mmol l^{-1}	14	Glucose concentration at which priming of unprimed granules is half-maximal
f_m	$[\text{ms}]^{-1}$	$1.6/6 \times 10^{-4}$	Maximum Ca^{2+}-induced fusion rate of primed mobilized granules
Ca_f	μmol l^{-1}	0.07	Intracellular Ca^{2+} concentration at which Ca^{2+} induced fusion of primed granules is activated
Ca_{f50}	μmol l^{-1}	0.4	Intracellular Ca^{2+} concentration at which fusion of primed granules is half-maximal
I_g	ng	8×10^{-6}	Insulin content of one granule
N	1	10^2–10^3	Number of cells per islet
n_β	1	10^2–10^3	Number of beta cells per islet

where s is previously defined in equations (6.7) and figure 6.5 right panel. The description and value ranges for all parameters in equation (6.10) can be found in table 6.2.

6.3.2 Model outcomes in comparison to experimental observations

The key assumptions made in the new model presented in equations (6.7)–(6.10) include: homogeneous beta cell activation, the rate of fusion within each beta cell is a graded increasing function of the glucose concentration, and fusion occurs from pre-docked granules as well as from newcomers. Numerical analysis shows that it is able to reproduce many of the reported insulin secretion dynamics [86], including the biphasic secretion after a step increase of glucose described in [106] (figure 6.6(A)), the potentiation phenomenon [106] (figure 6.6(B)), the staircase experiments [106] (figure 6.6(C)), and a number of other protocols [86, 107].

The model predicts that the strength of insulin granule mobilization, priming and fusion are critical limiting factors for the total amount of insulin release. The results indicate that the rapid spike in first-phase insulin secretion depends on rapid

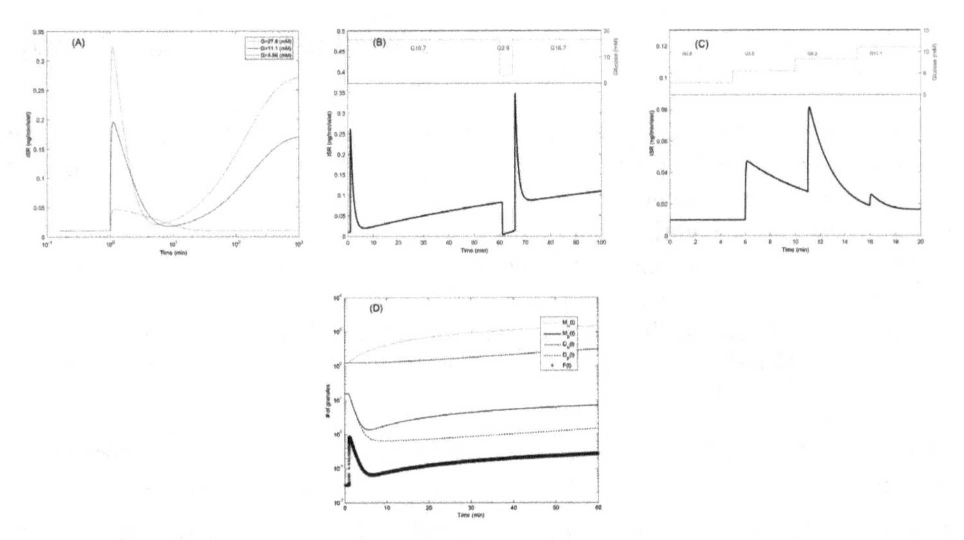

Figure 6.6. The mathematical model presented in figure 6.5 and equations (6.7) is able to reproduce the insulin secretion patterns observed in most experimental protocols, including (A) the biphasic secretion after a step increase of glucose described in [106], (B) the potentiation phenomenon (compare to figure 6.6 in [106]), (C) the staircase experiments (compare to figure 6.4 in [106]), and (D) the temporal change in the five granule pools after the step increase of extracellular glucose from basal to 16.7 mM.

depletion of the primed pools, while the second phase critically relies on granule mobilization, the strength of which determines the total amount of insulin released during that phase (figure 6.6(D)). Moreover, the strength of granule priming influences how fast the second phase increases and reaches steady state. The relative contribution of docked granules versus newcomers to the second phase depends on the relative strength of docking versus priming of unprimed mobilized granules.

The results indicate that heterogeneous recruitment is not necessary to reproduce observations obtained from the experimental protocols. For the staircase experiment in particular, heterogeneity has often been considered essential, as many previous mathematical models have indicated. The model described in equations (6.7) and (6.10) shows that a graded, glucose-dose dependent response from beta cells that activate at the same glucose concentration may also play a role. The amount of heterogeneity within an islet may be different between different species. Islet studies have shown that rat [108] and human islets [75, 109] may be less synchronized than mouse islets. However, studies of glucose-induced Ca^{2+} oscillations in beta cells within intact human islets have shown that signals may be synchronous within smaller clusters of beta cells grouped together [109]. Similarly, cytoarchitectural studies suggest that human and rat islets may contain multiple subcompartments of cell clusters. If recruitment heterogeneity is indeed low in each subcompartment, then the model used here would be appropriate for describing these islet subunits. Exactly how heterogeneity in glucose activation and a graded response contribute to the insulin secretion patterns in different species, and what the implications are for translational research, are promising areas for future study.

6.4 The role of islet network structure in insulin secretion rate and pulsatility

In this subsection, we present some of the latest work [110] to integrate the islet network electric oscillation model, presented in section 6.2 and the model of intracellular insulin granule trafficking and secretion, presented in section 6.3, to examine how islet network structure affects the islet insulin secretion rate and its pulsatility. We will investigate these questions through simulation of biphasic insulin secretion under a step increase of glucose concentration. Efforts to mathematically describe the biphasic insulin secretion are far from finished, with many complex mechanistic and technical issues still to be resolved [66, 79, 80, 103, 111–113].

6.4.1 A model of a network of coupled beta cells that incorporates glucose-dependent membrane potential, and intracellular Ca^{2+} and granule dynamics

In order to reduce the number of equations that needs to be solved for each beta cell, we simplify the model presented in section 6.3. By setting $\hat{d}_0 = \hat{d}(t) = p_{d0} = \hat{p}_d(t) = f_{d0} = \hat{f}_d(t) = 0$ in equations (6.3) that describe the evolution of the intracellular granule pools, we are left with only three equations for $\dot{M}_u$, $\dot{M}_p$ and $\dot{F}$, where M_u and M_p represent here unprimed and primed granule pools, respectively, each of which consists of both docked and undocked granules mobilized to the vicinity of the cell membrane. The simplified model with only three equations gives similar insulin secretion dynamics as the full system. We also assume that the time delays in the glucose-dependent rate constants are zero ($\delta_m = \delta_p = \delta_f = 0$). With the exception that there is no longer a delayed response, the solutions remain unaffected by this assumption.

Instead of assuming that each beta cell's level of granule fusion and insulin secretion are simply dependent on glucose, as was done previously, we now refine the model by using intracellular Ca^{2+} as the triggering factor. More specifically, in equation (6.9) we use the following fusion rate coefficient [110]

$$\hat{f}_m(t) = \frac{f_m\left[[\mathrm{Ca}^{2+}]_i(t) - \mathrm{Ca}_f\right]^{\kappa}}{\left[\mathrm{Ca}_{f50} - \mathrm{Ca}_f\right]^{\kappa} + \left[[\mathrm{Ca}^{2+}]_i(t) - \mathrm{Ca}_f\right]^{\kappa}} H([\mathrm{Ca}^{2+}]_i(t) - \mathrm{Ca}_f), \qquad (6.11)$$

where Ca_{f50} is the level of intracellular Ca^{2+} at which the Ca^{2+}-dependent rate of fusion of primed mobilized granules is half-maximal ($f_m/2$), and Ca_f is the activation threshold. By changing $\hat{f}_m(t)$ in this way, we are able to directly connect the model of granule trafficking given in equations (6.7) to that of beta-cell electrical activity given by equations (6.1)–(6.5) (in which the intracellular Ca^{2+}concentration is one of the dependent model variables).

In order to introduce into the model the dependence of electrical responses on glucose, we replace the constant $O_{\mathrm{K_{ATP}}}$, the fraction of open $\mathrm{K_{ATP}}$-channels (see chapter 2), with the following equation:

$$O_{K_{ATP}} = \frac{1}{1 + (G/G_{K_{ATP}50})^{\gamma}}, \tag{6.12}$$

where $G_{K_{ATP}50}$ is the glucose concentration at which $O_{K_{ATP}}$ is half-maximal and γ is a positive constant that determines the steepness of the channel open probability as a function of glucose (G). In this way, the conductance of the K_{ATP}-channel, given by $g_{K_{ATP}} O_{K_{ATP}}$ in equation (6.2), depends on glucose, G. In particular, as glucose is increased, the conductance is lowered due to fewer open K_{ATP}-channels.

6.4.2 Electrical response and insulin secretion as glucose is raised in a single step

We now present the results from simulating islets consisting of 587 beta cells arranged in a 3D HCP network. The results demonstrate how electrical oscillation and insulin secretion are altered when the fraction of beta cells, F_β, varies in the islet; this is done to mimic the compromised islet cell network integrity due to beta cell death under pathological conditions (such as during diabetes).

Figure 6.7, left panel, shows how the electrical response of all beta cells, in terms of the evolution of their membrane potential, varies with different glucose doses for a healthy functional islet with 10% non-beta cells (where $F_\beta = 0.9$). We see that the islet acts like a unit in which the cells are synchronized. The cells also exhibit a normal response to varying glucose doses. In particular, the cells are inactive at low glucose, and become active when the glucose concentration increases. For intermediate glucose concentrations, the cells are burster cells whose plateau fraction increases as the glucose concentration is increased. At high levels of glucose (>15 mM), all cells transform into asynchronous spiking cells which fire tonically due to the fact that their membrane potential is always depolarized. The results reproduce what we know about normal beta-cell function.

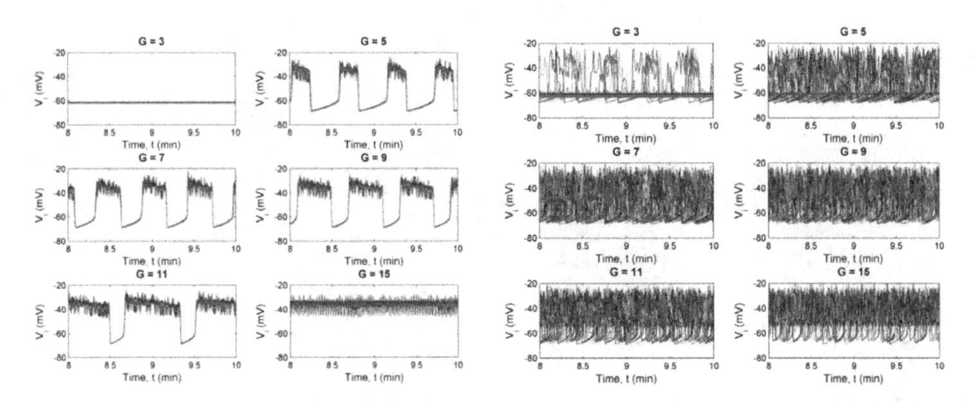

Figure 6.7. The role of islet beta-cell fraction in glucose dose dependent electrical oscillation and synchronization. Left: in a beta cell cluster with a low fraction of non-functional beta cells (10%), the response of different cells to glucose (whose does is specified on top of each subplot) is synchronized. At low glucose the islet is inactive. At intermediate glucose, the plateau fraction increases with increasing glucose concentration. Finally, at high glucose the islet cells fire constantly and asynchronously. Right: the same as for the left panels, but with high the fraction of non-functional beta cells (results from 90% is shown here); the response of different cells is unsynchronized, but a subset of beta cells already becomes active at very low glucose.

In an islet with extensive cell loss, however, such as the one shown in figure 6.7, right panel, where $F_\beta = 0.1$, the response is uncoordinated. Some cells activate even at very low glucose, with heterogeneous responses ranging from silent to bursting to tonically spiking. For intermediate glucose concentrations the oscillatory response is also much more robust when F_β is high.

One would naturally expect the insulin secretion rate, $I(t)$, given by equation (6.10), to be lower for islets with a small fraction of beta cells (F_β) than it is for islets with a high F_β due to fewer beta cells. Note that $n_\beta = n_{\text{islet}}F_\beta$, where n_{islet} is the total number of cells in an islet. Therefore, to examine the insulin secretion efficiency from each cell, we need to calculate the mean insulin secretion rate per beta cell: $ISR_{\text{cell}} = \frac{I}{n_{\text{islet}}F_\beta}$, and compare its values from islets with different F_β (90% versus 10%) as shown in figure 6.8, left panel. These simulation results reveal that, except for at the lowest glucose concentration, ISR_{cell} is higher in islets that have a low degree of cell loss.

When glucose is high, oscillations are suppressed even in the highly functional islets due to the asynchronous spiking nature of the cells. The oscillations in the insulin secretion rate depend on oscillations in the rate of granule fusion which in turn depends on oscillations in the intracellular Ca^{2+} concentration. At high glucose, the oscillations in the Ca^{2+} level are minor since the membrane potential of the cells stays constantly elevated.

Figure 6.8, left panels depicts the oscillations in the first phase of biphasic secretion. The right panel, however, shows how the total amount of insulin secreted (averaged per cell) during the first phase, i.e. the acute insulin response per cell AIR_{cell}, varies against plateau fraction. AIR_{cell} is determined using the trapezoid rule for approximating integrals:

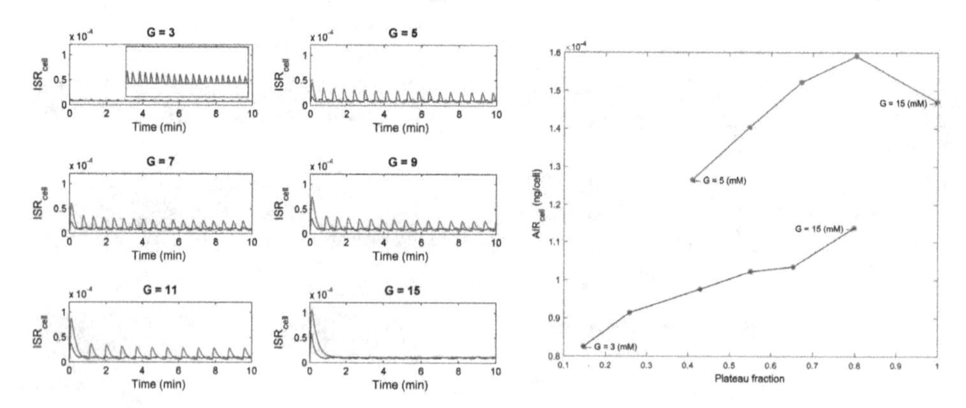

Figure 6.8. The role of beta cell fraction in glucose-dose dependent insulin secretion. Left: mean insulin secretion rate per beta cell, ISR_{cell}, for different glucose concentrations (specified on top of each panel) in islets where the amount of beta cell loss is low (red lines, $F_\beta = 0.9$) and high (blue lines, $F_\beta = 0.1$). Note how oscillations are dampened both by beta cell loss and high glucose. Right: the acute insulin response per cell, AIR_{cell}, versus plateau fraction in islets where the amount of beta cell loss is high (blue line, $F_\beta = 0.1$) and low (red line, $F_\beta = 0.9$).

$$AIR_{\text{cell}} = \frac{1}{n_\beta}\left(\frac{\Delta t}{2}\left(ISR_0 + \sum_{i=1}^{n-1} ISR_i + ISR_n\right)\right), \tag{6.13}$$

where $ISR_i = ISR(t_i)$, $t_i = i\Delta t$, $i = 0, \ldots, n$ and $\Delta t = T/n$ with T being the entire simulation time. For each glucose concentration, we calculate the islet plateau fraction as the mean of the fraction of time the membrane potential of each cell is above −50 mV.

We note that the total first phase per-cell insulin secretion is lower in islets with high cell loss. Comparing the two different islets, we observe that the data for the islet with low cell loss is shifted toward higher plateau fractions. We also point out that the total insulin secretion dips in the highly functional islet for the last data point. This point corresponds to the change from burster to spiker cells (i.e. $G = 15$). To try to understand why the burster cells are more efficient at insulin secretion, we calculated the area under the Ca^{2+} curve for all cells and then calculated the sum to get the total area under the curves. It turns out that the total area under the Ca^{2+} curves is higher for $G = 15$ than it is for $G = 11$, yet the total insulin secretion is lower. The insulin secretion rate is determined by the rate of fusion which is a product of an increasing function of intracellular Ca^{2+} and the number of primed granules. For the parameter values used in these simulations, the intracellular Ca^{2+} concentration in individual cells attains slightly larger maximum values when $G = 11$ than when $G = 15$, translating to a higher secretion rate. The secretion remains higher even during parts of the simulation where the level of primed granules becomes higher in the case of $G = 15$. We remark that when we use a lower sensitivity to Ca^{2+} (lower Hill coefficient) in the function determining the fusion rate, the advantage of the burster cells disappears.

Taking figures 6.7 and 6.8 together, in a coupled beta cell cluster, where cells burst synchronously, each cell generally secretes more insulin, and exhibits higher glucose dose sensitivity compared to spiking cells. This is consistent with observations from previous simulation studies where it was demonstrated that burster pancreatic islet beta cells exhibit a steeper glucose dose response than spiking ones [29], and that bursting endocrine pituitary cells are more effective at releasing hormones than spiking cells [114].

In summary, integration of the islet cell network model and the insulin granule trafficking model demonstrates that a compromised islet cell network will result in a poor glucose dose response, less efficient insulin secretion, and less robust oscillations.

6.5 Future work: application to diabetes and fundamental research of nonlinear dynamical systems

A quantitative, comprehensive understanding of the role of islet cell network in insulin secretion, and of the pulsatility of insulin secretion, is relevant to the study of many real-world problems in diabetes, as well as to the theoretical studies of nonlinear complex systems. Here we discuss several important examples.

6.5.1 Beta cell mass and beta cell function: islet is the basic unit

The nonlinear relationship between beta cell mass (BCM) and beta cell function (BCF) is a basic question in diabetes. Most pathological conditions that lead to diabetes involve defects in at least one of these three properties—BCM, BCF, and insulin action—though all three are ultimately compromised in diabetes [115, 116]. However, the relative contributions from the interplay between them, and from dysfunctions in each one of them, to the pathological development of diabetes, are not well understood, nor do we have an efficient means to dissect them [116, 117]. One major obstacle in addressing these challenges has been that, while there have been a number of clinical protocols (such as glucose tolerance tests) for evaluating BCF and insulin action, a means to directly and quantitatively assess BCM is still lacking. Although imaging technologies for measuring BCM are currently being actively pursued—for example, positron emission tomography (PET)—whether they will be able to offer quantitative information, or information relevant to function, remains to be seen [118]. Most often, variations in BCM are inferred qualitatively from BCF measurements, such as from the insulin appearance rate in blood during a glucose tolerance metabolic test [118–120]. However, the relationship between BCF and BCM is nonlinear, and has not been characterized to allow a quantitative estimation of BCM from BCF.

The modeling work presented in the previous subsections offers valuable insights into the origin of this nonlinearity. As depicted in figure 6.1, the pancreatic islet is a basic anatomical unit of both the BCM and the BCF. The total mass of beta cells is given by

$$BCM = \sum_{\text{all islet } i} n_{\beta,i} = n_{\text{islet}}\, n_{\text{beta}}, \tag{6.14}$$

where $n_{\beta,}$ i is the number of beta cells in islet i, and n_{islet} is the number of islets. BCM thus depends linearly on both n_{islet} and n_{β}. On the other hand, one can conceptually write:

$$\text{BCF} = \sum_{i} f_{\text{islet},i} = n_{\text{islet}} f_{\text{islet}}, \tag{6.15}$$

where f_{islet} is the insulin secretion from one islet. While BCF may be linear in n_{islet}, $f_{\text{islet}}(G, n_{\beta}, \ldots)$ is not linear in n_{β}, as evidenced in many laboratory and clinical results [121–123], as well as in mathematical modeling and numerical simulations presented in this section (figure 6.4, 6.7 and 6.8). Comparing the two equations, it is clear that the nonlinearity in the relationship between BCF and BCM is mainly contributed by the functional dependence of f_{islet} on n_{β}.

The modeling work presented in this section reveals that islet function f_{islet} not only depends on the properties of individual beta cells (ion channels, signaling pathways, granule trafficking, etc), but also on the islet network structure (size, number and strength of inter-beta cell couplings, etc), and the interplay between them [19]:

$$f_{\text{islet}} = f_{\text{islet network}}(G, n_{\beta}, n_{c}, g_{c})\, f_{\text{beta cells}}(G, g_{k_{\text{ATP}}}, g_{\text{Ca}}, \ldots). \tag{6.16}$$

This quick analysis highlight important avenues for dissecting the nonlinear relationship between BCM and BCF. While a considerable amount of effort has been devoted to mathematically and experimentally studying $f_{\text{beta cells}}$ [25, 29, 33, 35, 39, 55, 58, 62–67, 78, 99, 103, 113, 124], relatively little attention has been paid to $f_{\text{islet network}}$ until now. It is known that network structural properties impact system function more readily and more nonlinearly. Evidence suggests that the structural properties of BCM organization deserve more attention. Physiological (e.g. puberty, pregnancy, etc) and pathological (obesity, diabetes, etc) perturbations in BCM and BCF can occur at several different levels, affecting $f_{\text{islet network}}$, or $f_{\text{beta cells}}$, or both. These conditions can be utilized to model the nonlinear relationship between BCM and BCF, the results of which may in turn offer valuable insights into possible targets for clinical interventions.

6.5.2 Beta cell exhaustion

Normal beta cells contain a large insulin granule reserve (~10 000 insulin granules per cell), from which only a small fraction (5%–10%) is ever utilized [78, 124]. Experimental investigations using animal models indicate that with appropriate stimulation, one may be able to force the beta cells to inexhaustibly tap into this pool [125]. It is intriguing then what prevents beta cells to readily tap into this reserve under pathological conditions like obesity and type 2 diabetes, where there is increased demand of insulin due to insulin resistance. Mathematical modeling presented in this section revealed several potential rate-limiting steps during insulin granule trafficking [86]; additionally, the inter-beta cell connection and islet network integrity could offer avenues for the potential improvement of beta cell efficiency [110]. More experimental studies are needed to validate these predictions and to identify targets for augmenting insulin secretion when needed, such as in the presence of insulin resistance.

6.5.3 Mathematical modeling of glucose tolerance and disease risk prediction

A glucose tolerance test is a clinical protocol where a dose of glucose is given intravenously or orally, and changes in blood concentrations of glucose and insulin are measured for up to 3 h. Among various forms of this protocol, the oral glucose tolerance test (OGTT) has been part of the clinical standard to diagnose impaired glucose tolerance (IGT) and diabetes for several decades [126]. The temporal profiles of glucose and insulin obtained during a glucose tolerance test contain rich information on the BCF, BCM, and insulin action. A number of mathematical models have been proposed to simulate insulin and glucose dynamics during a glucose tolerance test in order to decode the information, to mechanistically understand the regulation of glucose homeostasis, and to identify early disease markers [106, 127–143]. Despite the extensive amount of work, we are still facing several outstanding gaps. For instance, dozens of indices have been defined to evaluate insulin sensitivity [144, 145], but few indices are available for evaluating

BCF and BCM [146]. Moreover, the various indices for measuring insulin action or BCF often show divergent patterns of association among themselves or with pathological conditions [147]. We do not know the total number of free indices needed to fully characterize the glycemic regulation[146]. Early disease risk prediction is still a challenge.

Figure 6.9 presents a schematic diagram of the most essential processes that need to be considered when modeling glucose tolerance: insulin clearance (f_1), glucose-stimulated insulin secretion (f_2), glucose-facilitated glucose disposal (f_3), and insulin action (f_4). While much attention has been devoted to describing insulin action (f_4), the insulin secretion component (f_2) tends to be under-modeled, and its dependence on BCM is either not considered or usually assumed to be linear [124]. This often renders the models inadequate at characterizing glucose tolerance for individuals with compromised BCM, BCF, or impaired insulin secretion.

The widely used MINMOD (standing for 'minimal model' described in detail in chapter 7) [148, 149] offers a good example for this point. It was one of the only two methods recommended by the ADA for assessing peripheral insulin resistance [150]. The model has demonstrated great success in assessing insulin sensitivity [151–153], and in utilizing model-derived insulin sensitivity as a quantitative trait in genetic studies of impaired glucose tolerance and type 2 diabetes [154]. However, there are also still several places in the model that leave room for improvement. MINMOD initially described insulin dynamics (f_2 in figure 6.9) using two terms: an exponential clearance term, and an insulin delivery term that is linear in time and glucose ($\dot{I}(t) = -nI(t) + \gamma(G(t) - h)t$ [148, 149]). This significant under-modeling not only

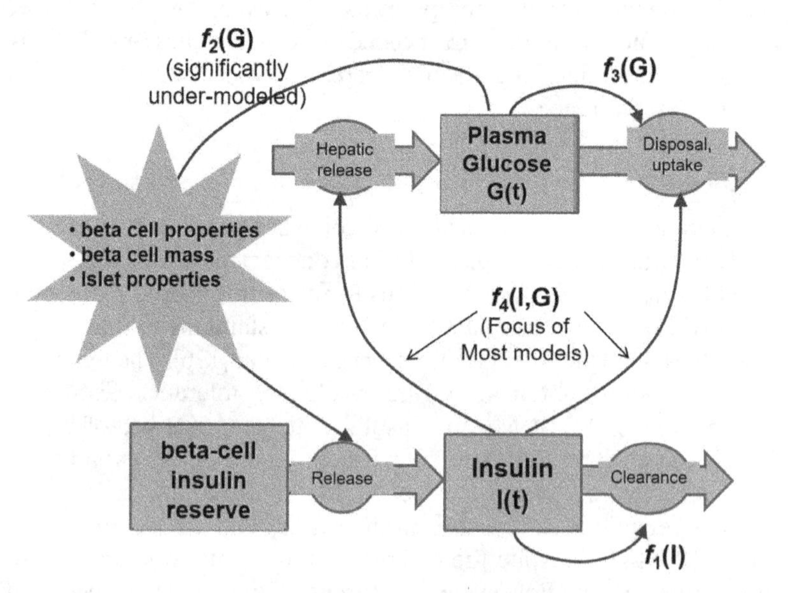

Figure 6.9. The minimal set of processes that need to be considered when modeling glucose tolerance: $f_1(I)$ clearance of insulin from blood circulation, $f_2(G)$ glucose stimulated insulin release, $f_3(G)$ glucose facilitated glucose disposal and uptake, $f_4(I, G)$ insulin suppressed hepatic glucose release, insulin stimulated glucose uptake.

limited this model's ability to assess BCF, but also affected its performance in evaluating insulin sensitivity [155, 156].

Glucose homeostasis is a complex nonlinear dynamic system, and a comprehensive characterization of all components and their intricate interactions is necessary. A quantitative appreciation of the BCM–BCF relationship, including consideration of beta cell network, is fundamental to developing a better model of the insulin secretion component (f_2 of figure 6.9). This is essential for a more efficient evaluation of glucose tolerance and for developing early predictive disease markers. On the laboratory side, more model systems with progressively controlled alteration of BCM (affecting n, g_c, and n_c) should be also developed.

6.5.4 Beta cell mass and beta cell function around disease onset: is T1D a geometric phase transition?

The modeling work presented in section 6.2 and 6.4 suggests that the structural integrity of the islet beta cell mass and network is critical for normal islet function. This prompted us to ask whether onset of Type 1 diabetes (T1D), which results from loss of pancreatic islet beta cells by infiltrating immune cells (to be discussed in more detail in chapters 8 and 9), could be due to compromised islet integrity [157]. This question was recently investigated [157] using percolation theory [158], a mathematical technique for studying network connectivity and its impact on network function and dynamics. A network is said to be percolated when there exists a large cluster of connected nodes that spans across the entire network from the boundary of one side to the opposite side, i.e. when a signal initiated from one end of the network is able to travel across the whole network to reach the other end. Percolation is a geometric phase transition that affects both network structure and emergent dynamics [158]. Percolation theory has been widely applied in a diverse range of fields of research; examples include characterizing the morphology, permittivity, mechanical and electrical properties of biological tissues [47, 159–161], tumor vascularization and growth, blood perfusion in a tumor vessel network [162, 163], the spread of infectious disease [164, 165] and of forest fires [166], emergent structures in nature [167], the search for oil fields [168], etc.

The primary and necessary condition for percolation to occur is a simple constraint on the number of connections each node has, namely the site open probability—the likelihood that any given node is available for coupling—has to be above a certain threshold, termed the critical site open probability (p_c). The value of p_c in an infinitely large network depends on the effective dimension of the network [158], and it is known for many frequently encountered network lattice structures. For a 3D HCP network p_c~0.199 [169], i.e. ~ 19.9% of the network nodes need to be available for connection with others. For networks of finite size, the probability for percolation to occur, often termed the spanning probability, is a continuous function of the site open probability that depends on the network size. Here we define effective critical site open probability $p_{c,\text{eff}}$ in networks of finite size to be the value at which the spanning probability reaches 50%. Table 6.3 lists $p_{c,\text{eff}}$ values for a HCP lattice network at different network sizes. For smaller networks the site open

Table 6.3. Effective critical site open probability $p_{c,\text{eff}}$ of 3D HCP networks. As the network size increases, $p_{c,\text{eff}}$ decreases and approaches the value for an infinitely large network. $p_{c,\text{eff}}$ is defined to be the site open probability at which the spanning probability is 50%.

Network size (N)	Effective percolation threshold ($p_{c,\text{eff}}$)
125	0.294 0
216	0.280 0
512	0.260 0
1000	0.247 0
3375	0.226 0
5832	0.219 0
∞	0.199

probability needs to be higher (more cells need to be viable) for the network to percolate.

As highlighted in the previous sections, majority of cell population in a human islet (figure 6.1), ~70%, is beta cells [75]; the other islet cell types mainly include the glucagon-secreting alpha cells and the somatostatin-secreting delta cells, which do not couple with the beta cells. This translates to a site open probability of $p \sim 0.70 \gg p_c$, implying that beta cell network is well percolated in a healthy normal islet.

Beta-cell destruction that occurs during the development of T1D is islet-wide, reducing the fraction of beta cells and altering the islet network structure. Clinical observations suggest that T1D onset occurs when ~70% of the beta cells have been lost [170–172]. Experimental studies of intercellular coupling and islet function also indicate that functional failure occurs at roughly the same threshold loss of BCM [26, 27]. It is currently unknown why disease occurs when there is still a significant amount (up to 30%) of viable beta cells and what determines this threshold. The ~70% loss of beta cells in a network with 30% non-beta cells translates to a site open probability of 0.21 (30% × 70%), a value very close to p_c. In fact, if we consider the effect of network size, the effective p_c is right around this value (see table 6.1). This quick analysis thus suggests that the clinically observed T1D onset and the experimentally studied loss of islet function both occur at the verge of islet beta-cell network losing its percolation. This analysis then raises an interesting possibility that T1D onset could be triggered by a geometric phase transition in the pancreatic islet beta-cell network.

While experimental validation of this idea is not feasible at this time, this was further analyzed from the perspective of beta-cell synchrony and percolation using numerical analysis. A close look at the results presented in figure 6.4 reveals that synchrony is indeed achieved only when the amount of beta cells renders the network above its critical site open probability, p_c (note that n_c in figure 6.4 is related to the site open probability p by $n_c = 12 * f_\beta * p$, where f_β is the beta cell fraction in an islet). In addition, figure 6.4 reveals that the relationship between synchrony and

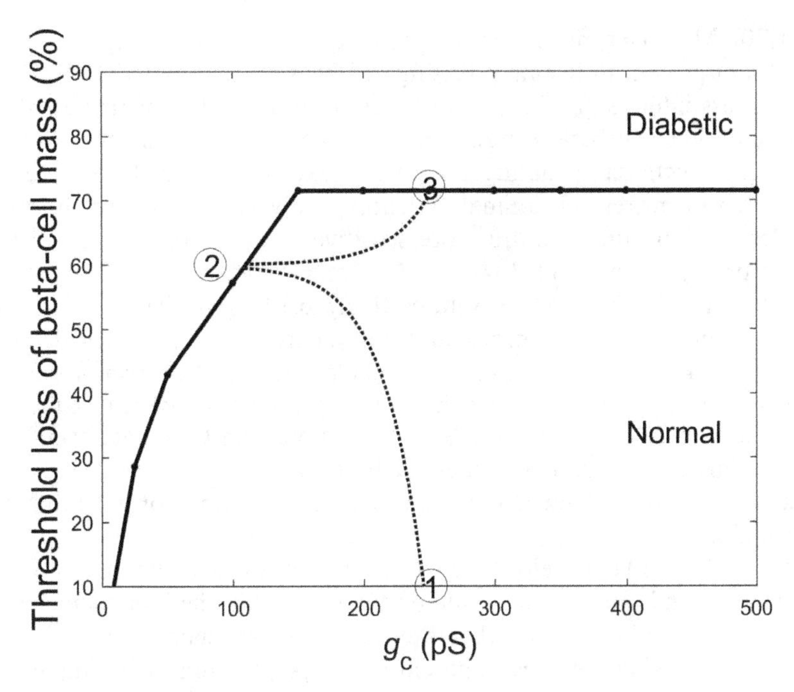

Figure 6.10. The critical residual beta-cell mass at the loss of synchronization depends on g_c. State 1: Normal; State 2 and 3: Diabetes onset. Transition 2 → 3: Honeymoon period. Plotted are results from HCP-587 cell clusters with 30% non-beta cells.

p also depends on inter-beta cell coupling strength, g_c. Figure 6.10, on the other hand, shows how the critical level of beta cell death associated with loss of synchrony depends on g_c.

Though not directly measured, existing evidence suggests that g_c is also likely compromised during T1D development. As depicted in figure 6.1, the beta cells are electrically coupled to each other via gap-junctions, through which ions and small molecules may be transferred between neighboring cells. The gap-junctional channels consist of a pair of so-called connexons [173]. In beta cells connexin-36 (Cx36) is the transmembrane protein building up the connexons (six connexins for each connexon; see figure 6.1, bottom right panel) [174–176]. Recently, the potentially significant role of connexin-dependent signaling in the early diabetes pathogenesis has been increasingly appreciated [177]. A study using a mice model of T2D, showed pre-diabetic animals with decreased levels of Cx36 protein, smaller gap-junctional plaques, and 30% lower inter-beta cell coupling compared to normal [178]. Also in human T2D, several authors have proposed that a substantial loss of Cx36 could occur [177, 179].

As we will see in chapters 8 and 9, T1D development is characterized by progressive loss of beta cell function, beta cell inflammation, oxidative stress, and chronic hyperglycemia. Ample evidence exists suggesting that they can suppress gap-junctional couplings. It has been reported that glucose represses expression of Cx36 transcript and protein in insulin-secreting cell lines and freshly isolated pancreatic rat islets [70]. Moreover, pro-inflammatory cytokines reduce the expression of Cx36 transcript and protein in insulin releasing cell lines (MIN6 and INS1E) [180], and oxidative stress inhibits gap-junctional function in a number of tissues [177, 181]. Transgenic Cx36 perturbation studies in animals also support the notion that inter-beta cell gap-junctional coupling may be weakened during T1D development. Removing Cx36 makes the animals phenotypically similar to prediabetics [182, 183]. Indeed, Cx36 null mice are more sensitive to cytotoxic assault whilst over-expression protects them [180, 184].

With this knowledge let us reexamine figure 6.10 again. The line represents a phase transition between synchrony and no-synchrony (see [157] for details of its determination). For each different g_c value, the phase transition occurs at a different threshold of beta cell loss, The islet is less tolerant to beta cell loss at weakened g_c. A disease scenario may involve the islet's starting at position 1 and finally ending up at position 3. This can be summarized as the follows.

Position 1: Normal islet with complete beta-cell mass and normal gap-junctional coupling.

Path 1 to 2: As T1D develops, beta cells are destroyed leading to a loss in beta-cell mass and a reduction in site occupancy within the islet cellular network. In addition to beta cell death, a number of factors, such as pro-inflammatory cytokines, oxidative stress, and chronic hyperglycemia, may impair the gap junctions [70, 71, 180, 181]. This would lead to a weakened bond strength, g_c [185], which together with the progressive loss of beta cells would cause the islet to head toward position 2, where a phase transition would occur, resulting in the lost of beta cell synchrony and function.

Position 2: Onset of T1D and start of therapeutic intervention.

Path 2 to 3: Treatment improves the local islet environment. Improved glycemic control, together with reduced levels of pro-inflammatory cytokines and oxidative stress, likely improves coupling between the beta cells [70, 71, 180, 181]. The gains in bond strength, g_c, cause the islet to reenter the region of the parameter space corresponding to normal synchrony and function. This means that the patient experiences improved production of endogenous insulin. However, with autoimmune destruction of beta cells still ongoing [186], the islet path ultimately curves up again toward position 3.

Position 3: Permanent loss of islet function.

This theoretical analysis thus predicts that the path taken by the islet during T1D development involves a therapy-induced transient remission after onset (from position 2 to 3). This is reminiscent of the poorly understood honeymoon phenomenon in T1D [187]. This phenomenon, which can last a few months up to a couple of years [187, 188], is often experienced by many T1D patients shortly after the start of their treatment. The patients endogenous insulin secretion capability is reestablished and smaller amounts of administered exogenous insulin is needed [189, 190]. Since the honeymoon phenomenon was first reported in 1940 by Jackson *et al* [191], several hypotheses to explain its underlying mechanisms have been proposed. They involve improved insulin secretion and reduced insulin resistance (alleviated by the removal of hyperglycemia) [192, 193], as well as beta cell regeneration and adaptive immune tolerance [186, 187]. However, the existing hypotheses do not fully explain the rapid and temporary dynamics of the honeymoon period. On the other hand, the scenario predicted by figure 6.10 describes well the salient features, such as rapid re-establishment of islet function upon treatment, and the one-off occurrence of remission, which can be explained by the rapid turnover dynamics of gap junctions [194, 195] and the interplay between the two beta cell network structural properties, including g_c and n_c, in determining islet function.

This example underscores the importance of mathematical modeling in understanding the dynamics and the nonlinearity of a biological process, which is essential for dissecting the major driving factors and in developing predictive biomarkers.

6.5.5 Islet as a model system to study nonlinear dynamical systems

In complex systems, structure defines function, and function (signaling, regulation, etc) is frequently carried out utilizing the emergent temporal orders. Synchrony through oscillation is a common way, and maybe the most efficient way, to coordinate signals and regulations in complex systems, particularly in living systems [1–3].

The relationship between structure and function is also an extremely challenging question, theoretically and mathematically, in the study of complex systems. The initial pioneering work of Wiener, Kuramoto, Peskin, and Winfree [196–203] is restricted to the simplest geometries of global all-to-all coupling, where each oscillator is coupled to all others and affects all others equally. Later, oscillators with local couplings were considered, but were arranged in regular, geometrically simple networks, such as a 1D rings and 2D planar lattices. It has been observed that new dynamic phenomena arise with increasing topological complexity. Presently, much attention is focused on networks constructed by planar graphs, and dynamics resulting from the nonlinearities of individual nodes or from external stimuli. Studies of high-dimension networks, the role of network topological structure and the changes in dynamics when the network topology is modified are just beginning [197, 204, 205].

In addition to the clinical significance discussed in previous subsections, the endocrine system (in particular the pancreatic islet) also offers an ideal model system to study structure and function in complex systems. The pancreatic islet is emblematic of many common features of living systems, including the hierarchical modular organization of its elements, the emergent spatial-temporal orders from inter-element interaction, the multi-stability, and the intrinsic stochastic processes, to name a few. In this section we modeled the role of cellular network structure in islet function; however, this effort is only a starting point. There are many open questions waiting to be answered; some examples were given in the previous subsections.

On the other hand, the islet is a relatively clean system with enough variations to enable experimental investigation of many fundamental questions in the structure-function relationship. For instance, one place is worthwhile investigating is to look into the qualitative and quantitative differences in islet cytoarchitectural organization across different mammal species [74, 75, 206] (see figures 2 and 4 of [74]). It has recently been appreciated that there may be increasing complexity going from

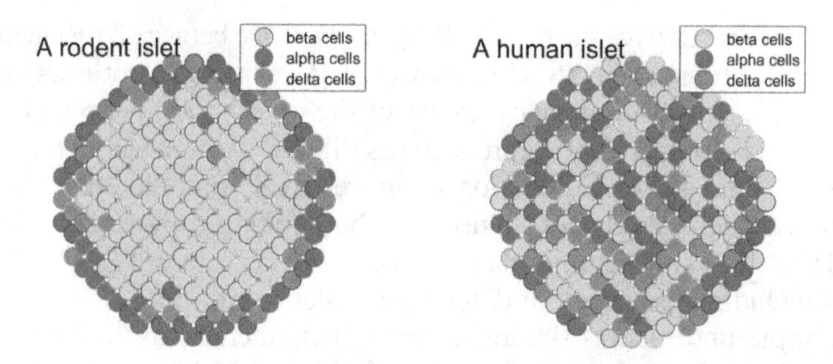

Figure 6.11. Schematic plots of the architecture of the islet cell network and the difference between rodent (left) and human (right) islets. In a rodent islet, beta cells (yellow) form a core surrounded by alpha (red) and delta (blue) cells. In a human islet, the proportion of non-beta cells is higher and all cell types are more intermingled.

rodents to larger mammals such as primates and humans [74, 75, 206]. Figure 6.11 presents a conceptual depiction of what was reported in these studies (for further detail see the reviews [207]). As mentioned in previous sections, a rodent islet has a well-defined beta cell core, surrounded by alpha and delta cells. In contrast, a human islet has proportionally fewer beta cells, and they are more intermingled with other cell types. The functional consequence of such differences is still not understood. These inter-species variations offer an ideal framework to study how changes in network structure affect function, conditions for multi-stability, and bifurcations in the network dynamics.

Exercises

1. Using the beta cell oscillation model presented in equations (6.1)–(6.6) to obtain phase diagrams and perform bifurcation analysis of membrane potential $V(t)$ and intracellular concentration of $Ca^{2+}(t)$. Examine conditions for the existence of oscillatory limit cycle solution (the active burst phase) and stable steady states how they depend on the fraction of functional beta cells and the inter-beta cell coupling strength g_c.
2. One ongoing discussion in the field is, whether there exist a small number of hub beta cells in islets, which are responsible for oscillation synchronize of all beta cells. Using the model presented in the section, implement simulations to investigate if there are such cells, and whether their hub-ness is a permanent or emergent transient property of the cells.
3. In figure 6.9, explore additional components that could be added to the four processes to model glucose tolerance, and propose and refine their mathematical formulation.
4. From the literature, gather data of the following in rat, mouse, and humans: (1) Average body weights for adults. (2) Average number of islets and its distribution. (3) Average islet size (in terms of number of cells) and its distribution. Examine the relationship between the three measures.

5. From the literature, identify three other examples in nature where network structure of a complex system affects function. Your examples can be either living or inanimate systems.
6. In addition to the HCP lattice, are there any other lattice types where the percolation threshold agrees with existing knowledge of the critical level of beta cell loss that causes islet functional failure?
7. How do the infiltrating immune cells affect islet beta cell network integrity?
8. In humans, if islet cells could release all of their insulin granules, how many islets would it take to equal the normal 1st phase insulin release?
9. In addition to the number and strength of intercellular couplings, the synchronization of a network of coupled electrical oscillators also depends on the network size. This has largely to do with the fraction of boundary cells for which the number of connections is lower. Discuss which network is affected more by this problem, the HCP or the SCP, and why.

References

[1] Tiana G, Krishna S, Pigolotti S, Jensen M H and Sneppen K 2007 Oscillations and temporal signalling in cells *Phys. Biol.* **4** R1–17

[2] Kim J R, Shin D, Jung S H, Heslop-Harrison P and Cho K H 2010 A design principle underlying the synchronization of oscillations in cellular systems *J. Cell Sci.* **123** 537–43

[3] Klevecz R R, Bolen J, Forrest G and Murray D B 2004 A genomewide oscillation in transcription gates DNA replication and cell cycle *Proc. Natl Acad. Sci. USA* **101** 1200–5

[4] Haus E 2007 Chronobiology in the endocrine system *Adv. Drug Deliv. Rev.* **59** 985–1014

[5] Bergsten P 2000 Pathophysiology of impaired pulsatile insulin release *Diabetes Metab. Res. Rev.* **16** 179–91

[6] Ward G M 1987 The insulin receptor concept and its relation to the treatment of diabetes *Drugs* **33** 156–70

[7] Matthews D R, Naylor B A, Jones R G, Ward G M and Turner R C 1983 Pulsatile insulin has greater hypoglycemic effect than continuous delivery *Diabetes* **32** 617–21

[8] Marsh B D, Marsh D J and Bergman R N 1986 Oscillations enhance the efficiency and stability of glucose disposal *Am. J. Physiol.* **250** E576–82

[9] Bratusch-Marrain P R, Komjati M and Waldhausl W K 1986 Efficacy of pulsatile versus continuous insulin administration on hepatic glucose production and glucose utilization in type I diabetic humans *Diabetes* **35** 922–6

[10] O'Meara N M, Sturis J, Van Cauter E and Polonsky K S 1993 Lack of control by glucose of ultradian insulin secretory oscillations in impaired glucose tolerance and in non-insulin-dependent diabetes mellitus *J. Clin. Invest.* **92** 262–71

[11] Polonsky K S, Given B D, Hirsch L J, Tillil H, Shapiro E T, Beebe C, Frank B H, Galloway J A and Van Cauter E 1988 Abnormal patterns of insulin secretion in non-insulin-dependent diabetes mellitus *N. Engl. J. Med.* **318** 1231–9

[12] Simon C, Brandenberger G, Follenius M and Schlienger J L 1991 Alteration in the temporal organisation of insulin secretion in type 2 (non-insulin-dependent) diabetic patients under continuous enteral nutrition *Diabetologia* **34** 435–40

[13] Sturis J, Scheen A J, Leproult R, Polonsky K S and van Cauter E 1995 24 -hour glucose profiles during continuous or oscillatory insulin infusion. demonstration of the functional significance of ultradian insulin oscillations *J. Clin. Invest.* **95** 1464–71

[14] Paolisso G, Scheen A J, Giugliano D, Sgambato S, Albert A, Varricchio M, D'Onofrio F and Lefebvre P J 1991 Pulsatile insulin delivery has greater metabolic effects than continuous hormone administration in man: importance of pulse frequency *J. Clin. Endocrinol. Metab.* **72** 607–15

[15] Lefebvre P J, Paolisso G, Scheen A J and Henquin J C 1987 Pulsatility of insulin and glucagon release: physiological significance and pharmacological implications *Diabetologia* **30** 443–52

[16] Ahlersova E, Ahlers I, Smajda B and Kassayova M 1992 The effect of various photoperiods on daily oscillations of serum corticosterone and insulin in rats *Physiol. Res.* **41** 315–21

[17] Berman N, Chou H F, Berman A and Ipp E 1993 A mathematical model of oscillatory insulin secretion *Am. J. Physiol.* **264** R839–51

[18] Nittala A, Ghosh S and Wang X 2007 Investigating the role of islet cytoarchitecture in its oscillation using a new beta-cell cluster model *PLoS One* **2** e983

[19] Rhodes C J 2005 Type 2 diabetes-a matter of beta-cell life and death? *Science* **307** 380–4

[20] Ionescu-Tirgoviste C, Gagniuc P A, Gubceac E, Mardare L, Popescu I, Dima S and Militaru M 2015 A 3D map of the islet routes throughout the healthy human pancreas *Sci. Rep.* **5** 14634

[21] Carter J D, Dula S B, Corbin K L, Wu R and Nunemaker C S 2009 A practical guide to rodent islet isolation and assessment *Biol. Proced. Online* **11** 3–31

[22] Bavamian S, Klee P, Britan A, Populaire C, Caille D, Cancela J, Charollais A and Meda P 2007 Islet-cell-to-cell communication as basis for normal insulin secretion *Diabetes Obes. Metab.* **9** 118–32

[23] Nlend R N *et al* 2006 Connexin36 and pancreatic beta-cell functions *Arch. Physiol. Biochem.* **112** 74–81

[24] Ravier M A, Guldenagel M, Charollais A, Gjinovci A, Caille D, Sohl G, Wollheim C B, Willecke K, Henquin J C and Meda P 2005 Loss of connexin36 channels alters beta-cell coupling, islet synchronization of glucose-induced Ca^{2+} and insulin oscillations, and basal insulin release *Diabetes* **54** 1798–807

[25] Sherman A and Rinzel J 1991 Model for synchronization of pancreatic beta-cells by gap junction coupling *Biophys. J.* **59** 547–59

[26] Rocheleau J V, Remedi M S, Granada B, Head W S, Koster J C, Nichols C G and Piston D W 2006 Critical role of gap junction coupled katp channel activity for regulated insulin secretion *PLoS Biol* **4** e26

[27] Le Gurun S, Martin D, Formenton A, Maechler P, Caille D, Waeber G, Meda P and Haefliger J A 2003 Connexin-36 contributes to control function of insulin-producing cells *J. Biol. Chem.* **278** 37690–7

[28] Hopcroft D W, Mason D R and Scott R S 1985 Structure-function relationships in pancreatic islets: support for intraislet modulation of insulin secretion *Endocrinology* **117** 2073–80

[29] Smolen P, Rinzel J and Sherman A 1993 Why pancreatic islets burst but single beta cells do not. The heterogeneity hypothesis *Biophys. J.* **64** 1668–80

[30] Salomon D and Meda P 1986 Heterogeneity and contact-dependent regulation of hormone secretion by individual B cells *Exp. Cell Res.* **162** 507–20

[31] Jonkers F C, Jonas J C, Gilon P and Henquin J C 1999 Influence of cell number on the characteristics and synchrony of Ca^2+ oscillations in clusters of mouse pancreatic islet cells *J. Physiol.* **520** 839–49

[32] Nittala A and Wang X 2008 The hyperbolic effect of density and strength of inter-cell coupling in islet bursting *Theor. Biol. Med. Model.* **5** 17

[33] Chay T R and Kang H S 1988 Role of single-channel stochastic noise on bursting clusters of pancreatic beta-cells *Biophys. J.* **54** 427–35

[34] Aguirre J, Mosekilde E and Sanjuan M A 2004 Analysis of the noise-induced bursting-spiking transition in a pancreatic beta-cell model *Phys. Rev. E Stat. Nonlinear Soft Matter Phys.* **69** 041910

[35] Sherman A, Rinzel J and Keizer J 1988 Emergence of organized bursting in clusters of pancreatic beta-cells by channel sharing *Biophys. J.* **54** 411–25

[36] Pedersen M G 2005 A comment on noise enhanced bursting in pancreatic beta-cells *J. Theor. Biol.* **235** 1–3

[37] MacDonald P E and Rorsman P 2006 Oscillations, intercellular coupling, and insulin secretion in pancreatic beta cells *PLoS Biol* **4** e49

[38] Hellman B, Gylfe E, Bergsten P, Grapengiesser E, Lund P E, Berts A, Tengholm A, Pipeleers D G and Ling Z 1994 Glucose induces oscillatory Ca^{2+} signalling and insulin release in human pancreatic beta cells *Diabetologia* **37** S11–20

[39] Porksen N, Hollingdal M, Juhl C, Butler P, Veldhuis J D and Schmitz O 2002 Pulsatile insulin secretion: detection, regulation, and role in diabetes *Diabetes* **51** S245–54

[40] Paolisso G, Sgambato S, Torella R, Varricchio M, Scheen A, D'Onofrio F and Lefebvre P J 1988 Pulsatile insulin delivery is more efficient than continuous infusion in modulating islet cell function in normal subjects and patients with type 1 diabetes *J. Clin. Endocrinol. Metab.* **66** 1220–6

[41] Paolisso G, Sgambato S, Scheen A J, Torella R and Lefebvre P J 1988 Advantages of pulsatile administration of human insulin on endogenous pancreatic secretion in the normal subject and in type 1 diabetics *Ann. Med. Interne* **139** 144–5

[42] Zarkovic M, Ciric J, Stojanovic M, Penezic Z, Trbojevic B, Dresgic M and Nesovic M 1999 Effect of insulin sensitivity on pulsatile insulin secretion *Eur. J. Endocrinol.* **141** 494–501

[43] Bingley P J, Matthews D R, Williams A J, Bottazzo G F and Gale E A 1992 Loss of regular oscillatory insulin secretion in islet cell antibody positive non-diabetic subjects *Diabetologia* **35** 32–8

[44] O'Rahilly S, Turner R C and Matthews D R 1988 Impaired pulsatile secretion of insulin in relatives of patients with non-insulin-dependent diabetes *N. Engl. J. Med.* **318** 1225–30

[45] Schmitz O, Porksen N, Nyholm B, Skjaerbaek C, Butler P C, Veldhuis J D and Pincus S M 1997 Disorderly and nonstationary insulin secretion in relatives of patients with NIDDM *Am. J. Physiol.* **272** E218–26

[46] Bertuzzi F and Ricordi C 2007 Prediction of clinical outcome in islet allotransplantation *Diabetes Care* **30** 410–7

[47] Dissado L A 1990 A fractal interpretation of the dielectric response of animal tissues *Phys. Med. Biol.* **35** 1487–503

[48] Hodgkin A L and Huxley A F 1952 A quantitative description of membrane current and its application to conduction and excitation in nerve *J. Physiol.* **117** 500–44

[49] Meissner H P and Schmelz H 1974 Membrane potential of beta-cells in pancreatic islets *Pflugers Arch.* **351** 195–206

[50] Matthews E K and Sakamoto Y 1975 Electrical characteristics of pancreatic islet cells *J. Physiol.* **246** 421–37

[51] Tarasov A, Dusonchet J and Ashcroft F 2004 Metabolic regulation of the pancreatic beta-cell ATP-sensitive K+ channel: a pas de deux *Diabetes* **53** S113–22

[52] Atwater I L, Mears D and Rojas E 1996 *Electrophysiology of the pancreatic β-cell* (Philadelphia, PA: Lippincott-Raven)

[53] MacDonald P E, Joseph J W and Rorsman P 2005 Glucose-sensing mechanisms in pancreatic beta-cells *Philos. Trans. R. Soc. Lond. B Biol. Sci.* **360** 2211–25

[54] Atwater I, Carroll P and Li M X 1989 *Electrophysiology of the Pancreatic Beta-cell* (New York: Alan R. Liss)

[55] Sherman A 1996 Contributions of modeling to understanding stimulus-secretion coupling in pancreatic beta-cells *Am. J. Physiol.* **271** E362–72

[56] Barg S *et al* 2001 Fast exocytosis with few Ca^{2+} channels in insulin-secreting mouse pancreatic b cells *Biophys. J.* **81** 3308–23

[57] Wiser O, Trus M, Hernandez A, Renstrom E, Barg S, Rorsman P and Atlas D 1999 The voltage sensitive lc-type Ca^{2+} channel is functionally coupled to the exocytotic machinery *Proc. Natl Acad. Sci. USA* **96** 248–53

[58] Jo J, Kang H, Choi M Y and Koh D S 2005 How noise and coupling induce bursting action potentials in pancreatic beta-cells *Biophys. J.* **89** 1534–42

[59] Meissner H P 1976 Electrophysiological evidence for coupling between beta cells of pancreatic islets *Nature* **262** 502–4

[60] Mears D, Sheppard N F Jr, Atwater I and Rojas E 1995 Magnitude and modulation of pancreatic beta-cell gap junction electrical conductance *in situ J. Membr. Biol.* **146** 163–76

[61] Perez-Armendariz M, Roy C, Spray D C and Bennett M V 1991 Biophysical properties of gap junctions between freshly dispersed pairs of mouse pancreatic beta cells *Biophys. J.* **59** 76–92

[62] Chay T R and Keizer J 1983 Minimal model for membrane oscillations in the pancreatic beta-cell *Biophys. J.* **42** 181–90

[63] Bertram R, Previte J, Sherman A, Kinard T A and Satin L S 2000 The phantom burster model for pancreatic beta-cells *Biophys. J.* **79** 2880–92

[64] de Vries G and Sherman A 2001 From spikers to bursters via coupling: help from heterogeneity *Bull. Math. Biol.* **63** 371–91

[65] Zimliki C L, Mears D and Sherman A 2004 Three roads to islet bursting: emergent oscillations in coupled phantom bursters *Biophys. J.* **87** 193–206

[66] Pedersen M G, Bertram M and Sherman A 2005 Intra- and inter-islet synchronization of metabolically driven insulin secretion *Biophys. J.* **89** 107–19

[67] Zhang M, Fendler B, Peercy B, Goel P, Bertram R, Sherman A and Satin L 2008 Long lasting synchronization of calcium oscillations by cholinergic stimulation in isolated pancreatic islets *Biophys. J.* **95** 4676–88

[68] Fridlyand L E, Tamarina N A and Philipson L H 2010 Bursting and calcium oscillations in pancreatic beta cells: specific pacemakers for specific mechanisms *Am. J. Physiol. Endocrinol. Metab.* **299** E517–32

[69] Nadal A, Quesada I and Soria B 1999 Homologous and heterologous asynchronicity between identified alpha-, beta- and delta-cells within intact islets of Langerhans in the mouse *J. Physiol.* **517** 85–93

[70] Allagnat F, Martin D, Condorelli D F, Waeber G and Haefliger J A 2005 Glucose represses connexin36 in insulin-secreting cells *J. Cell Sci.* **118** 5335–44
[71] In't Veld P A, Pipeleers D G and Gepts W 1986 Glucose alters configuration of gap junctions between pancreatic islet cells *Am. J. Physiol.* **251** C191–6
[72] Giugliano M, Bove M and Grattarola M 2000 Insulin release at the molecular level: metabolic-electrophysiological modeling of the pancreatic beta-cells *IEEE Trans. Biomed. Eng.* **47** 611–23
[73] Gylfe E *et al* 2000 Signaling underlying pulsatile insulin secretion *Ups. J. Med. Sci.* **105** 35–51
[74] Brissova M, Fowler M J, Nicholson W E, Chu A, Hirshberg B, Harlan D M and Powers A C 2005 Assessment of human pancreatic islet architecture and composition by laser scanning confocal microscopy *J. Histochem. Cytochem.* **53** 1087–97
[75] Cabrera O, Berman D M, Kenyon N S, Ricordi C, Berggren P-O and Alejandro Caicedo A 2006 The unique cytoarchitecture of human pancreatic islets has implications for islet cell function *Proc. Natl Acad. Sci. USA* **103** 2334–9
[76] Charollais A *et al* 2000 Junctional communication of pancreatic beta cells contributes to the control of insulin secretion and glucose tolerance *J. Clin. Invest.* **106** 235–43
[77] Straub S G and Sharp G W 2002 Glucose-stimulated signaling pathways in biphasic insulin secretion *Diabetes Metab. Res. Rev.* **18** 451–63
[78] Rorsman P and Renstrom E 2003 Insulin granule dynamics in pancreatic beta cells *Diabetologia* **46** 1029–45
[79] Rorsman P, Eliasson L, Renstrom E, Gromada J, Barg S and Gopel S 2000 The cell physiology of biphasic insulin secretion *News Physiol. Sci.* **15** 72–7
[80] Curry D L, Bennett L L and Grodsky G M 1968 Dynamics of insulin secretion by the perfused rat pancreas *Endocrinology* **83** 572–84
[81] Grodsky G M 1972 A threshold distribution hypothesis for packet storage of insulin and its mathematical modeling *J. Clin. Invest.* **51** 2047–59
[82] Grodsky G M, Bennett L L, Smith D F and Schmid F G 1967 Effect of pulse administration of glucose or glucagon on insulin secretion *in vitro Metabolism* **16** 222–33
[83] Porte D Jr and Pupo A A 1969 Insulin responses to glucose: evidence for a two pool system in man *J. Clin. Invest.* **48** 2309–19
[84] Wang Z and Thurmond D C 2009 Mechanisms of biphasic insulin-granule exocytosis—roles of the cytoskeleton, small gtpases and snare proteins *J. Cell Sci.* **122** 893–903
[85] Ohara-Imaizumi M, Nishiwaki C, Kikuta T, Nagai S, Nakamichi Y and Nagamatsu S 2004 TIRF imaging of docking and fusion of single insulin granule motion in primary rat pancreatic beta-cells: different behaviour of granule motion between normal and Goto-Kakizaki diabetic rat beta-cells *Biochem. J.* **381** 13–8
[86] Stamper I J and Wang X 2013 Mathematical modeling of insulin secretion and the role of glucose-dependent mobilization and priming of insulin granules *J. Theor. Biol.* **318** 210–25
[87] Allersma M W, Bittner M A, Axelrod D and Holz R W 2006 Motion matters: secretory granule motion adjacent to the plasma membrane and exocytosis *Mol. Biol. Cell* **17** 2424–38
[88] Burchfield J G, Lopez J A, Mele K, Vallotton P and Hughes W E 2010 Exocytotic vesicle behaviour assessed by total internal reflection fluorescence microscopy *Traffic* **11** 429–39
[89] Holz R W and Axelrod D 2008 Secretory granule behaviour adjacent to the plasma membrane before and during exocytosis: total internal reflection fluorescence microscopy studies *Acta Physiol.* **192** 303–7

[90] Kasai K, Fujita T, Gomi H and Izumi T 2008 Docking is not a prerequisite but a temporal constraint for fusion of secretory granules *Traffic* **9** 1191–203

[91] Shibasaki T *et al* 2007 Essential role of Epac2/Rap1 signaling in regulation of insulin granule dynamics by cAMP *Proc. Natl Acad. Sci. USA* **104** 19333–8

[92] Degtyar V E, Allersma M W, Axelrod D and Holz R W 2007 Increased motion and travel, rather than stable docking, characterize the last moments before secretory granule fusion *Proc. Natl Acad. Sci. USA* **04** 15929–34

[93] Nagamatsu S, Ohara-Imaizumi M, Nakamichi Y, Kikuta T and Nishiwaki C 2006 Imaging docking and fusion of insulin granules induced by antidiabetes agents: sulfonylurea and glinide drugs preferentially mediate the fusion of newcomer, but not previously docked, insulin granules *Diabetes* **55** 2819–25

[94] Verhage M and Sorensen J B 2008 Vesicle docking in regulated exocytosis *Traffic* **9** 1414–24

[95] Ohara-Imaizumi M *et al* 2007 Imaging analysis reveals mechanistic differences between first- and second-phase insulin exocytosis *J. Cell Biol.* **177** 695–705

[96] Grodsky G M, Curry D, Landahl H and Bennett L 1969 Further studies on the dynamic aspects of insulin release *in vitro* with evidence for a two-compartmental storage system *Acta Diabetol. Lat.* **6** 554–78

[97] Nesher R and Cerasi E 2002 Modeling phasic insulin release: immediate and time-dependent effects of glucose *Diabetes* **51** S53–9

[98] Straub S G and Sharp G W 2004 Hypothesis: one rate-limiting step controls the magnitude of both phases of glucose-stimulated insulin secretion *Am. J. Physiol. Cell Physiol.* **287** C565–71

[99] Straub S G, Shanmugam G and Sharp G W 2004 Stimulation of insulin release by glucose is associated with an increase in the number of docked granules in the beta-cells of rat pancreatic islets *Diabetes* **53** 3179–83

[100] Barg S, Huang P, Eliasson L, Nelson D J, Obermuller S, Rorsman P, Thevenod F and Renstrom E 2001 Priming of insulin granules for exocytosis by granular Cl^- uptake and acidification *J. Cell Sci.* **114** 2145–54

[101] Hill A V 1910 The possible effects of the aggregation of the molecules of haemoglobin on its dissociation curves *J. Physiol.* **40** iv–vii

[102] Murray J D 2002 *Mathematical Biology: I. An Introduction Interdisciplinary Applied Mathematics* 3rd edn (New York: Springer)

[103] Pedersen M G, Corradin A, Toffolo G M and Cobelli C 2008 A subcellular model of glucose-stimulated pancreatic insulin secretion *Philos. Trans. A Math. Phys. Eng. Sci.* **366** 3525–43

[104] Salgado A P, Santos R M, Fernandes A P, Tome A R, Flatt P R and Rosario L M 2000 Glucose-mediated Ca^{2+} signalling in single clonal insulin-secreting cells: evidence for a mixed model of cellular activation *Int. J. Biochem. Cell Biol.* **32** 557–69

[105] Jonkers F C and Henquin J C 2001 Measurements of cytoplasmic Ca^{2+} in islet cell clusters show that glucose rapidly recruits beta-cells and gradually increases the individual cell response *Diabetes* **50** 540–50

[106] O'Connor M D, Landahl H and Grodsky G M 1980 Comparison of storage- and signal-limited models of pancreatic insulin secretion *Am. J. Physiol.* **238** R378–89

[107] Henquin J C, Dufrane D and Nenquin M 2006 Nutrient control of insulin secretion in isolated normal human islets *Diabetes* **55** 3470–7

[108] Manning Fox J E, Gyulkhandanyan A V, Satin L S and Wheeler M B 2006 Oscillatory membrane potential response to glucose in islet beta-cells: a comparison of islet-cell electrical activity in mouse and rat *Endocrinology* **147** 4655–63

[109] Quesada I, Todorova M G, Alonso-Magdalena P, Beltra M, Carneiro E M, Martin F, Nadal A and Soria B 2006 Glucose induces opposite intracellular Ca^{2+} concentration oscillatory patterns in identified alpha- and beta-cells within intact human islets of Langerhans *Diabetes* **55** 2463–9

[110] Stamper I J and Wang X 2019 Integrated multiscale mathematical modeling of insulin secretion reveals the role of islet network integrity for proper oscillatory glucose-dose response *J. Theor. Biol.* **21** 1–24

[111] Taguchi N, Aizawa T, Sato Y, Ishihara F and Hashizume K 1995 Mechanism of glucose-induced biphasic insulin release: physiological role of adenosine triphosphate-sensitive K+ channel-independent glucose action *Endocrinology* **136** 3942–8

[112] Bratanova-Tochkova T K, Cheng H, Daniel S, Gunawardana S, Liu Y J, Mulvaney-Musa J, Schermerhorn T, Straub S G, Yajima H and Sharp G W 2002 Triggering and augmentation mechanisms, granule pools, and biphasic insulin secretion *Diabetes* **51** S83–90

[113] Pedersen M G, Toffolo G M and Cobelli C 2010 Cellular modeling: insight into oral minimal models of insulin secretion *Am. J. Physiol. Endocrinol. Metab.* **298** E597–601

[114] Tagliavini A, Tabak J, Bertram R and Pedersen M G 2016 Is bursting more effective than spiking in evoking pituitary hormone secretion? a spatiotemporal simulation study of calcium and granule dynamics *Am. J. Physiol. Endocrinol. Metab.* **310** E515–25

[115] De Gaetano A, Hardy T, Beck B, Abu-Raddad E, Palumbo P, Bue-Valleskey J and Porksen N 2008 Mathematical models of diabetes progression *Am. J. Physiol. Endocrinol. Metab.* **295** E1462–79

[116] Cho J H, Kim J W, Shin J A, Shin J and Yoon K H 2011 Beta-cell mass in people with type 2 diabetes *J. Diabetes Invest.* **2** 6–17

[117] Tuomilehto J 2004 Who is at risk of type 2 diabetes? *Diabetes Prim. Care*

[118] Robertson R P 2007 Estimation of beta-cell mass by metabolic tests: necessary, but how sufficient? *Diabetes* **56** 2420–4

[119] Larsen M O, Rolin B, Sturis J, Wilken M, Carr R D, Porksen N and Gotfredsen C F 2006 Measurements of insulin responses as predictive markers of pancreatic beta-cell mass in normal and beta-cell-reduced lean and obese Gottingen minipigs *in vivo Am. J. Physiol. Endocrinol. Metab.* **290** E670–7

[120] Larsen M O, Rolin B, Wilken M, Carr R D and Gotfredsen C F 2003 Measurements of insulin secretory capacity and glucose tolerance to predict pancreatic beta-cell mass *in vivo* in the nicotinamide/streptozotocin Gottingen minipig, a model of moderate insulin deficiency and diabetes *Diabetes* **52** 118–23

[121] Aizawa T *et al* 2001 Size-related and size-unrelated functional heterogeneity among pancreatic islets *Life Sci.* **69** 2627–39

[122] Jo J, Choi M Y and Koh D S 2007 Size distribution of mouse Langerhans islets *Biophys. J.* **93** 2655–66

[123] Hellman B 1959 Actual distribution of the number and volume of the islets of Langerhans in different size classes in non-diabetic humans of varying ages *Nature* **184** 1498–9

[124] Bertuzzi A, Salinari S and Mingrone G 2007 Insulin granule trafficking in beta-cells: mathematical model of glucose-induced insulin secretion *Am. J. Physiol. Endocrinol. Metab.* **293** E396–409

[125] Chen J, Nittala A, Gao S, Ghosh S, Wang X and Patel S 2010 Kinetics of insulin secretion to acute, repetitive stimulation of islets *in vivo* in Sprague Dawley rats *Islet* **2** 10–7

[126] Bartoli E, Fra G P and Carnevale 2011 Schianca G P The oral glucose tolerance test (OGTT) revisited *Eur. J. Intern. Med.* **22** 8–12
[127] Ackerman E, Gatewood L C, Rosevear J W and Molnar G D 1965 Model studies of blood-glucose regulation *Bull. Math. Biophys.* **27** 21–37
[128] Bolie V W 1961 Coefficients of normal blood glucose regulation *J. Appl. Physiol.* **16** 783–8
[129] Subba Rao G, Bajaj J S and Subba Rao J 1990 A mathematical model for insulin kinetics. II. Extension of the model to include response to oral glucose administration and application to insulin-dependent diabetes mellitus (IDDM) *J. Theor. Biol.* **142** 473–83
[130] Turner R C, Rudenski A S, Matthews D R, Levy J C, O'Rahilly S P and Hosker J P 1990 Application of structural model of glucose-insulin relations to assess beta-cell function and insulin sensitivity *Horm. Metab. Res. Suppl.* **24** 66–71
[131] Proietto J 1990 Estimation of glucose kinetics following an oral glucose load. methods and applications *Horm. Metab. Res. Suppl.* **24** 25–30
[132] Nomura M, Shichiri M, Kawamori R, Yamasaki Y, Iwama N and Abe H 1984 A mathematical insulin-secretion model and its validation in isolated rat pancreatic islets perifusion *Comput. Biomed. Res.* **17** 570–9
[133] Sluiter W J, Erkelens D W, Terpstra P, Reitsma W D and Doorenbos H 1976 Glucose tolerance and insulin release, a mathematical approach. II. Approximation of the peripheral insulin resistance after oral glucose loading *Diabetes* **25** 245–9
[134] Gatewood L C, Ackerman E, Rosevear J W, Molnar G D and Burns T W 1968 Tests of a mathematical model of the blood-glucose regulatory system *Comput. Biomed. Res.* **2** 1–14
[135] Ceresa F, Ghemi F, Martini P F, Martino P, Segre G and Vitelli A 1968 Control of blood glucose in normal and in diabetic subjects Studies by compartmental analysis and digital computer technics *Diabetes* **17** 570–8
[136] Jansson L, Lindskog L and Norden N E 1980 Diagnostic value of the oral glucose tolerance test evaluated with a mathematical model *Comput. Biomed. Res.* **13** 512–21
[137] Sturis J, Polonsky K S, Mosekilde E and Van Cauter E 1991 Computer model for mechanisms underlying ultradian oscillations of insulin and glucose *Am. J. Physiol.* **260** E801–9
[138] Topp B, Promislow K, deVries G, Miura R M and Finegood D T 2000 A model of beta-cell mass, insulin, and glucose kinetics: pathways to diabetes *J. Theor. Biol.* **206** 605–19
[139] Tolic I M, Mosekilde E and Sturis J 2000 Modeling the insulin-glucose feedback system: the significance of pulsatile insulin secretion *J. Theor. Biol.* **207** 361–75
[140] Makroglou A J L and Kuang Y 2005 Mathematical models and software tools for the glucose-insulin regulatory system and diabetes: an overview *Appl. Numer. Math.* **56** 559–73
[141] Bennett L S G 2004 Asymptotic properties of a delay differential equation model for the interaction of glucose with plasma and interstitial insulin *Appl. Math. Comput.* **151** 189–207
[142] Prager R, Wallace P and Olefsky J M 1986 *In vivo* kinetics of insulin action on peripheral glucose disposal and hepatic glucose output in normal and obese subjects *J. Clin. Invest.* **78** 472–81
[143] Shapiro E T, Tillil H, Polonsky K S, Fang V S, Rubenstein A H and Van Cauter E 1988 Oscillations in insulin secretion during constant glucose infusion in normal man: relationship to changes in plasma glucose *J. Clin. Endocrinol. Metab.* **67** 307–14
[144] Radikova Z 2003 Assessment of insulin sensitivity/resistance in epidemiological studies *Endocr. Regul.* **37** 189–94
[145] Monzillo L U and Hamdy O 2003 Evaluation of insulin sensitivity in clinical practice and in research settings *Nutr. Rev.* **61** 397–412

[146] Ferrannini E and Mari A 2004 Beta cell function and its relation to insulin action in humans: a critical appraisal *Diabetologia* **47** 943–56

[147] Ferrannini E, Gastaldelli A, Miyazaki Y, Matsuda M, Mari A and DeFronzo R A 2005 beta-cell function in subjects spanning the range from normal glucose tolerance to overt diabetes: a new analysis *J. Clin. Endocrinol. Metab.* **90** 493–500

[148] Bergman R N, Ider Y Z, Bowden C R and Cobelli C 1979 Quantitative estimation of insulin sensitivity *Am. J. Physiol.* **236** E667–77

[149] Toffolo G, Bergman R N, Finegood D T, Bowden C R and Cobelli C 1980 Quantitative estimation of beta cell sensitivity to glucose in the intact organism: a minimal model of insulin kinetics in the dog *Diabetes* **29** 979–90

[150] American Diabetes Association 1998 Consensus development conference on insulin resistance. 5–6 November 1997 *Diabetes Care* **21** 310–4

[151] Bergman R N 1989 Lilly lecture 1989. toward physiological understanding of glucose tolerance. minimal-model approach *Diabetes* **38** 1512–27

[152] Bergman R N 2003 The minimal model of glucose regulation: a biography *Adv. Exp. Med. Biol.* **537** 1–19

[153] Bergman R N 2005 Minimal model: perspective from 2005 *Horm. Res.* **64** 8–15

[154] Knowler W C, Pettitt D J, Savage P J and Bennett P H 1981 Diabetes incidence in Pima Indians: contributions of obesity and parental diabetes *Am. J. Epidemiol.* **113** 144–56

[155] Caumo A, Vicini P and Cobelli C 1996 Is the minimal model too minimal? *Diabetologia* **39** 997–1000

[156] Caumo A, Vicini P, Zachwieja J J, Avogaro A, Yarasheski K, Bier D M and Cobelli C 1999 Undermodeling affects minimal model indexes: insights from a two-compartment model *Am. J. Physiol.* **276** E1171–93

[157] Stamper I J, Jackson E and Wang X 2014 Phase transitions in pancreatic islet cellular networks and implications for type-1 diabetes *Phys. Rev. E Stat. Nonlinear Soft Matter Phys.* **89** 012719

[158] Stauffer D and Aharony A 1994 *Introduction to Percolation Theory* (Boca Raton, FL: CRC Press)

[159] Sahimi M 2003 *Heterogeneous Materials: I. Linear Transport and Optical Properties* (New York: Springer) pp 162–4

[160] Smye S W, Evans C J, Robinson M P and Sleeman B D 2007 Modelling the electrical properties of tissue as a porous medium *Phys. Med. Biol.* **52** 7007–22

[161] Tommasini S M, Wearne S L, Hof P R and Jepsen K J 2008 Percolation theory relates corticocancellous architecture to mechanical function in vertebrae of inbred mouse strains *Bone* **42** 743–50

[162] Craciunescu O I, Das S K, Poulson J M and Samulski T V 2001 Three-dimensional tumor perfusion reconstruction using fractal interpolation functions *IEEE Trans. Biomed. Eng.* **48** 462–73

[163] Baish J W, Gazit Y, Berk D A, Nozue M, Baxter L T and Jain R K 1996 Role of tumor vascular architecture in nutrient and drug delivery: an invasion percolation-based network model *Microvasc. Res.* **51** 327–46

[164] Davis S, Trapman P, Leirs H, Begon M and Heesterbeek J A 2008 The abundance threshold for plague as a critical percolation phenomenon *Nature* **454** 634–7

[165] Meyers L A 2007 Contact network epidemiology: Bond percolation applied to infectious disease prediction and control *Bull. Amer. Math. Soc.* **44** 63–86

[166] Beer T and Enting I G 1990 Fire spread and percolation modeling *Math. Comput. Model.* **13** 77–96

[167] Saberi A A 2015 Recent advances in percolation theory and its applications *Phys. Rep.-Rev. Sect. Phys. Lett.* **578** 1–32

[168] King P R, Buldyrev S V, Dokholyan N V, Havlin S, Lee Y, Paul G, Stanley H E and Vandesteeg N 2001 Predicting oil recovery using percolation theory *Petrol. Geosci.* **7** S105–7

[169] Lorenz C D, May R and Ziff R M 2000 Similarity of percolation thresholds on the HCP and FCC lattices *J. Stat. Phys.* **98** 961–70

[170] Gepts W 1965 Pathologic anatomy of the pancreas in juvenile diabetes mellitus *Diabetes* **14** 619–33

[171] Matveyenko A V and Butler P C 2008 Relationship between beta-cell mass and diabetes onset *Diabetes Obes. Metab.* **10** 23–31

[172] van Belle T L, Coppieters K T and von Herrath M G 2011 Type 1 diabetes: etiology, immunology, and therapeutic strategies *Physiol. Rev.* **91** 79–118

[173] Sohl G and Willecke K 2004 Gap junctions and the connexin protein family *Cardiovasc. Res.* **62** 228–32

[174] Serre-Beinier V, Le Gurun S, Belluardo N, Trovato-Salinaro A, Charollais A, Haefliger J A, Condorelli D F and Meda P 2000 Cx36 preferentially connects beta-cells within pancreatic islets *Diabetes* **49** 727–34

[175] Moreno A P, Berthoud V M, Perez-Palacios G and Perez-Armendariz E M 2005 Biophysical evidence that connexin-36 forms functional gap junction channels between pancreatic mouse beta-cells *Am. J. Physiol. Endocrinol. Metab.* **288** E948–56

[176] Serre-Beinier V *et al* 2009 Cx36 makes channels coupling human pancreatic beta-cells, and correlates with insulin expression *Hum. Mol. Genet.* **18** 428–39

[177] Hamelin R, Allagnat F, Haefliger J A and Meda P 2009 Connexins, diabetes and the metabolic syndrome *Curr. Protein Peptide Sci.* **10** 18–29

[178] Carvalho C P, Oliveira R B, Britan A, Santos-Silva J C, Boschero A C, Meda P and Collares-Buzato C B 2012 Impaired beta-cell-beta-cell coupling mediated by cx36 gap junctions in prediabetic mice *Am. J. Physiol. Endocrinol. Metab.* **303** E144–51

[179] Wright J A, Richards T and Becker D L 2012 Connexins and diabetes *Cardiol. Res. Pract* **2012** 496904

[180] Klee P, Allagnat F, Pontes H, Cederroth M, Charollais A, Caille D, Britan A, Haefliger J A and Meda P 2011 Connexins protect mouse pancreatic beta cells against apoptosis *J. Clin. Invest.* **121** 4870–9

[181] Upham B L and Trosko J E 2009 Oxidative-dependent integration of signal transduction with intercellular gap junctional communication in the control of gene expression *Antioxid. Redox Signal.* **11** 297–307

[182] Meda P 2012 The *in vivo* beta-to-beta-cell chat room: connexin connections matter *Diabetes* **61** 1656–8

[183] Head W S, Orseth M L, Nunemaker C S, Satin L S, Piston D W and Benninger R K 2012 Connexin-36 gap junctions regulate *in vivo* first- and second-phase insulin secretion dynamics and glucose tolerance in the conscious mouse *Diabetes* **61** 1700–7

[184] Klee P, Bavamian S, Charollais A, Caille D, Cancela J, Peyrou M and Meda P 2008 *Gap Junctions and Insulin Secretion* (Tokyo: Springer) pp 111–32

[185] Speier S, Gjinovci A, Charollais A, Meda P and Rupnik M 2007 Cx36-mediated coupling reduces beta-cell heterogeneity, confines the stimulating glucose concentration range, and affects insulin release kinetics *Diabetes* **56** 1078–86

[186] Akirav E, Kushner J A and Herold K C 2008 Beta-cell mass and type 1 diabetes: going, going, gone? *Diabetes* **57** 2883–8
[187] Aly H and Gottlieb P 2009 The honeymoon phase: intersection of metabolism and immunology *Curr. Opin. Endocrinol. Diabetes Obes.* **16** 286–92
[188] Abdul-Rasoul M, Habib H and Al-Khouly M 2006 'the honeymoon phase' in children with type 1 diabetes mellitus: frequency, duration, and influential factors *Pediatr. Diabetes* **7** 101–7
[189] Heinze E and Thon A 1983 Honeymoon period in insulin-dependent diabetes mellitus *Pediatrician* **12** 208–12
[190] Lombardo F, Valenzise M, Wasniewska M, Messina M F, Ruggeri C, Arrigo T and De Luca F 2002 Two-year prospective evaluation of the factors affecting honeymoon frequency and duration in children with insulin dependent diabetes mellitus: the key-role of age at diagnosis *Diabetes Nutr. Metab.* **15** 246–51
[191] Jackson R L, Boyd J D and Smith T E 1940 Stabilization of the diabetic child *Am. J. Dis. Child.* **59** 332–41
[192] Unger R H and Grundy S 1985 Hyperglycaemia as an inducer as well as a consequence of impaired islet cell function and insulin resistance: implications for the management of diabetes *Diabetologia* **28** 119–21
[193] Rossetti L, Giaccari A and DeFronzo R A 1990 Glucose toxicity *Diabetes Care* **13** 610–30
[194] Hesketh G G, Van Eyk J E and Tomaselli G F 2009 Mechanisms of gap junction traffic in health and disease *J. Cardiovasc. Pharmacol.* **54** 263–72
[195] Laird D W 2006 Life cycle of connexins in health and disease *Biochem. J.* **394** 527–43
[196] Strogatz S H 2003 *Sync: The Emerging Science of Spontaneous Order* 1st edn (Theia, New York: Penguin)
[197] Strogatz S 2000 From Kuramoto to Crawford: exploring the onset of synchronization in populations of coupled oscillators *Physica* D **143** 1–20
[198] Wiener N 1958 *Nonlinear Problems in Random Theory* (Cambridge, MA: MIT Press)
[199] Wiener N 1961 *Cybernetics* 2nd edn (Cambridge, MA: MIT Press)
[200] Kuramoto Y 1984 *Chemical Oscillations, Waves, and Turbulence* (Berlin: Springer)
[201] Kuramoto Y 1975 *International Symposium on Mathematical Problems in Theoretical Physics* (Lecture Notes in Physics vol 39) (New York: Springer) pp 420–2
[202] Winfree A T 1967 Biological rhythms and the behavior of populations of coupled oscillators *J. Theor. Biol.* **16** 15
[203] Winfree A T 2001 *The Geometry of Biological Time. Interdisciplinary Applied Mathematics* vol 12 2nd edn (New York: Springer)
[204] Coombes S, Doiron B, Josi K and Shea-Brown E 2006 Towards blueprints for network architecture, biophysical dynamics and signal transduction *Philos. Trans. R. Soc. A: Math. Phys. Eng. Sci.* **364** 3301–18
[205] Arenas A, Diaz-Guilera A and Perez-Vicente C J 2006 Synchronization reveals topological scales in complex networks *Phys. Rev. Lett.* **96** 114102
[206] Erlandsen S L, Hegre O D, Parsons J A, McEvoy R C and Elde R P 1976 Pancreatic islet cell hormones distribution of cell types in the islet and evidence for the presence of somatostatin and gastrin within the D cell *J. Histochem. Cytochem.* **24** 883–97
[207] Khadra A and Schnell S 2015 Development, growth and maintenance of beta-cell mass: models are also part of the story *Mol. Aspects Med.* **42** 78–90

Part III

Insulin release in health and disease

IOP Publishing

Diabetes Systems Biology
Quantitative methods for understanding beta-cell dynamics and function
Anmar Khadra

Chapter 7

Insulin release in health and disease

Chiara Dalla Man and Morten Gram Pedersen

In this chapter we will present minimal and multiscale models used to assess beta-cell function in humans. Minimal models are parsimonious descriptions of the key components of a system's functionality, capable of measuring crucial processes of glucose metabolism. They are characterized by a relatively small number of parameters, so that these can be estimated from the data, providing insights about dynamics of variables not easily accessible by measurements. Multiscale modeling aims to couple events at different physiological levels using modelling and mathematical analysis. For this purpose, appropriate cellular models, built from a mechanistic description of well-characterized subcellular events, must be analysed and simplified.

Minimal models describing insulin secretion will be presented first. These models will be linked to a mechanistic mathematical model of insulin granule dynamics that incorporates cell-to-cell heterogeneity in the glucose threshold for cell activation and secretion with a multiscale approach. We analyse this model to get insight into the biological meaning of minimal model parameters. Finally, since the beta-cell function should always be evaluated in light of the prevailing insulin sensitivity to assess its contribution to the development of diabetes, models to estimate insulin sensitivity will be also described and the concept of disposition index will be introduced.

7.1 Minimal and multiscale models of insulin release

Peripheral insulin concentrations reflect the net effect of insulin secretion and hepatic extraction of portal insulin [1]. While C-peptide and insulin are secreted into the portal vein in equimolar concentrations, plasma insulin and C-peptide concentrations do not preserve equimolarity, since the liver extracts insulin but not C-peptide. Prior studies have suggested that hepatic extraction (HE) removes about 50% of insulin appearing in the portal circulation; however, this process appears to be dynamic and is affected by the amplitude of portal insulin pulses [2], circulating free

doi:10.1088/978-0-7503-3739-7ch7

fatty acids [3], and hyperglycemia [4]. Therefore, the best marker of insulin secretion by the beta cell is not plasma insulin but plasma C-peptide concentration. All the models presented below exploit this knowledge, that C-peptide and insulin are secreted in equimolar concentrations and that C-peptide, at variance with insulin, is not extracted by the liver, so that they use C-peptide concentration data either during an intravenous glucose tolerance test (IVGTT), an oral glucose tolerance test (OGTT), or meal tolerance test (MTT).

The standard IVGTT consists of injecting, say at time 0, glucose over a period of 30–60 s and measuring the resulting plasma glucose, insulin and C-peptide concentrations. The full sampling schedule of the IVGTT usually consists of three pre-test samples taken at (−30, −20, −10) min, and 21 test samples taken at (0, 2, 4, 6, 8, 10, 15, 20, 22, 25, 26, 28, 31, 35, 45, 60, 75, 90, 120, 180, 240) min for measuring glucose and hormone concentrations. A reduced sampling schedule has also been used [5].

7.1.1 The C-peptide minimal model during IVGTT

The first models describing insulin during an IVGTT were those of Licko [6] and Hagander [7]. Starting from the cellular model of insulin secretion by Grodsky [8] (discussed later), Licko developed a whole-body IVGTT model by making a number of assumptions and proposing for the first time beta-cell function indices, consisting of 1st and 2nd phase responsivity.

A few years later, the minimal modeling methodology was applied to the C-peptide subsystem (figure 7.1). Plasma glucose is the 'input' (assumed known), and C-peptide is the 'output' [9] of the model. The kinetics of C-peptide is described in the model by two compartments as proposed by Eaton [10]. Secretion, on the other hand, is modeled with two components (see figure 7.2):

1. First phase secretion, likely representing exocytosis of previously primed insulin secretory granules, is portrayed as the release of insulin from a rapidly

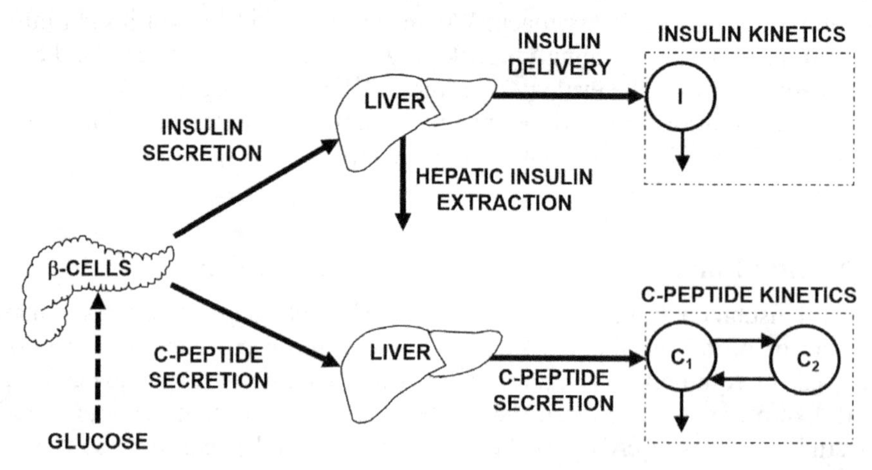

Figure 7.1. Insulin and C-peptide are secreted equimolarly by the pancreatic beta cells, then the liver extracts part of the secreted insulin but not C-peptide, making the latter a better marker of insulin secretion *in vivo*.

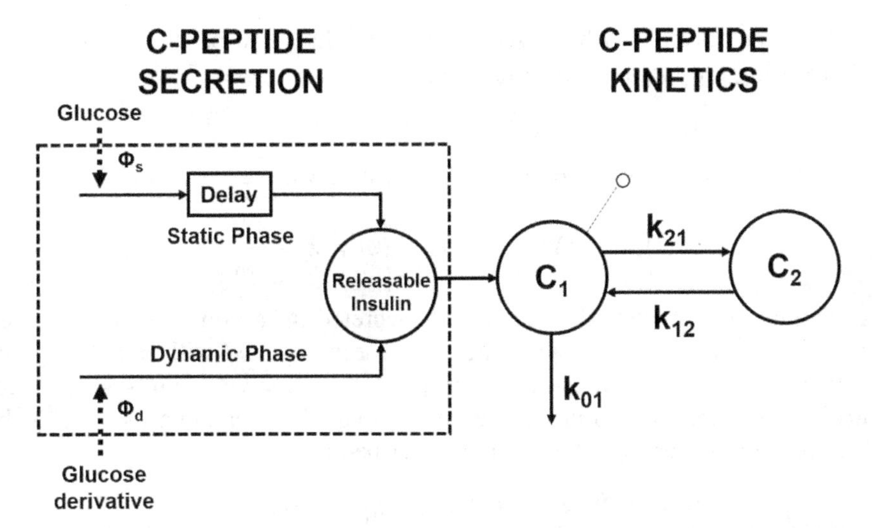

Figure 7.2. The minimal model of C-peptide kinetics during IVGTT [9]: glucose and its derivative stimulate C-peptide secretion, thus filling the releasable pool; C-peptide enters then in the circulation (compartment C_1) and distributes into tissues (compartment C_2).

turning-over compartment (2 min). This process is controlled by the glucose derivative, since first phase secretion is assumed to be proportional to the increase of glucose from its basal level all the way to its maximum level, through a parameter that defines the first phase responsivity.

2. Second phase insulin secretion is believed to be derived from the provision and/or docking of new insulin secretory granules, and is assumed to be proportional to glucose concentration through a parameter that defines the second phase responsivity. This second phase secretion term includes a delay, presumably representing the time required for new granules to dock, be primed and then exocytosed.

Later studies by Toffolo *et al* assumed that C-peptide secretion rate (S) is linked to plasma C-peptide concentration by the two-compartment model of C-peptide kinetics originally proposed by Eaton *et al* [10]. In particular, the C-peptide kinetics model was given by

$$\begin{cases} \dot{C}_1(t) = -(k_{01} + k_{21})C_1(t) + k_{12}C_2(t) + S(t) & C_1(0) = C_b \\ \dot{C}_2(t) = k_{21}C_1(t) - k_{12}C_2(t) & C_2(0) = C_b\dfrac{k_{21}}{k_{12}}, \end{cases} \tag{7.1}$$

where C_1(t) and C_2(t) are the C-peptide concentration in the accessible and peripheral compartments respectively (both in pmol l^{-1}), C_b is the basal C-peptide concentration, k_{01}, k_{12} and k_{21} (in min^{-1}) are C-peptide kinetic parameters. The secretion rate, $S(t)$, normalized to the C-peptide volume of distribution V, was defined as

$$S(t) = S_b + mX(t), \tag{7.2}$$

where S_b is the normalized basal secretion rate, which can be derived from model equations using the steady-state constraint

$$S_b = k_{01}C_b. \tag{7.3}$$

$X(t)$ represents the releasable pool which is refilled via the provision $Y(t)$

$$\dot{X}(t) = -mX(t) + Y(t), X(0) = X_0 + \frac{S_b}{m}, \tag{7.4}$$

X_0 is the amount of C-peptide immediately secreted in response of the super rapid glucose rise. As a matter of fact, when glucose is injected with a bolus, glucose derivative can be represented as an impulse. As known from system theory, perturbing a system with an impulsive input, is equal to changing its initial value.

The provision of new insulin, $Y(t)$, is described as

$$\dot{Y}(t) = \begin{cases} -\alpha \cdot Y(t) + \alpha \cdot S_b + \alpha \cdot \beta \cdot [G(t) - h] & if[G(t) - h] \geqslant 0 \\ -\alpha \cdot Y(t) + \alpha \cdot S_b & if[G(t) - h] < 0 \end{cases}, Y(0) = S_b, \tag{7.5}$$

where the deviation of glucose $G(t)$ (in mg dl^{-1}) from the threshold value h is a forcing function. The absolute total secretion rate, expressed in pmol min^{-1}, was given by $S_t(t) = S(t) \cdot V$.

The model parameters are given by:

1. the secretion rate constant m (measured in min^{-1});
2. the provision rate constant α (measured in min^{-1});
3. the second-phase sensitivity to glucose β (measured in dl mg^{-1} pmol l^{-1} min^{-1});
4. the threshold level for glucose concentration h (measured in mg dl^{-1}). As shown in [9], h is expected to be estimated close to the basal glucose value.
5. the incremental amount of C-peptide secreted during the first phase X_0 (measured in pmol l^{-1}).

The measurement equation was

$$Cpep(t) = C_1(t), \tag{7.6}$$

where $Cpep$ is the measured C-peptide data. An example of C-peptide concentration measured during IVGTT is reported in figure 7.3 (upper panel).

Model identification

The parameters of the model defined by equations (7.1)–(7.6) were estimated from plasma glucose and C-peptide concentrations measured during an IVGTT, by fitting model prediction ($C_1(t)$ of equation (7.1)) to C-peptide data. This was done by using weighted nonlinear least square fitting [11] by assuming that the error on C-peptide measurement is uncorrelated, with zero mean and known standard deviation as reported in [12]. Very briefly, the weighted nonlinear least square method finds the optimal combination of model parameters which makes the squared distance

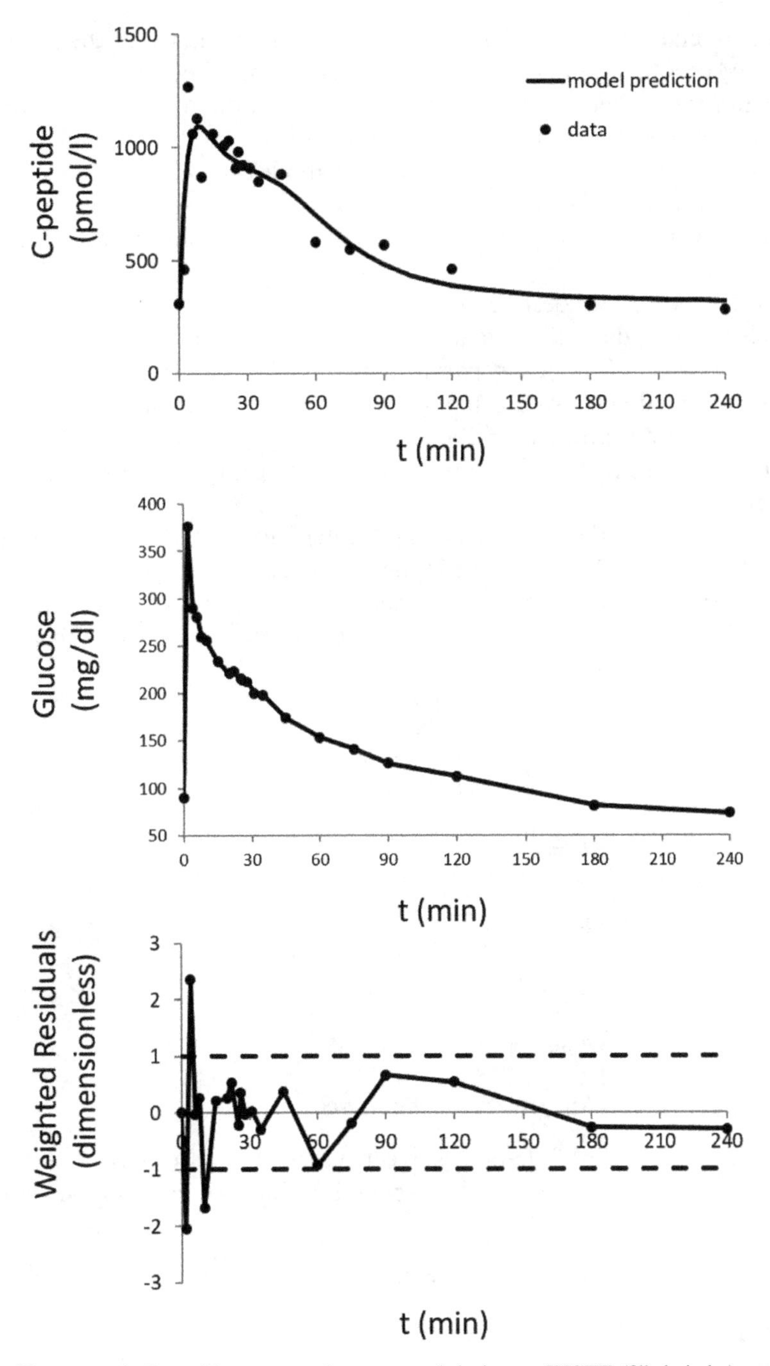

Figure 7.3. Upper panel: C-peptide concentration measured during an IVGTT (filled circles) against model prediction (continuous line). Middle panel: glucose concentration measured during an IVGTT and used as forcing function in the model. Lower panel: weighted residuals. If model fit is good, weighted residuals should be as sequence of random noise mostly lying in the [−1, +1].

between data and model prediction, weighted by the variance of the measurement error, (called cost function) as close as possible to zero. When the model is not linear in the parameters, there is not a closed form solution to this problem and the minimum of the cost function is achieved with an iterative procedure. The difference between each datum and the corresponding model prediction, weighted by the standard deviation of the measurement error, is called weighted residual. If the model is good, this quantity should resemble a white noise with zero mean and variance equal to 1.

It is worth noting that, despite in theory (*a priori*) all model parameters should be identifiable from the data, they are usually estimated with poor precision. One would obtain better results if kinetics parameters (i.e. the rates k_{21}, k_{12}, k_{01}) could be otherwise estimated. This would require performing a separate kinetic experiment, e.g. infusing somatostatin (to inhibit endogenous C-peptide secretion), injecting a C-peptide bolus intravenously and measuring C-peptide concentration for the following 3 h.

That was done in the study by Van Cauter and co-workers in 200 subjects, including 11 normal, 53 obese and 36 type 2 diabetic subjects [13]. The authors then set up a regression model among each kinetic parameter and subject anthropometric characteristics, so that, the C-peptide kinetics parameters can be derived offline from using the following formulas:

$$\begin{cases} A = \begin{cases} 0.140 & \text{Normal} \\ 0.152 & \text{Obese} \\ 0.153 & \text{2DM} \end{cases} \\ B = \dfrac{0.693\,15}{0.14 \cdot \text{age} + 29.16} \\ \text{FRA} = \begin{cases} 0.76 & \text{Normal} \\ 0.78 & \text{Obese \& 2DM} \end{cases} \\ k_{12} = \text{FRA} \cdot B + (1 - \text{FRA}) \cdot A \\ k_{01} = \dfrac{A \cdot B}{k_{12}} \\ k_{21} = A + B - k_{12} - k_{01} \end{cases}, \tag{7.7}$$

where 2DM denote a type 2 diabetic subject. Van Cauter also presented expressions that allowed for the derivation of the C-peptide volume of distribution, V, given by

$$V \begin{cases} BSA \cdot 1.11 + 2.04 & \text{for women} \\ BSA \cdot 1.92 + 0.64 & \text{for men} \end{cases}, \tag{7.8}$$

where BSA is the body surface area calculated as

$$\text{BSA (m}^2) = 0.007184 \cdot \text{Height (cm)}^{0.725} \cdot \text{Weight (kg)}^{0.425}. \tag{7.9}$$

An example of a model fit against data and weighted residuals is shown in figure 7.3.

Definition of beta-cell responsivity index

One of the interesting features of the C-peptide minimal model described above is that it put in a 'causal' relationship the measured C-peptide (the output of the model) with the measured glucose concentration (the input of the model). It is thus possible to derive, from model parameters, some indices of the efficiency of the system (the beta-cell responsivity), quantifying how well the pancreas respond to the glucose stimulus.

The beta-cell responsivity indices can be derived from model parameters as follows:

1. **the first phase beta-cell responsivity to glucose**

$$\Phi_1 = \frac{X_0}{\Delta G}, \tag{7.10}$$

where ΔG is the maximal excursion of glucose above basal;

2. **the second phase beta-cell responsivity to glucose**

$$\Phi_2 = \beta, \tag{7.11}$$

3. **the basal beta-cell responsivity to glucose**

$$\Phi_b = \frac{S_b}{G_b} = \frac{k_{01}C_b}{G_b}, \tag{7.12}$$

4. **the total beta-cell responsivity to glucose**

$$\Phi_{\mathrm{Tot}} = \Phi_2 + \frac{\Phi_1 \cdot \Delta G}{\int_0^{\infty} (G(t) - h)dt}, \tag{7.13}$$

representing the average increase of the above basal pancreatic secretion over the average glucose stimulus.

7.1.2 The C-peptide minimal model during OGTT

In follow up studies, beta-cell function was assessed from a more physiological oral test, i.e. meal tolerance test (MTT) or oral glucose tolerance test (OGTT), by properly adapting the IVGTT minimal model to the more gradual changes in glucose, insulin and C-peptide concentrations.

Also in this case, the C-peptide kinetics was still described with Eaton's model and Van Cauter formulas, described above. The model assumed that S is made up of a static (S_s) and a dynamic (S_d) component [14], i.e.

$$S(t) = S_d(t) + S_s(t), \tag{7.14}$$

S_s was assumed to be equal to the provision of releasable insulin by beta cells (Y), controlled with some delay ($T = 1/\alpha$) by glucose concentration above a threshold level h, through the parameter β. In other words,

$$\dot{S}_s(t) = \dot{Y}(t) = \begin{cases} -\alpha \cdot Y(t) + \alpha \cdot S_b + \alpha \cdot \beta \cdot [G(t) - h] & if[G(t) - h] \geqslant 0 \\ -\alpha \cdot Y(t) + \alpha \cdot S_b & if[G(t) - h] < 0 \end{cases} \quad Y(0) = S_b, \tag{7.15}$$

It is worth noting that, even in this context, the parameter h should be very close to the basal glucose value. S_b is the normalized basal secretion rate, which can be derived from equation (7.1) using steady state constraint. S_d represents the secretion of insulin from the readily releasable pool and is proportional to glucose derivative, which has been shown to be important for the description of insulin secretion during an OGTT/MTT, where glucose rise last 30–40 min and cannot be represented with a step or an impulsive increase, as during the IVGTT [14, 15]. This means that

$$S_d(t) = \begin{cases} K \cdot \dfrac{dG(t)}{dt} & if \dfrac{dG(t)}{dt} \geqslant \text{ and } G(t) \geqslant G_b \\ 0 & \text{otherwise} \end{cases}. \tag{7.16}$$

Definition of beta-cell responsivity index

As stated for the IVGTT C-peptide minimal model, also one of the interesting features of the oral C-peptide minimal model described above is that it put in a 'causal' relationship the measured C-peptide (the output of the model) with the measured glucose concentration (the input of the model). It is thus possible to derive, from model parameters, some indices of the efficiency of the system (the beta-cell responsivity), quantifying how well the pancreas responds to the glucose stimulus.

According to [14] beta-cell responsivity index can be derived from model parameters as follows:

1. **Dynamic beta-cell responsivity**

$$\Phi_d = \frac{\int_0^\infty S_d(t)dt}{\int_0^\infty \frac{dG(t)}{dt}dt} = \frac{\int_{G_b}^{G_{\max}} KdG}{G_{\max} - G_b} = K, \tag{7.17}$$

2. **Static beta-cell responsivity**

$$\Phi_s = \frac{\int_0^\infty [S_s(t) - S_b]dt}{\int_0^\infty [G(t) - h]dt} = \beta, \tag{7.18}$$

3. **Basal beta-cell responsivity**

$$\Phi_b = \frac{S_b}{G_b} = \frac{k_{01}C_b}{G_b}, \tag{7.19}$$

4. **Overall or total beta-cell responsivity**

$$\Phi_{tot} = \frac{\int_0^\infty [S(t) - S_b]dt}{\int_0^\infty [G(t) - h]dt} = \Phi_s + \frac{\Phi_d \cdot (G_{\max} - G_b)}{AUC(G) - h \cdot T}, \tag{7.20}$$

where AUC indicates the area under the curve and T is the duration of the experiment.

An example of model fit against data and weighted residuals is shown in figure 7.4.

7.2 Mechanistic modelling of pancreatic insulin secretion

Sections 7.2 and 7.3 feature excerpts from [11].

7.2.1 Phenomenological modelling

In the 1970s, Grodsky and co-workers [8, 16] and Cerasi *et al* [17] developed models of the pancreatic insulin response to various patterns of glucose stimuli. Because of the limited knowledge of beta-cell biology at that time, these early models were phenomenological. Only recently has our knowledge of the events leading to exocytosis of insulin granules reached a level that allows us to formulate mechanistically based models, as explained in chapter 3. In this section, the aim is to present modelling of secretion from the entire pancreatic beta-cell population where cell-to-cell and islet-to-islet heterogeneity plays an important role.

Cerasi and co-workers [17] suggested that the dynamics of insulin secretion was due to largely unidentified time-dependent inhibitory and potentiating signals [18], which act on insulin secretion at different times. For biphasic insulin secretion, inhibition is responsible for creating the nadir after the first phase peak, while potentiation acts later to produce the second phase. An alternative view on the signal hypothesis is that the dynamics of intracellular Ca^{2+}, the triggering signal for insulin release, dictates the dynamics of insulin secretion [19]. In particular, Ca^{2+} shows a phasic pattern in response to a step in glucose concentration [20, 21], which could drive biphasic insulin secretion. However, it should be noted that when Ca^{2+} is kept at a constant and high level, insulin secretion remains biphasic [22]. Thus, other signals or mechanisms appear to be operating in addition to Ca^{2+}. Future research may provide insight into these hypothetical signals as well as into the control and role of Ca^{2+} dynamics in biphasic insulin release.

As an alternative to Cerasi's signal hypothesis, Grodsky and colleagues [8, 23] proposed that insulin was located in 'packets', plausibly the insulin containing granules, but Grodsky suggested that the packets could also possibly be identified with the individual beta cells. In this model, part of the insulin is stored in a reserve pool, while other insulin packets belong to a labile and releasable pool. The rapid release of the labile pool results in the first phase of insulin secretion [8, 23], while the reserve pool is responsible for the sustained second phase. The distinction between reserve and 'readily releasable' insulin has been at least partly confirmed when the

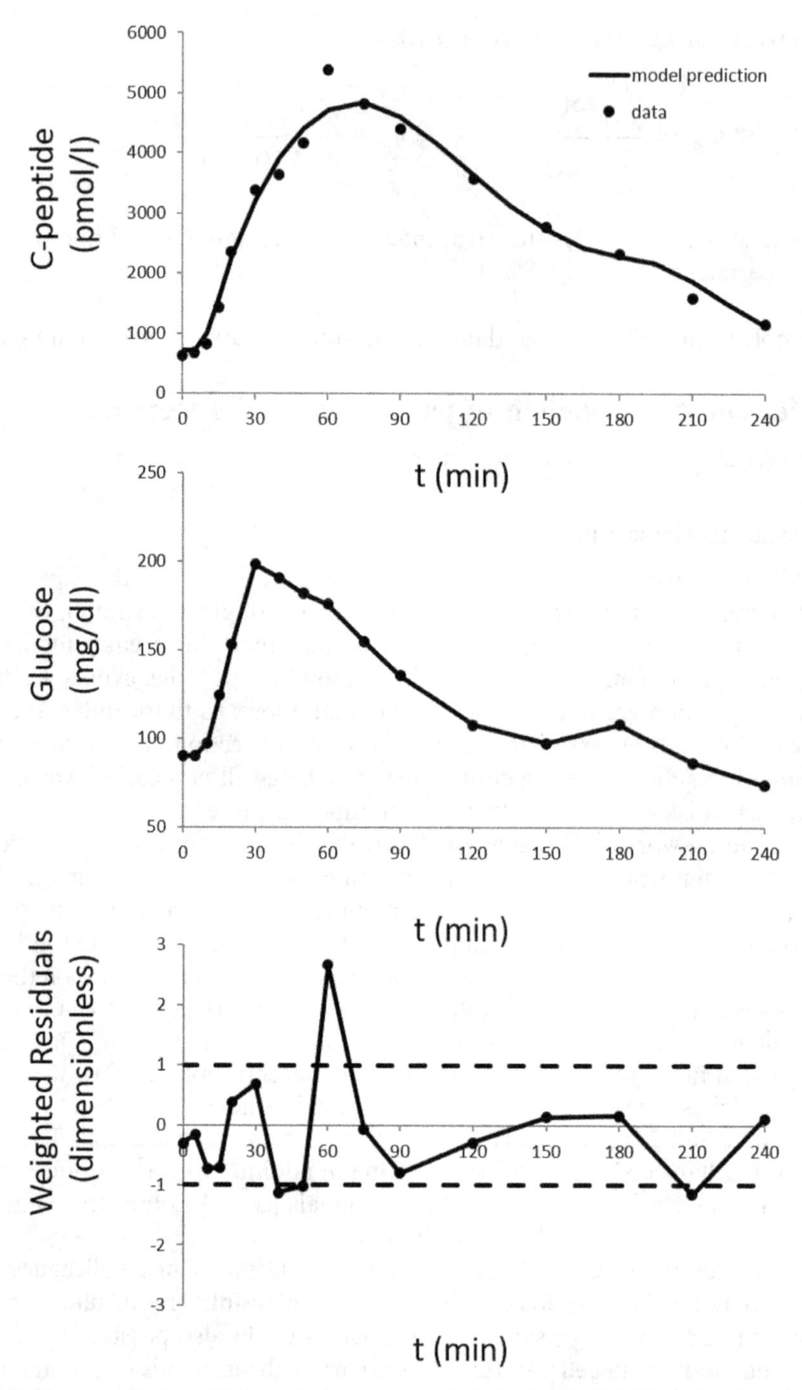

Figure 7.4. Upper panel: C-peptide concentration measured during an OGTT (filled circles) against model prediction (continuous line). Middle panel: glucose concentration measured during an OGTT and used as forcing function in the model. Lower panel: weighted residuals. If model fit is good, weighted residuals should be as sequence of random noise mostly lying in the [−1, +1].

packets are identified with granules [24, 25]. Mathematically, this model is essentially as explained in chapter 2, exercise 5.

However, a modification of this conceptually simple model is needed to explain the so-called staircase experiment where the glucose concentration is increased in consecutive steps, each step giving rise to a peak of insulin. Grodsky [8] assumed that the labile pool is heterogeneous in the sense that the packets in the pool have different thresholds with respect to glucose beyond which they release their content. The resulting mathematical model describing this process was able to reproduce the staircase and many other experiments. Although there has been no support of granules having different thresholds [18], Grodsky [8] mentioned that cells apparently have different thresholds based on electrophysiological measurements [26]. In the following, we provide an update of Grodsky's model based on the idea and recent data of cell-to-cell heterogeneity with respect to their activation threshold.

7.2.2 A mechanistic pancreatic model of insulin secretion

A mathematical model of beta-cell function [27] starting from the idea that the insulin-containing granules belong to different pools [28] has been developed based on several biological experiments [21, 24, 25]. Insulin is synthesized in the endoplasmic reticulum and packed in secretory granules. These newly prepared insulin granules enter a large internal 'reserve pool' from which the granules are mobilized to the cell membrane. This is likely to happen via messengers such as Ca^{2+} [29], as well as ATP which drives the motor protein myosin V [30]. Mobilized granules are either reinternalized or proceed to dock to the membrane. Once docked, the granules can then undergo priming, a process that involves acidification of the granules [31]. The primed granules belong to the readily releasable pool (RRP) and can fuse with the cell membrane in response to raised intracellular Ca^{2+} level when calcium channels open at stimulatory glucose levels. In the model, glucose controls fusion (which is triggered by calcium influx not included in the model) and regulates mobilization from the reserve pool to the cell membrane. In some sense, the model is a simplified representation of the processes described in detail in chapter 3.

The model was simplified by considering the reserve pool constant (infinite). To do so, synthesis of new granules and changes in the size of the reserve pool, due to mobilization, reinternalization and crinophagy, were neglected. This hypothesis is supported by the observation that more than 80 per cent of the granules are located in the centre of the cell [28], and the fact that modifying the biosynthesis rate appears to have little effect on secretion over the first 2 h after an increased glucose stimulus [32]. In addition, the mobilized and docked granules were merged into a single 'docked pool'. The details of the model are summarized in figure 7.5.

Numerical simulations of the resulting model were then confronted with experimental data obtained from the intact pancreas. All beta cells in the pancreas were considered. In agreement with earlier electrophysiological findings [26], recent Ca^{2+} imaging experiments [33, 34] showed that the beta cells have different glucose thresholds for triggering Ca^{2+} influx. Above this threshold, the intracellular Ca^{2+} concentration changes little with glucose. Of interest, very recent imaging studies of

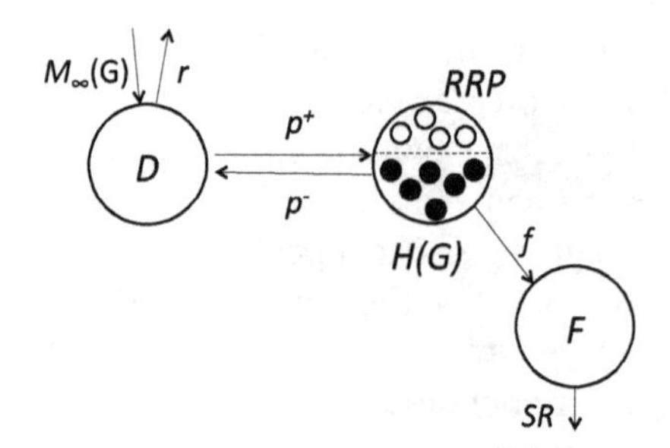

Figure 7.5. Overview of the mechanistic pancreatic model [27].

exocytotic events in intact islets show that beta cells are recruited to release insulin at different glucose concentrations [35].

Based on these observations, one can distinguish between silent and active beta cells. By denoting the fraction of cells that are active at a given extracellular glucose concentration (g) by $\Phi(g)$, which is a sigmoidal function of g [26, 33, 34], the density function describing the fraction of cells with threshold between g and $g + dg$ is given by $\phi(g) = d\Phi/dg$. All rate constants modelling the granule pool dynamics are assumed identical in all cells. Thus the only difference between cells is whether they are active and release insulin as a result of exocytosis, or are silent and, hence, their readily releasable granules do not fuse.

This distinction therefore leads to a heterogeneous total pancreatic RRP, where a part of the readily releasable granules are located in silent cells and another part is located in active cells and undergoes exocytosis. We denote the part of the RRP in active cells by H.

As in [36], mobilization was assumed to occur with no delay at a rate

$$M_\infty(G) = M_0 + c\frac{G^n}{G^n + K_M^n}. \tag{7.21}$$

The delay in mobilization was not needed to reproduce the characteristic biphasic profile in response to a step in glucose concentration. Indeed, a very short delay in M was used [27] in order to reproduce the data from O'Connor *et al* [16], further confirming that the assumption of removing the delay in M was valid.

To derive the equations describing the various pools, one needs to account for the fluxes in and out of each pool. The docked pool D is refilled by mobilization of new granules with rate $M_\infty(G)$, and by 'unpriming', i.e. a process where readily releasable granules lose their release capacity, with a rate constant p^-. Granules then leave the docked pool, due either to undocking and reinternalization occurring at a rate constant r, and due to priming occurring at a rate p^+. This produces the equation

$$\frac{dD}{dt} = M_\infty(G) + p^-R - (r + p^+)D. \tag{7.22}$$

The dynamics of the granules for the entire RRP R are thus given by

$$\frac{dR}{dt} = p^{+}D - p^{-}R - fH, \tag{7.23}$$

where the last term describes fusion of granules located in the RRP in active cells, i.e. in H, and f is the fusion constant.

The modelling of the part of the RRP in active cells (H) is slightly trickier. Unpriming occurs with rate $p^{-}H$. The priming flux is assumed identical in all cells. This implies that the flux of granules entering $H(G)$ due to priming is proportional to $\Phi(G)$, and equal to $p^{+}D\Phi(G)$. H also changes when G changes due to activation or deactivation of cells. Intuitively, this term should be proportional to dG/dt, because the faster G increases, the more rapidly do new cells activate, and their RRPs enter H at the same rate. The total RRP in cells with activation threshold G is given by $h(G) = dH/dG$. Based on this, one can arrive at the following equation (see also exercise 6),

$$\frac{dH(G)}{dt} = p^{+}D\Phi(G) - (f + p^{-})H(G) + h(G)\frac{dG}{dt}. \tag{7.24}$$

The pool of fused granules F follows

$$\frac{dF}{dt} = fH - mF, \tag{7.25}$$

where m is the rate constant of insulin release. Finally, the secretion rate is given by

$$S = mF + S_b, \tag{7.26}$$

where SR_b is the basal secretion rate. As shown in figure 7.6, the model is able to satisfactorily reproduce both the biphasic secretion after a step in the glucose concentration, as well as the three peaks in response to the staircase protocol. The latter is due to the recruitment of more cells at each step of glucose, which then causes the RRPs of these cells to be released.

7.3 Multiscale modelling of insulin secretion

7.3.1 Multiscale analysis of the IVGTT

It would be interesting to compare the two expressions for the secretion rate, equation (7.2) for the minimal model, and equation (7.26), for the mechanistic model during an IVGTT. Inspecting the two equations reveals that one needs to relate F to X. In the following, this coupling is done by approximating F by a variable X^{*} with kinetics similar to that of X.

First, it is important to point out that the last term in equation (7.24) describes derivative control, i.e. that insulin secretion does not only depend on glucose level, but also on its rate of change. As discussed previously, this fact is important during oral and meal tolerance tests and included in the oral minimal secretion model. However, during an IVGTT, the glucose concentration decreases monotonically (figure 7.7(A)), in contrast to the rising glucose level during the first phase of an oral test. It was previously argued that in the mechanistic model presented above (see

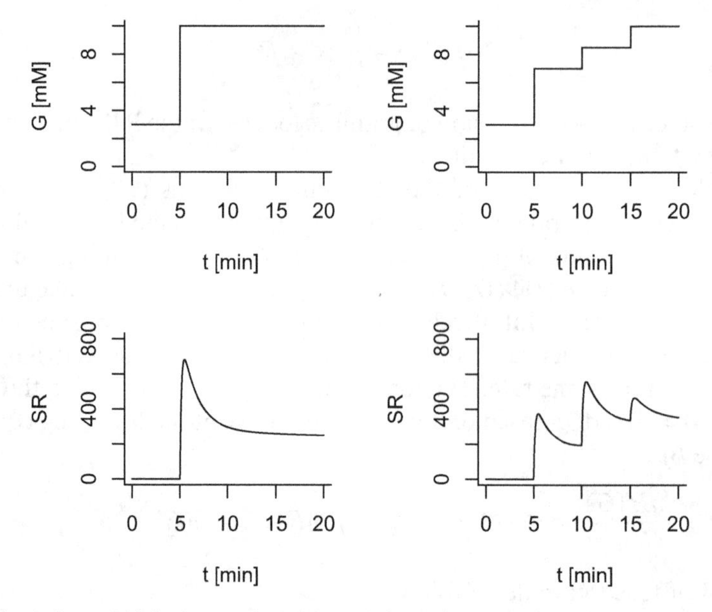

Figure 7.6. Simulations of biphasic secretion (left) and the staircase protocol (right) using the mechanistic model [27] shown in figure 7.5. Upper panel indicates the glucose protocol, lower panels show the resulting simulated secretion rates. Parameters as in [36]: $r = 0.08$ min^{-1}, $p^+ = 0.003$ min^{-1}, $p^- = 0.01$ min^{-1}, $f = 6.2$ min^{-1}, $m = 0.62$ min^{-1}, $K_M = 10$ mM, $n = 4$, $c = 200$ µg min^{-1}, $M_0 = 14$ µg min^{-1}, $K_\Phi = 7.22$ mM.

equations (4.21)–(7.26)), derivative control is negligible when dG/dt is negative [37], since almost all RRP granules with threshold G would have been released, i.e. $h(G) \approx 0$. For the IVGTT, this claim was further supported by a more careful analysis as shown in [36]. Hence, derivative control can be neglected during an IVGTT because of the decreasing glucose profile, and the fraction of the RRP located in active cells can be described by

$$\frac{dH}{dt} = p^+ D\Phi(G) - (f + p^-)H(G). \tag{7.27}$$

In order to derive the large-scale minimal model, QSS approximation can be invoked based on an evaluation of the time-scales of the different processes. The timescale for H is $1/(f + p^-)$, which is of the order of seconds due to rapid fusion. This makes it much faster than other time-scales of the model and of the glucose dynamics during the IVGTT. Hence once can apply QSS approximation on H to obtain

$$H(G) = p^+ D\Phi(G)/(f + p^-). \tag{7.28}$$

Similarly, we can assume that the entire RRP is in a steady-state, which after some algebra (see exercise 7) leads to

$$\frac{dD}{dt} = M_\infty(G) - \rho(G)D, \tag{7.29}$$

and

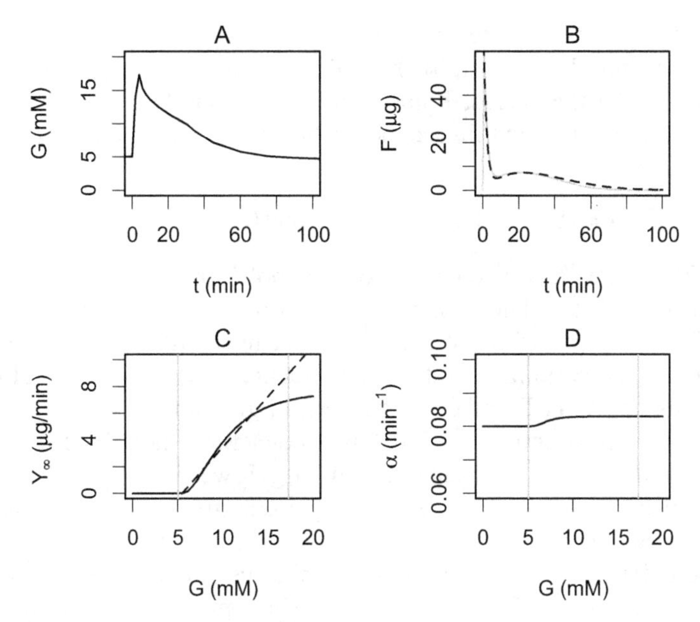

Figure 7.7. (A) Mean average plasma glucose profile in a cohort of 204 healthy subjects during an IVGTT. (B) The fused pool F as a function of time (full grey curve) and the approximation X^* (dashed black curve) in response to the glucose profile in panel A. Parameters as in [36]: $r = 0.08$ min^{-1}, $p^+ = 0.003$ min^{-1}, $p^- = 0.01$ min^{-1}, $f = 6.2$ min^{-1}, $m = 0.62$ min^{-1}, $K_M = 10$ mM, $n = 4$, $c = 200$ µg min^{-1}, $M_0 = 14$ µg min^{-1}, $K_\Phi = 7.22$ mM. (C) The steady-state mobilization rate $Y^*_\infty(G)$ (full curve) is compared with the linear function $\beta(G - h)$ with $h = G_b = 5$ mM. The vertical grey lines indicate the basal and maximal glucose levels from panel A. (D) The delay parameter T^* as a function of G. The vertical grey lines indicate the basal and maximal glucose levels from panel A.

$$\frac{dF}{dt} = \frac{fp^+ D\Phi(G)}{f + p^-} - mF, \tag{7.30}$$

where

$$\rho(G) = r + p^+\Phi(G)\frac{f}{f + p^-}.$$

By defining $X^* := F$ and $Y^* := [fp^+ \Phi(G)/(f + p^-)] \cdot D$, we obtain (see exercise 7)

$$\frac{dX^*}{dt} = Y^* - mX^*, \tag{7.31}$$

$$\frac{dY^*}{dt} = (Y^*_\infty(G) - Y^*)/T^*, \tag{7.32}$$

with $T^* = 1/\rho(G)$ and

$$Y^*_\infty(G) = \frac{fp^+ \Phi(G)}{f + p^-} \times \frac{M_\infty(G)}{\rho(G)}.$$

Note the analogy between the models in equations (7.4)–(7.5) and (7.31)–(7.32) (exercise 7). The initial condition for Y^* is $Y^* = 0$, because $\Phi(G_b) = 0$. When G rises to $G_{\max}$ rapidly after the glucose bolus, a part of the RRP that is equal to $H\,(G_{\max})$ fuses rapidly and enters F within seconds. This means that the initial condition for $X^* = F$ is

$$X^*(0) = X_0^* = H(G_{\max}) = p^+ D(0)\Phi(G_{\max})/(f + p^-). \tag{7.33}$$

Figure 7.7(B) shows a typical pattern of the pool F and its approximation X^* with excellent correspondence. The asymptotic function for mobilization $Y_\infty^*(G)$ is plotted in figure 7.7(C) together with the corresponding linear function $\max\{0, \beta(G - h)\}$ from equation (7.5). Note the good correspondence over most of the glucose range attained during an IVGTT (indicated by vertical grey lines).

The minimal model equation (7.5) has a constant delay $T = 1/\alpha$, whereas the delay T^* in equation (7.32) depends on G. However, T^* is nearly constant (figure 7.7(D)) because r is an order of magnitude greater than p^+, justifying the minimal model assumption of constant delay. With the parameters used here, we find that the delay T^* is of the order of 12 min, an outcome that is in reasonable agreement with results in Toffolo *et al* [38], who found a delay of approximately 15 min.

7.3.2 Multiscale analysis of the MTT/OGTT

Similar to the analysis presented above for the IVGTT, it was investigated how the three main components of oral minimal models, derivative control, proportional control, and delay, are related to the subcellular events [37]. As for the IVGTT, the simplification of the mechanistic model [27] was based on QSS assumptions using timescale analysis. Specifically, it was found that during the MTT/OGTT, release is rapid (timescale $1/m \sim 1.5$ min) compared to glucose dynamics (timescale of tens of minutes), in contrast to the beginning of the IVGTT where the glucose levels change on a timescale comparable to release. Hence, during the MTT/OGTT, the pool of fused granules F is at QSS and its dynamics cannot be inferred. The overall result of the analysis (exercise 8) is that during an MTT/OGTT, the secretion rate is approximately equal to

$$SR(t) \approx \frac{f}{f + p^-}\left(p^+ D(t)\, \Phi(G(t)) + h(G, t)\, \frac{dG}{dt}\right) + SR_b. \tag{7.34}$$

Hence secretion is, besides basal secretion SR_b, composed of a static and a dynamic term, in close correspondence with the minimal oral model (7.14). Static secretion includes a delay because of the time needed for mobilization and docking. In addition it depends on priming and the number of active cells.

Dynamic secretion is related to $h(G, t)$, which in the minimal model is described by the parameter K in equation (7.16). As for the IVGTT, it can be argued that h is negligible when G decreases [37], which supports the minimal model assumption of a contribution from dG/dt only when G increases equation (7.16). Of note, Breda *et al* [14] and Toffolo *et al* [38] considered a variant of the model where K decreased with increasing glucose concentrations, and in some cases this extended model was needed to fit the data satisfactorily [38]. Interestingly, the glucose-dependence of K corresponds well to the mechanistic model since h is bell-shaped and peaks near the basal glucose level and then decreases [37].

7.3.3 Confronting multiscale analyses of insulin release

The previous analysis shows that it is possible to link the mathematical model of insulin secretion from groups of islets, or isolated pancreas, to minimal models of *in vivo* insulin secretion. This multiscale connection suggests that it should be possible to investigate beta-cell function mechanistically based on data obtained in a clinical setting. In other words, the minimal models can be given mechanistic underpinning by the use of multiscale modelling. In addition, a mechanistic model was recently suggested as the core of a bio-inspired artificial pancreas [39], further underlining the need for a thorough multiscale understanding of the cellular model.

The biology underlying insulin secretion during an IVGTT or an MTT/OGTT is the same for both, while the minimal secretion models, although similar, are not identical. It was shown (see equations (7.31)–(7.32)) that the model description of the cellular events underlying glucose-stimulated insulin secretion [27] simplifies to the IVGTT minimal model [9] when beta cells respond to an IVGTT glucose profile [36]. It was similarly shown (equation (7.34); reference [37]) that the beta-cell model [27] reduces to the MTT/OGTT minimal secretion model [14] when subjected to a typical glucose stimulus seen during an MTT/OGTT. Thus, depending on the clinical setting, a single mechanistic beta-cell model simplifies to either the IVGTT or the MTT/OGTT minimal secretion models needed for parameter identifiability in tests of beta-cell function. This fact justifies on the one hand the differences between the two minimal models that represent the same underlying biology but under different conditions, and on the other hand highlights why the two minimal models have a structural similarity.

In the IVGTT multiscale analysis, the delay parameter T^* is nearly constant and, surprisingly, approximately equal to the inverse of the reinternalization rate $1/r$. The parameter $T^* = 1/\rho(G)$ reflects the time needed for the docked pool D to respond to changes in G (see equation (7.29)), and when granule movement to and from the membrane is substantial, this time-constant is mainly controlled by the reinternalization rate. A recent study using total internal reflection fluorescence imaging experiments has indeed suggested such frequent movement [40].

The IVGTT minimal model parameter X_0 corresponds to the amount of the RRP that is released when glucose increases to $G_{\max}$ (see equation (7.33)), and the first phase index $\Phi_1 = X_0/(G_{\max} - G_b)$ is hence related to the function $h = \partial H/\partial G$. For the MTT/OGTT minimal model, a similar relation between the dynamic index Φ_D and the function h exists, again reflecting that the two minimal models share similarities because they are reflecting the same underlying biology. However, note that during the rising phase, which lasts ~60 min during an MTT/OGTT (figure 4.4), mobilization, docking and priming increases the RRP. This means that the dynamic term described by $h(G, t)$ in equation (7.34) increases over time. Hence, the dynamic index Φ_D reflects not only the size of the RRP at the basal state, but also how it changes because of granule maturation.

The second phase index $\Phi_2 = \beta$ in equations (7.5) is approximately equal to $dY^*_\infty(G)/dG$ (figure 7.7(C)). Using that $f \gg p^-$ and $\rho(G) \approx r$, we find that

$$\beta \approx \frac{dY^*_\infty(G)}{dG} = \frac{p^+}{r} \cdot \frac{d}{dG}(\Phi(G)\, M_\infty(G)). \tag{7.35}$$

Hence, the second phase index $\Phi_2 = \beta$ reflects the combined effect of mobilization, cell recruitment and the strength of priming versus reinternalization, i.e. the net effect of the processes that lead to an increased amount of readily releasable insulin. The same processes influence the static index Φ_S of the minimal OGTT model, as discussed above.

Recent work has addressed how dynamic signals such as Ca^{2+} interact with granule pool dynamics to shape the patterns of insulin secretion [41]. The insight obtained from these studies should be linked to the minimal models with multiscale modelling as outlined here. Multiscale modelling of glucagon secretion and incretins will also be important future subjects to address. Finally, the cellular models should be built as far as possible based on data from human islets and beta cells [42–44].

7.4 Type 2 diabetes and insulin release

Beta cell responsivity should also be evaluated in relation to prevailing insulin sensitivity [45].

This important index, measuring the ability of insulin to inhibit glucose production and promote glucose disposal, can also be estimated from either IVGTT or OGTT tests, as described below.

7.4.1 The glucose minimal model during IVGTT

The IVGTT glucose minimal model is a compartmental model allowing the quantitative description of glucose metabolism and insulin control, initially proposed in [46]. It is described by the following equations:

$$\begin{aligned} \dot{Q}(t) &= -(p_1 + X(t)) \cdot Q(t) + p_1 \cdot \mathrm{V} \cdot \mathrm{G_b} \quad && Q(0) = G_b \cdot V + D \\ \dot{X}(t) &= -p_2 \cdot X(t) + p_3 \cdot (I(t) - I_b) && X(0) = 0, \end{aligned} \tag{7.36}$$

and measurement equation:

$$G(t) = \frac{Q(t)}{V}, \tag{7.37}$$

where G is plasma glucose concentration, I is plasma insulin concentration, suffix ‘b’ denotes basal values, D the intravenous glucose dose, X is insulin action, V is the distribution volume, p_1, p_2 and p_3 are model parameters. Specifically, p_1 is the fractional glucose effectiveness, S_G, p_2 is the rate constant describing the dynamics of insulin action, and p_3 is the parameter governing the magnitude of insulin action (figure 7.8).

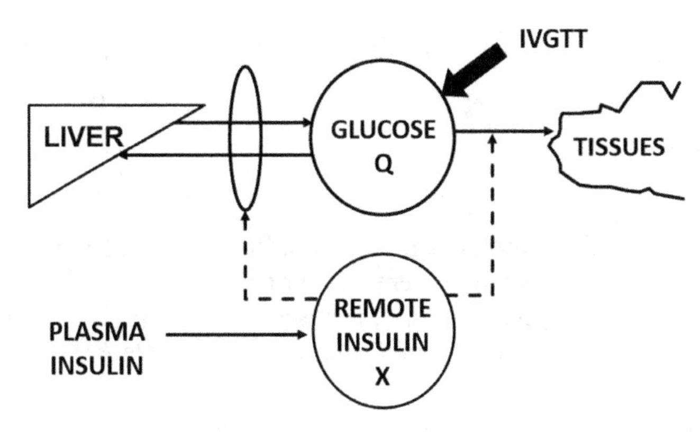

Figure 7.8. The minimal model of glucose kinetics during IVGTT [46]. Glucose is assumed to distribute almost uniformly in the body, so that its kinetics can be represented by a single compartment (Q). The liver produces and utilizes glucose while the tissues only utilize glucose (thin black arrows). The black tick arrow represents the glucose income due to the IVGTT bolus. Insulin in a compartment remote from plasma (X), and not plasma insulin, is the one which actually controls both glucose production and utilization (dashed arrows).

Two indices can be derived from model parameters. These are:

1. **fractional glucose effectiveness** S_G, which measures the ability of glucose per se, at basal insulin, to stimulate glucose disappearance and to inhibit endogenous production:

$$S_G = p_1, \tag{7.38}$$

2. **insulin sensitivity** S_I, which measures the ability of insulin to enhance glucose disappearance and inhibit glucose endogenous production; it can be described by the equation

$$S_I = \frac{p_3}{p_2}. \tag{7.39}$$

Of note, if G is measured in (mg dl^{-1}), D in (mg kg^{-1}) and I in (unit of insulin concentration, for instance μU ml^{-1}), then Q is measured in (mg kg^{-1}), S_G in (min^{-1}), S_I in (min^{-1} (μU ml)$^{-1}$), p_2 in (min^{-1}) and V in (dl kg^{-1}).

This model is widely used to investigate glucose metabolism from a standard intravenous glucose tolerance test (IVGTT) or insulin-modified IVGTT.

7.4.2 The glucose minimal model during OGTT

The oral glucose minimal model (OGMM) [47, 48] is described by the following equations (figure 7.9)

$$\begin{cases} \dot{Q}(t) = -(p_1 + X(t)) \cdot Q(t) + p_1 \cdot V \cdot G_b + Ra(t) & Q(0) = G_b \cdot V \\ \dot{X}(t) = -p_2 \cdot X + p_3 \cdot (I(t) - I_b) & X(0) = 0 \end{cases}$$
$$G(t) = \frac{Q(t)}{V}, \tag{7.40}$$

where Ra is the rate of glucose appearance in plasma coming from the meal/OGTT and the previously defined symbols remain before.

The parametric description of Ra proposed in [47, 48] is a piecewise-linear function with known break-point t_i and unknown amplitude k_i:

$$Ra(t) = \begin{cases} k_{i-1} + \dfrac{k_i - k_{i-1}}{t_i - t_{i-1}} \cdot (t - t_{i-1}) & for\ \ t_{i-1} \leqslant t \leqslant t_i \\ 0 & \text{otherwise} \end{cases}. \tag{7.41}$$

OGMM identification requires a number of assumptions which were discussed in detail in [47, 48]. Briefly, to ensure *a priori* identifiability of the model, one has to assume values for V and p_1 (= S_G), which cannot be estimated from the data. They

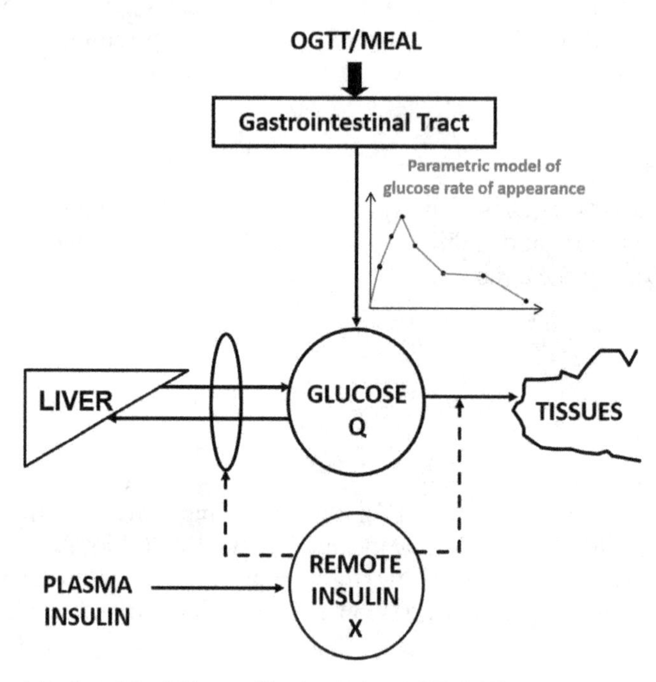

Figure 7.9. The minimal model of Glucose kinetics during OGTT/MTT [47, 48]. Glucose is assumed to distribute almost uniformly in the body, so that its kinetics can be represented by a single compartment (Q). The liver produces and utilizes glucose while the tissues only utilize glucose (thin black arrows). At variance with the IVGTT, here the exogenous input is the rate of appearance of glucose coming from the gastro-intestinal tract. Similarly to the IVGTT case, insulin in a compartment remote from plasma (X), and not plasma insulin, is the one which actually controls both glucose production and utilization (dashed arrows).

were fixed to population values provided in [44] ($V = 1.45$ dl kg^{-1}; $p_1 = 0.025$ min^{-1}). To improve numerical identifiability of the remaining parameters p_2, p_3, k_i ($i = 1...8$), a Gaussian Bayesian prior was considered on the p_2 (with $\sqrt{p_2} \in N(0.1, 0.01)$). In other words, the cost function to be minimized to obtain the optimal parameter vector does not only include the squared distance between data and model prediction (like the weighted nonlinear least squares), but also the squared distance between the estimated parameter (here $\sqrt{p_2}$) and its expected value (the mean of the distribution) weighted by the variance. Finally, a constraint was imposed to guarantee that the area under the estimated *Ra* equals the total amount of ingested glucose, *D*, multiplied by the fraction of the ingested dose that is actually absorbed, f (fixed to population value: $f = 0.9$), and normalized to body weight (BW). This constraint provides an additional relationship among the unknown parameters k_i, thus reducing the number of unknowns by one (see exercise 9).

Of note, since the model has been developed on a database where plasma samples were collected very frequently (at $t = -120, -30, -20, -10, 0, 5, 10, 15, 20, 30, 40, 50, 60, 75, 90, 120, 150, 180, 210, 240, 260, 280, 300, 360, 420$ min), $Ra(t)$ could be described with eight parameters. In case of less frequent sampling schedule, it is convenient to appropriately reduce the number of parameters describing *Ra* (see below).

Measurement error on glucose data is usually assumed to be independent, Gaussian, with zero mean and known standard deviation (CV = 2%). Insulin concentration, the model forcing function, is assumed to be known without error.

Alternative parameterization

When one has to deal with data of type 2 diabetics, or very insulin resistant individuals, estimating S_I with precision can be difficult. To improve numerical identifiability of the model one can use the constraints

$$p_1 = \frac{GEZI + S_I \cdot I_b}{V}, \tag{7.42}$$

to link S_I to p_1 through the parameter GEZI, representing glucose effectiveness at zero insulin [45].

GEZI can be fixed to 0.036 dl kg^{-1} min^{-1} in healthy individuals and 0.025 dl kg^{-1} min^{-1} in type 2 diabetic patients. These values were derived in [49]. In fact, in that study, the meal rate of glucose appearance was estimated using a sophisticated tracer experiment, so that parameters p_1 and V could be identified in each individual. This allowed them to calculate parameter GEZI in each subject and derive a population average. Interestingly, if S_I is well identifiable, the two parameterizations provide virtually the same results, while in the case of too low or very imprecise S_I, this last parameterization is more robust.

7.4.3 The concept of disposition index

It is an accepted notion that beta-cell function needs to be interpreted in light of the prevailing insulin sensitivity. One possibility is to resort to a normalization of beta-cell function based on the disposition index (DI) paradigm, first introduced in 1981 [45], and recently revisited in [50] where beta-cell function is multiplied by insulin sensitivity. This concept is illustrated in figure 7.10. If in a subject, insulin sensitivity is reduced but beta-cell responsivity compensates for that impairment, that subject remains on the same hyperbola, maintaining normal tolerance (left panel). If in a subject, insulin sensitivity is reduced and beta-cell responsivity does not compensate for that impairment, that subject moves to a different hyperbola, and the glucose tolerance of this individual becomes impaired (right panel). In other words, different tolerances are represented by different hyperbolas, given by

$$\mathrm{DI} = S_I \cdot \Phi = \text{constant}, \tag{7.43}$$

and the individual's beta cells ability to respond to a decrease in insulin sensitivity, by adequately increasing insulin secretion, can be assessed by measuring the product of beta-cell function and insulin sensitivity. This measure of beta-cell function, was first introduced for IVGTT and has been used also for MTT/OGTT. Thus, disposition indices can be calculated by multiplying responsivity indices Φ_d, Φ_s, Φ_{tot} by S_I to determine if the first phase, second phase and global beta-cell function are appropriate in light of the prevailing insulin sensitivity. Interestingly, while S_I was found to be significantly lower in MTT than OGTT and Φ_d, Φ_s, Φ_{tot} significantly higher in MTT than OGTT, the corresponding DIs are the same with the two tests, making it a good marker of glucose tolerance [51].

Another important use of the DI paradigm is the monitoring in time of the individual components of tolerance and the assessment of different treatment strategies. Referring to figure 7.11, if a subject lies in the orange zone, this subject needs to increase insulin sensitivity, e.g. by taking an insulin sensitizer, if it lies in the left upper part of the graph (red circle), while if it lies in the right lower part of the

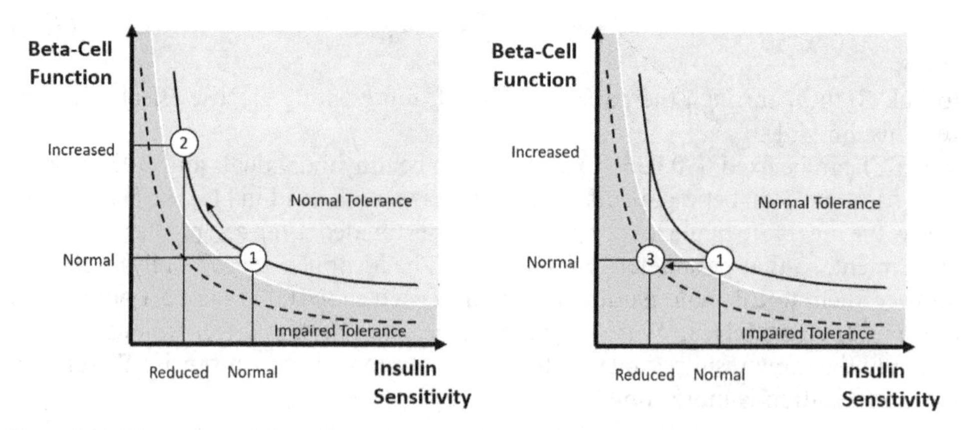

Figure 7.10. The concept of disposition index [45]. If in a subject insulin sensitivity is reduced but beta-cell responsivity compensates for that impairment, that subject remains on the same hyperbola, maintaining normal tolerance (left); if in a subject insulin sensitivity is reduced and beta-cell responsivity does not compensate for that impairment, that subject moves to a different hyperbola, and the glucose tolerance becomes impaired (right).

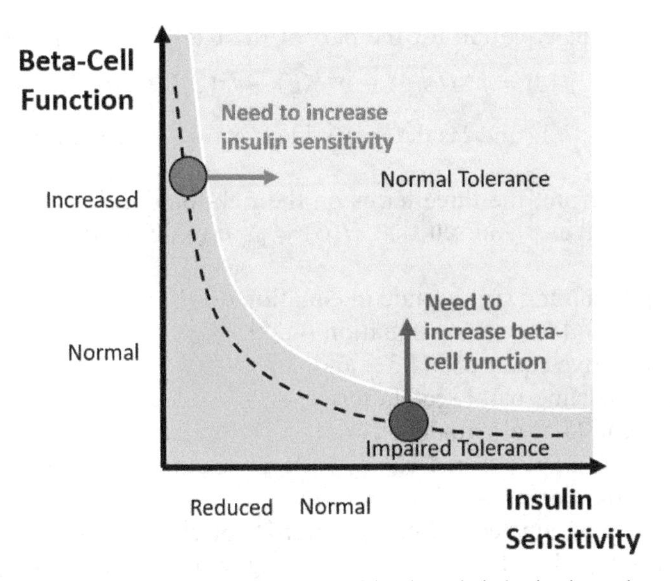

Figure 7.11. Treatment strategies: the subject represented by the red circle clearly needs to increase insulin sensitivity to recover a normal glucose tolerance, e.g. by taking an insulin sensitizer; conversely, the subject represented by the blue circle needs to improve beta-cell responsiveness, likely taking an insulin secretagogue drug.

graph (blue circle), the subject needs to improve beta-cell responsiveness, likely taking an insulin secretagogue drug (e.g. a DPP4-inhibitor [52]).

A final note concerns the often forgotten role of insulin hepatic extraction in the above picture. In fact, since the effect of insulin on peripheral tissues is also determined by the amount of insulin to which the tissue is exposed, hepatic insulin extraction should also be taken into account, thus adding a further dimension to the relationship between insulin secretion and action shown in figure 7.10.

Exercises

1. Derive the C-peptide kinetics, using the Van Cauter formulas for an individual with the following anthropometric characteristics: age = 34 years, weight = 78 kg, height = 1.82 m, sex = M; status: Healthy.
2. Calculate the basal insulin secretion rate (SR_b) and the basal concentration of CP in the second compartment, given that $CP_b = 500$ pmol l^{-1} in the subject described in exercise 1.
3. Calculate Φ_b in a subject with $CP_b = 620$ pmol l^{-1}, $G_b = 155$ mg dl^{-1}, age = 65 years, weight = 96 kg, height = 1.75 m, sex = M; status: 2DM.
4. Starting from equations (7.1) and (7.15)–(7.16), derive equation (7.20) defining parameter Φ_{tot} as function of Φ_d and Φ_s.
5. Calculate Φ_{tot} during a meal when $\Phi_s = 35 \times 10^{-9}$ min^{-1}, $\Phi_d = 455 \times 10^{-9}$, $G_b = 90$ mg dl^{-1}, $G_{\max} = 184$ mg dl^{-1}, $h = 95$ mg dl^{-1}, $T = 420$ min, $AUC(G) = 50540$ mg dl^{-1} min^{-1} (be careful with the measurement unit of G).

6. Consider the equation for the part of the RRP $h(g)$ with threshold g,

$$\mathrm{d}h(g)/\ \mathrm{dt} = p^{+}D\varphi(g) - p^{-}h(g) - fh(g)?\ (G - g),\ (*)$$

where $\varphi = \mathrm{d}?/\mathrm{d}g$ and ? is the Heaviside function, $?(x) = 1$ for $x > 0$, $?(x) = 0$ for $x < 0$.
   (a) Interpret the three terms on the right-hand side of (*).
   (b) Convince yourself that $H(G) = {}_0{}^G\ h(g)\ \mathrm{d}g$. Derive equation (4.24).
7.
   (a) Assuming steady-state in equation (4.23), and using equations (4.28) and (4.31), derive equation (4.29).
   (b) Derive equations (4.32) and (4.33).
   (c) Redefine basal secretion appropriately to show that equations (4.4) and (4.5) are analogous to equations (4.32) and (4.33).
8. For the OGTT, assume steady-state in equations (4.24) and (4.25), and derive equation (4.36).
9. Calculate parameter k_3 (mg kg^{-1} min^{-1}) of glucose rate of appearance Ra (equation (7.41)), knowing that $D = 75$ g, weight = 70 kg, $k_1 = 1.5$ mg kg^{-1} min^{-1}, $k_2 = 2.9$ mg kg^{-1} min^{-1}, $k_4 = 4.3$ mg kg^{-1} min^{-1}, $k_5 = 2.1$ mg kg^{-1} min^{-1}, $k_6 = 1.1$ mg kg^{-1} min^{-1}, $k_7 = 0.5$ mg kg^{-1} min^{-1}, $k_8 = 0$ mg kg^{-1} min^{-1} and the breakpoints are fixed in [0, 10, 20, 30, 60, 90, 120, 180, 420] min.
10. Given the model in equation (7.40), determine the steady state value of G (G_{ss}), as function of G_b if $Ra = 0$, $p_1 = 0.02$ min^{-1}, $p_2 = 0.01$ min^{-1}; $p_3 = 10^{-5}$ min^{-1} per μU ml^{-1}, $I_b = 10$ μU ml^{-1} and I brought constant to 1/2 I_b.
11. Suppose that a subject has initially $S_I = 10^{-3}$ dl kg^{-1} min^{-1} per μU ml^{-1} and $\Phi_{\mathrm{tot}} = 550 \times 10^{-9}$ min^{-1} and, due to a bad life style his S_I decreases to 5×10^{-4} dl kg^{-1} min^{-1} per μU ml^{-1}. How much has the beta cell function increase to maintain the same level of glucose tolerance?

References

[1] Polonsky K S and Rubenstein A H 1984 C-peptide as a measure of the secretion and hepatic extraction of insulin. Pitfalls and limitations *Diabetes* **33** 486–94

[2] Meier J J, Veldhuis J D and Butler P C 2005 Pulsatile insulin secretion dictates systemic insulin delivery by regulating hepatic insulin extraction in humans *Diabetes* **54** 1649–56

[3] Wiesenthal S R, Sandhu H, McCall R H, Tchipashvili V, Yoshii H, Polonsky K, Shi Z Q, Lewis G F, Mari A and Giacca A 1999 Free fatty acids impair hepatic insulin extraction *in vivo Diabetes* **48** 766–74

[4] Duckworth W C, Bennett R G and Hamel F G 1998 Insulin degradation: progress and potential *Endocr. Rev.* **19** 608–24

[5] Magni P, Sparacino G, Bellazzi R and Cobelli C 2006 Reduced sampling schedule for the glucose minimal model: importance of Bayesian estimation *Am. J. Physiol. Endocrinol. Metab.* **290** E177–84

[6] Licko V and Silver A 1975 Open-loop glucose-insulin control with threshold secretory mechanism: analysis of intravenous glucose tolerance tests in man *Math. Biosci.* **27** 319–32

[7] Hagander P, Tranberg K G, Thorell J and Distefano J III 1978 Models for the insulin response to intravenous glucose *Math. Biosci.* **42** 15–29

[8] Grodsky G M 1972 A threshold distribution hypothesis for packet storage of insulin and its mathematical modeling *J. Clin. Invest.* **51** 2047–59

[9] Toffolo G, De Grandi F and Cobelli C 1995 Estimation of beta-cell sensitivity from intravenous glucose tolerance test C-peptide data. Knowledge of the kinetics avoids errors in modeling the secretion *Diabetes* **44** 845–54

[10] Eaton R P, Allen R C, Schade D S, Erickson K M and Standefer J 1980 Prehepatic insulin production in man: kinetic analysis using peripheral connecting peptide behavior *J. Clin. Endocrinol. Metab.* **51** 520–8

[11] Cobelli C and Carson E R 2008 *Introduction to Modelling in Physiology and Medicine* (San Diego, CA: Academic)

[12] Toffolo G, Campioni M, Basu R, Rizza R A and Cobelli C 2006 A minimal model of insulin secretion and kinetics to assess hepatic insulin extraction *Am. J. Physiol. Endocrinol. Metab.* **290** E169–76

[13] Van Cauter E, Mestrez F, Sturis J and Polonsky K S 1992 Estimation of insulin secretion rates from C-peptide levels. Comparison of individual and standard kinetic parameters for C-peptide clearance *Diabetes* **41** 368

[14] Breda E, Cavaghan M K, Toffolo G, Polonsky K S and Cobelli C 2001 Oral glucose tolerance test minimal model indexes of beta-cell function and insulin sensitivity *Diabetes* **50** 150–8

[15] Breda E, Toffolo G, Polonsky K S and Cobelli C 2002 Insulin release in impaired glucose tolerance: oral minimal model predicts normal sensitivity to glucose but defective response times *Diabetes* **51** S227–33

[16] O'Connor M D, Landahl H and Grodsky G M 1980 Comparison of storage- and signal-limited models of pancreatic insulin secretion *Am. J. Physiol.* **238** R378–89

[17] Cerasi E, Fick G and Rudemo M 1974 A mathematical model for the glucose induced insulin release in man *Eur. J. Clin. Invest.* **4** 267–78

[18] Nesher R and Cerasi E 2002 Modeling phasic insulin release: immediate and time-dependent effects of glucose *Diabetes* **51** S53–9

[19] Eliasson L, Abdulkader F, Braun M, Galvanovskis J, Hoppa M B and Rorsman P 2008 Novel aspects of the molecular mechanisms controlling insulin secretion *J. Physiol.* **586** 3313–24

[20] Valdeolmillos M, Santos R M, Contreras D, Soria B and Rosario L M 1989 Glucose-induced oscillations of intracellular Ca^{2+} concentration resembling bursting electrical activity in single mouse islets of Langerhans *FEBS Lett.* **259** 19–23

[21] Henquin J C, Nenquin M, Stiernet P and Ahren B 2006 *In vivo* and *in vitro* glucose-induced biphasic insulin secretion in the mouse: pattern and role of cytoplasmic Ca^{2+} and amplification signals in beta-cells *Diabetes* **55** 441–51

[22] Henquin J C, Ishiyama N, Nenquin M, Ravier M A and Jonas J C 2002 Signals and pools underlying biphasic insulin secretion *Diabetes* **51** S60–7

[23] Grodsky G M, Curry D, Landahl H and Bennett L 1969 [Further studies on the dynamic aspects of insulin release *in vitro* with evidence for a two-compartmental storage system] *Acta Diabetol. Lat.* **6** 554–78

[24] Daniel S, Noda M, Straub S G and Sharp G W 1999 Identification of the docked granule pool responsible for the first phase of glucose-stimulated insulin secretion *Diabetes* **48** 1686–90

[25] Olofsson C S, Göpel S O, Barg S, Galvanovskis J, Ma X, Salehi A, Rorsman P and Eliasson L 2002 Fast insulin secretion reflects exocytosis of docked granules in mouse pancreatic B-cells *Pflugers Arch.* **444** 43–51

[26] Dean P M and Matthews E K 1970 Glucose-induced electrical activity in pancreatic islet cells *J. Physiol.* **210** 255–64

[27] Pedersen M G, Corradin A, Toffolo G M and Cobelli C 2008 A subcellular model of glucose-stimulated pancreatic insulin secretion *Philos. Trans. A Math. Phys. Eng. Sci.* **366** 3525–43

[28] Rorsman P and Renström E 2003 Insulin granule dynamics in pancreatic beta cells *Diabetologia* **46** 1029–45

[29] Jing X *et al* 2005 CaV2.3 calcium channels control second-phase insulin release *J. Clin. Invest.* **115** 146–54

[30] Ivarsson R, Jing X, Waselle L, Regazzi R and Renström E 2005 Myosin 5a controls insulin granule recruitment during late-phase secretion *Traffic* **6** 1027–35

[31] Barg S, Huang P, Eliasson L, Nelson D J, Obermüller S, Rorsman P, Thevenod F and Renström E 2001 Priming of insulin granules for exocytosis by granular Cl^- uptake and acidification *J. Cell. Sci.* **114** 2145–54 https://jcs.biologists.org/content/114/11/2145.long

[32] Bertuzzi A, Salinari S and Mingrone G 2007 Insulin granule trafficking in beta-cells: mathematical model of glucose-induced insulin secretion *Am. J. Physiol. Endocrinol. Metab.* **293** 396–409

[33] Jonkers F C and Henquin J C 2001 Measurements of cytoplasmic Ca^{2+} in islet cell clusters show that glucose rapidly recruits beta-cells and gradually increases the individual cell response *Diabetes* **50** 540–50

[34] Heart E, Corkey R F, Wikstrom J D, Shirihai O S and Corkey B E 2006 Glucose-dependent increase in mitochondrial membrane potential, but not cytoplasmic calcium, correlates with insulin secretion in single islet cells *Am. J. Physiol. Endocrinol. Metab.* **290** E143–8

[35] Low J T, Mitchell J M, Do O H, Bax J, Rawlings A, Zavortink M, Morgan G, Parton R G, Gaisano H Y and Thorn P 2013 Glucose principally regulates insulin secretion in mouse islets by controlling the numbers of granule fusion events per cell *Diabetologia* **56** 2629–37

[36] Pedersen M G and Cobelli C 2013 Multiscale modelling of insulin secretion during an intravenous glucose tolerance test *Interface Focus* **3** 20120085

[37] Pedersen M G, Toffolo G M and Cobelli C 2010 Cellular modeling: insight into oral minimal models of insulin secretion *Am. J. Physiol. Endocrinol. Metab.* **298** 597–601

[38] Toffolo G, Breda E, Cavaghan M K, Ehrmann D A, Polonsky K S and Cobelli C 2001 Quantitative indexes of beta-cell function during graded up&down glucose infusion from C-peptide minimal models *Am. J. Physiol. Endocrinol. Metab.* **280** 2–10

[39] Herrero P, Georgiou P, Oliver N, Johnston D G and Toumazou C 2012 A bio-inspired glucose controller based on pancreatic β-cell physiology *J. Diabetes Sci. Technol.* **6** 606–16

[40] Hatlapatka K, Matz M, Schumacher K, Baumann K and Rustenbeck I 2011 Bidirectional insulin granule turnover in the submembrane space during K^+ depolarization-induced secretion *Traffic* **12** 1166–78

[41] Pedersen M G, Tagliavini A and Henquin J-C 2019 Calcium signaling and secretory granule pool dynamics underlie biphasic insulin secretion and its amplification by glucose: experiments and modeling *Am. J. Physiol.-Endocrinol. Metab.* **316** E475–86

[42] Pedersen M G 2010 A biophysical model of electrical activity in human β-cells *Biophys. J.* **99** 3200–7

[43] Riz M, Braun M and Pedersen M G 2014 Mathematical modeling of heterogeneous electrophysiological responses in human β-cells *PLoS Comput. Biol.* **10** e1003389
[44] Gandasi N R *et al* 2017 Ca^{2+} channel clustering with insulin-containing granules is disturbed in type 2 diabetes *J. Clin. Invest.* **127** 2353–64
[45] Bergman R N, Phillips L S and Cobelli C 1981 Physiologic evaluation of factors controlling glucose tolerance in man: measurement of insulin sensitivity and beta-cell glucose sensitivity from the response to intravenous glucose *J. Clin. Invest.* **68** 1456–67
[46] Bergman R N, Ider Y Z, Bowden C R and Cobelli C 1979 Quantitative estimation of insulin sensitivity *Am. J. Physiol.* **236** E667–77
[47] Dalla Man C, Caumo A and Cobelli C 2002 The oral glucose minimal model: estimation of insulin sensitivity from a meal test *IEEE Trans. Biomed. Eng.* **49** 419–29
[48] Dalla Man C, Caumo A, Basu R, Rizza R, Toffolo G and Cobelli C 2004 Minimal model estimation of glucose absorption and insulin sensitivity from oral test: validation with a tracer method *Am. J. Physiol. Endocrinol. Metab.* **287** E637–43
[49] Basu A, Dalla Man C, Basu R, Toffolo G, Cobelli C and Rizza R A 2009 Effects of type 2 diabetes on insulin secretion, insulin action, glucose effectiveness, and postprandial glucose metabolism *Diabetes Care* **32** 866–72
[50] Denti P, Toffolo G M and Cobelli C 2012 The disposition index: from individual to population approach *Am. J. Physiol. Endocrinol. Metab.* **303** E576–86
[51] Bock G, Dalla Man C, Campioni M, Chittilapilly E, Basu R, Toffolo G, Cobelli C and Rizza R 2007 Effects of nonglucose nutrients on insulin secretion and action in people with pre-diabetes *Diabetes* **56** 1113–9
[52] Drucker D J and Nauck M A 2006 The incretin system: glucagon-like peptide-1 receptor agonists and dipeptidyl peptidase-4 inhibitors in type 2 diabetes *Lancet* **368** 1696–705

Part IV

Autoimmune type 1 diabetes

IOP Publishing

Diabetes Systems Biology
Quantitative methods for understanding beta-cell dynamics and function
Anmar Khadra

Chapter 8

Applying systems biology to the genetics of age-of-onset dependent heterogeneity in type 1 diabetes

Shouguo Gao, Nathaniel Wolanyk, Soumitra Ghosh and Xujing Wang

8.1 The design of an integrative genomics approach to study T1D age-at-onset (AAO)

8.1.1 Challenges in the genetic study of T1D

Type 1 diabetes (T1D) is a complex genetic disease that results from the autoimmune destruction of the insulin-producing pancreatic islet beta cells [1–3]. It currently affects about 0.5% of the population in the US, and is one of the most common chronic diseases in children and adolescents. Patients with T1D depend on insulin for life, with an estimated 10–15 years decrease in life span even following a rigorous disease management protocol. The risk for disease is attributed to a combination of genetic variants, environmental factors, and interactions between genes and environment [4, 5]. The fact that concordance rates in monozygotic twins is significantly higher than that in dizygotic twins, ~50% or higher by the age of 40 [6] versus ~5%–6% [7], suggests a much stronger contribution from genetic disposition than from environmental factors. Indeed it is believed that genetic disposition determines about 80% of an individual's risk of developing T1D [8, 9]. The sibling recurrence risk ratio (λ_S, the ratio of the disease prevalence among siblings of affected individuals, over that in the general population) is estimated to be at 15, one of the highest among common complex diseases [10].

Given the high contribution to disease risk, much effort has been devoted toward dissecting the genetic architecture of T1D [11–13]. Initial works mainly included linkage mapping and candidate gene association analysis. Linkage analysis examines the co-segregation in families of a disease trait locus with a genomic region tagged by polymorphic markers, through the relationships between disease phenotype and marker variations. It returns odds scores (termed LOD, standing for

doi:10.1088/978-0-7503-3739-7ch8

Logarithm of the odds) of how likely a disease locus exists in the proximity of each marker [14]. The linkage studies started in the 1970s and continued for the next two to three decades, revealing a major contribution of the major histocompatibility complex (MHC) to the genetic disposition [15] and close to 20 other chromosomal regions that potentially house disease genes. Except for the strong linkage to the HLA locus, clear consistent evidence for linkage was not always observed at other loci [16–20], indicating extensive locus heterogeneity. The signal-to-noise ratio of the linkage mapping results was low; the majority of genes in these regions (totaling several thousands) were not expected to be disease genes. Disease risk association analysis of candidate genes subsequently yielded a number of positive results, including the insulin gene INS, CTLA-4 PTPN22, and IL2RA.

Association analysis, on the other hand, directly examined the correlation between each genetic variant and trait variation in a population of subjects, thus identifying genetic variants that might play a causal role in the creation of the trait [21]. Before genotyping and sequencing cost came down in the past decades, association analysis had been typically restricted to a list of likely candidates. The challenge then was to inclusively collect potential candidates, which is not feasible when disease etiology is not understood well, as it has been the case for T1D. The recent two decades have seen unprecedented advancements in high-throughput genomic technologies, which enabled more comprehensive, larger-scale whole genome genetic screening studies. Since the mid 2000s several genome-wide association studies (GWAS) of T1D have together genotyped over 30 000 subjects [22]. However, till now, the drastically increased coverage and sample size did not solve the genetics of any complex disease, as scientists originally hoped. Several challenges remain. Typically, the new genome wide scans confirmed the major loci previously known, and uncovered a number of new ones that contribute moderately to weakly. Presently, over 50 loci are purported to be associated with disease risk [10, 22–24]. These regions together host more than 300 protein coding genes, and a large number of potential regulatory elements. Only a small fraction of the loci have been attributed to particular genes [25], and even for them, the functional role of the attributed genes in disease pathogenesis remain mostly unclear [18].

The challenges faced by T1D are shared by most common complex diseases [26, 27]. After decades of efforts and several generations of technology revolutions, we are still faced with the same old questions. What are all the functional variants? How do they lead to disease? How can we translate the genetic results to predicting power for disease risk and disease prevention strategies?

The failure has been mostly attributed to the intrinsic inadequacies of the two approaches to mapping trait relevant genes; that includes (i) the limitations of linkage analysis in addressing the complexities like imprinting and age-dependent genetic effect of complex traits, and (ii) the limitations of association analysis in covering all functional SNPs within Linkage Disequilibrium of typed markers and in covering rare variants [28–31]. In the coming years, the technology will continue to advance to bring the cost down, and deeper genome wide screening including whole genome sequencing of large, more diverse populations will become increasingly more feasible. While these will help us to overcome some of these limitations, several more fundamental questions emerged from the failures of the past decades may remain: can we ever fully and automatically dissect the genetics of complex disease without understanding the disease biology? Do we have the right analytical model to dissect the phenotype–genotype relationship for complex traits? With the current conceptual framework, do we have enough statistical power to dissect genetics of complex traits even if we can screen all humans on Earth?

Conceptually, conventional approaches to complex trait genetics mostly adopted similar way of thinking to those employed in previous efforts that successfully mapped numerous Mendelian disorders, namely, testing one trait locus or one gene at a time. However, biological functions arises from interactions among cellular macromolecules and are shaped within given environment backgrounds [32]. The importance of interactions of genes and of gene-environments in forming complex traits have long been appreciated, but development of analytical models and methods to dissect their contribution is lagging behind. Clearly a brute force approach to assess all possible interactions is not feasible due to the astronomical number of them. Biology of the traits needs to be brought in [28, 30] to narrow the search space to the relevant pathways, and to model the interactions structures.

8.1.2 The design of an integrative genomics approach to study a focused problem in T1D: the age-at-onset heterogeneity in T1D

In this section, we present an integrated approach, combining mathematical biology, bioinformatics, and systems biology to model the disease initiation and progression of T1D, and subsequently utilize the results to identify and prioritize candidate disease genes. Figure 8.1 depicts the overall design. Given the complexity of T1D disease etiology and locus heterogeneity and the many technical challenges to overcome when developing a new method, this work focuses on a simplified question, the age-at-onset (AAO) heterogeneity in T1D, rather than attempts to solve the overall genetic architecture of T1D.

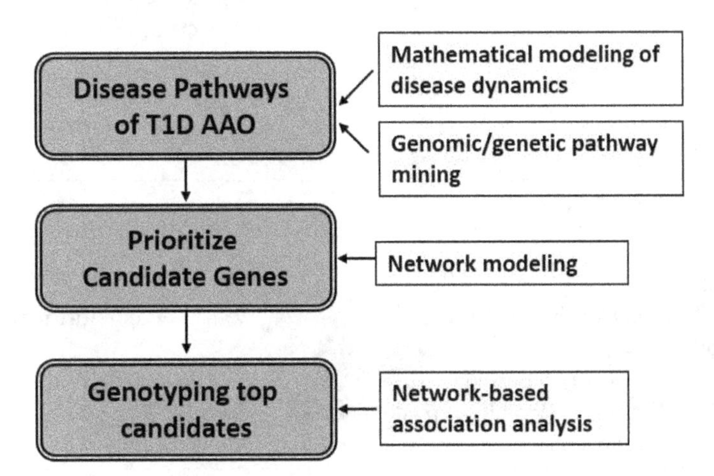

Figure 8.1. The scheme of the integrative genomics framework to dissect the genetics of T1D AAO.

A complex disease is often defined by a cluster of subtypes caused by different combinations of genetic variants and environmental factors, with both overlapping and heterogeneous traits, that lead to failure of a given function. In T1D, extensive genetic, epidemiological, physiological and phenotypical heterogeneities have been observed and several distinct subtypes have been proposed [11, 33]. One major contributing factor to the heterogeneities is AAO [17, 34–38]. Although sometimes referred to as juvenile diabetes, over 50% of T1D cases onset after 20 years of age [39, 40], with much different disease dynamics, phenotypes and genetic architecture from the pre-adolescent patients [41–43], as summarized in table 8.1. In the study of other complex diseases, the value of locus heterogeneity reduction in disease gene cloning through AAO based cohort stratification has been demonstrated [44]. Examples include the *BRCA1* (OMIM No. 113705, [45]) and *BRCA2* (OMIM No. 600185), responsible for a proportion of early onset breast cancer, the *APP* (OMIM No. 104760), *PSEN1* (OMIM No. 104311) and *PSEN2* (OMIM No. 600759) in Alzheimer's disease, and *SNCA* (OMIM No. 163890) in Parkinson's disease. In T1D, within the HLA locus, an association with AAO has been observed for both class II (including HLA-DRB1 and DQB1) [46] and class I alleles [46, 47]. Presently it is not understood why in some patients the disease starts later, nor how to delay disease onset; indeed, few epidemiological or genetic studies have compared patients with different ages of onset or investigated factors that regulate AAO of T1D [35, 46, 47].

This more focused AAO problem is likely more feasible to solve than the general T1D genetic landscape, with potentially more clinical relevance. Understanding why and how the disease starts late in some can shed light to developing new interventional methods that delay the onset, ideally beyond one's normal life span. Even just to delay T1D onset beyond adolescence would be of great significance considering the difficulty of diabetes management in the young. Adolescent patients face unique added challenges, such as hormonal changes, elevated growth rates, increased insulin resistance, as well as psychological and social behavioral changes, which can adversely affect efforts to tightly control blood glucose levels [48, 49]. Poorly controlled T1D leads to significantly increased risk for complications [50].

Table 8.1. The heterogeneity between young- and adult-onset T1D.

	Young-onset	Adult-onset
Disease progression	Rapid	Slow
Beta cell destruction	Complete within 1 year	Can take up to decades
Insulitis	Frequently observed	Less likely observed
Genetics	High-risk HLA alleles predominate the genetic risk	Non HLA genes also contribute significantly

The framework of the integrative approach presented in this section contains three major steps (figure 8.1): first, the identification of functional pathways important for T1D AAO. This is accomplished by mathematical modeling of the AAO dependent disease dynamic differences [51], assisted by disease pathway mining through integration of multiple genomics/genetic data types. Secondly, prioritization of genes in candidate disease pathways. This is achieved through integrating properties and prior knowledge of the candidate genes, as well as their topological properties in context-specific gene interaction networks. Lastly, validation of the predictions by association analysis of the top candidates in large cohorts of adult- (onset ⩾19 years) versus young-onset (<19 years) T1D. This is a modular design, where each module can be readily updated and improved separately as the technology and our understanding of disease biology advance. In the next subsections we will describe each step in more detail and the findings obtained to date.

8.2 Mathematical modeling of AAO heterogeneity of T1D and disease pathway identification

8.2.1 The Copenhagen model of T1D pathogenesis and its extension

In T1D, overt disease occurs after a pre-diabetic period of destruction of the pancreatic islet beta cells. The disease etiology is complex and not well understood [51]. However, it is generally believed that the following sequence of events occur during disease initiation: a local insult (viral infection, for example) causes beta cell damage and death, leading to the release of beta cell antigens. The residing antigen presenting cells (APCs) in islets, mainly macrophages in T1D, become activated upon uptake of beta cell autoantigens [52, 53]. Activated macrophages will migrate to pancreatic lymph nodes and recruit T cells to islets. Both the infiltrating T cells and activated macrophages release cytokines that cause further beta cell destruction. T cells can also kill β cells directly through effector cell mechanism [54]. Damaged beta cells will release more autoantigens that will be presented to the immune system, thereby amplifying and closing the perpetuating loop of beta cell destruction. From this scenario, one can see that the minimal set of the most essential components involved in beta cell killing at the site of pancreatic islet include (figure 8.2):

1. beta cells and beta cell autoantigens;
2. resting and activated tissue macrophages and infiltrating T cells.

De Blasio *et al* were the first to mathematically investigate the dynamics of the cellular interactions and termed it the Copenhagen model [53]. Using M, M_A, A, and T to denote the number of resting and activated macrophages, beta cell autoantigens, and infiltrating T cells, the model proposed that the kinetics of each cell populations given by [53]

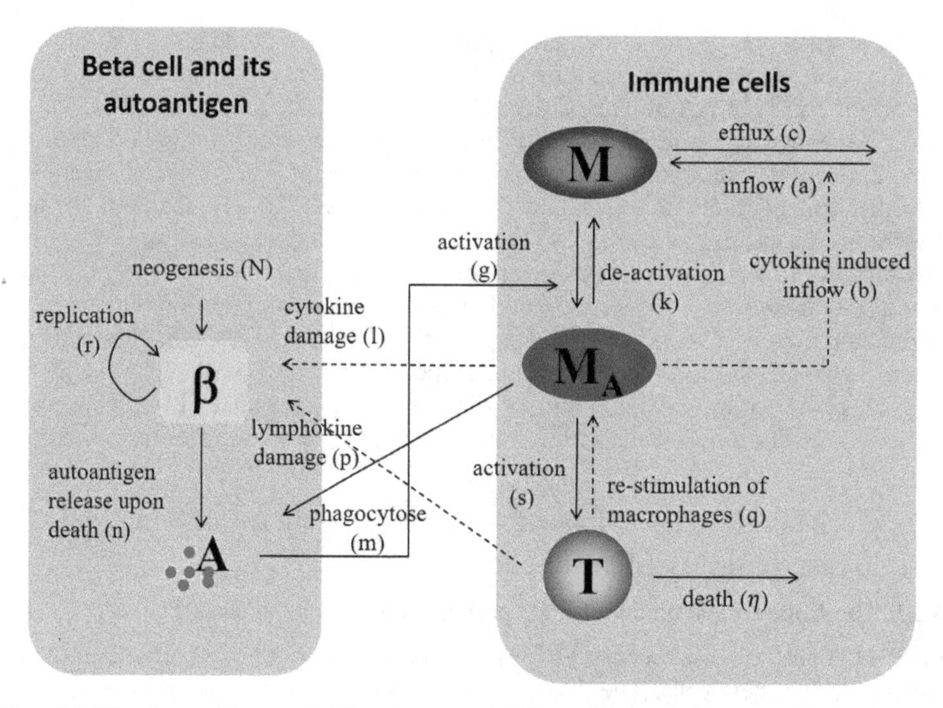

Figure 8.2. The scheme of the extended Copenhagen model. The model proposed two essential components in beta cell destruction: (i) the beta cell and its autoantigen; and (ii) the immune cells. The critical cellular interactions that initiate the T1D disease process are depicted.

$$\begin{aligned} \dot{M} &= a + (k + b)M - cM_A - gMA \\ \dot{M}_A &= gMA - kM_A \\ \dot{T} &= sM_AT - \eta T \\ \dot{A} &= lM_A + pT + qM_AT - mA \end{aligned}, \tag{8.1}$$

where the coefficients are the rate constants of the corresponding processes: a is macrophage inflow rate to islets; b is macrophage recruitment rate to the islet; k is rate of M_A decay back to M; c is rate of macrophage efflux; g is macrophage activation rate by uptake of autoantigen A; s is induced T-cell proliferation rate; η is death rate of activated T-cells; l is rate of beta cell death induced by M_A released cytokines IL-1 and TNF-α; p is rate of beta cell death induced by lymphokines such as IFN-γ; q is beta cell death rate caused by effector macrophages; m is clearing rate of autoantigens A.

The study found that the model predicted the existence of two equilibrium states when there is no infiltrating T cells: one without and one with activated macrophages M_A, representing the healthy state and the slow onset T1D, respectively [53]. It is important to point out that there were infiltrating T cells, a third equilibrium state exists, but it is not stable and hence did not correspond to any chronic state.

The Copenhagen model as a pioneer work in the field still left room for improvement. The dynamics of the immune cells-autoantigen system was not fully analyzed, the beta cell kinetics were not directly modeled, and it was not able to reproduce the AAO dependent heterogeneity. Therefore, the Copenhagen model was subsequently extended by including an explicit description of the beta-cell dynamics, as described by the scheme given in figure 8.2. The pancreatic islet beta cells actively participate in the process of their destruction [55, 56], and its own turnover (cell renewal and loss) continues after birth with drastically different rates between young and adult ages [57, 58]. This beta cell turnover is given by

$$\dot{\beta}_{\text{turnover}} = r(t)\beta - e(t)\beta + N(t), \tag{8.2}$$

where $r(t)$ and $e(t)$ are the per cell rates of beta cell replication and death, respectively, and $N(t)$ is the rate of beta cell neogenesis. When the immune cells are activated to infiltrate islets, induce beta cell death mediated by cytokines and effector T-cells [53]. As a first order approximation, each killing term is assumed to be proportional to the amount of cytokines or effector cells (consequently linear with respect to the number of the corresponding source of cell types), with

$$\dot{\beta}_{\text{induced death}} = lM_A + pT + qM_AT, \tag{8.3}$$

where l, p and q are the rate constants of beta cell death due to the cytokines secreted by macrophage and T cells, and effector cells, respectively. The complete extended model that includes a description of beta cell dynamics is thus given by [34]

$$\begin{aligned}
\dot{M} &= a + (k + b)M_A - cM - gMA \\
\dot{M_A} &= gMA - kM_A \\
\dot{T} &= sM_AT - \eta T \\
\dot{\beta} &= r\beta - e\beta + N - lM_A - pT - qM_AT \\
\dot{A} &= n(\beta + lM_A + pT + qM_AT) - mA
\end{aligned}, \tag{8.4}$$

8.2.2 The extended model is able to reproduce the AAO-dependent heterogeneity

The investigation of the dynamics of an ODE-model of a dynamics system, the first steps of analysis typically involves two steps: (1) dimensional analysis to understand model structure, determine degree of freedom (DOF) and simplify the mathematics; and (2) dynamic stability analysis, to identify equilibrium states of the system and examine the key factors affecting their stability. The stable equilibrium states usually describe chronic states in physiology, such as a healthy state. Examining conditions under which the equilibrium states lose stability will shed light on how a disease process is initiated. Details of such analyses were presented in [34]; here the major findings are summarized.

Among the interactions between the two compartments presented in figure 8.2 (beta cells and autoantigens along with immune cells), the key feedback that perpetuates immune destruction is the macrophage activation by beta cell auto-antigens A. When $A = 0$, the two components completely decouple from each other,

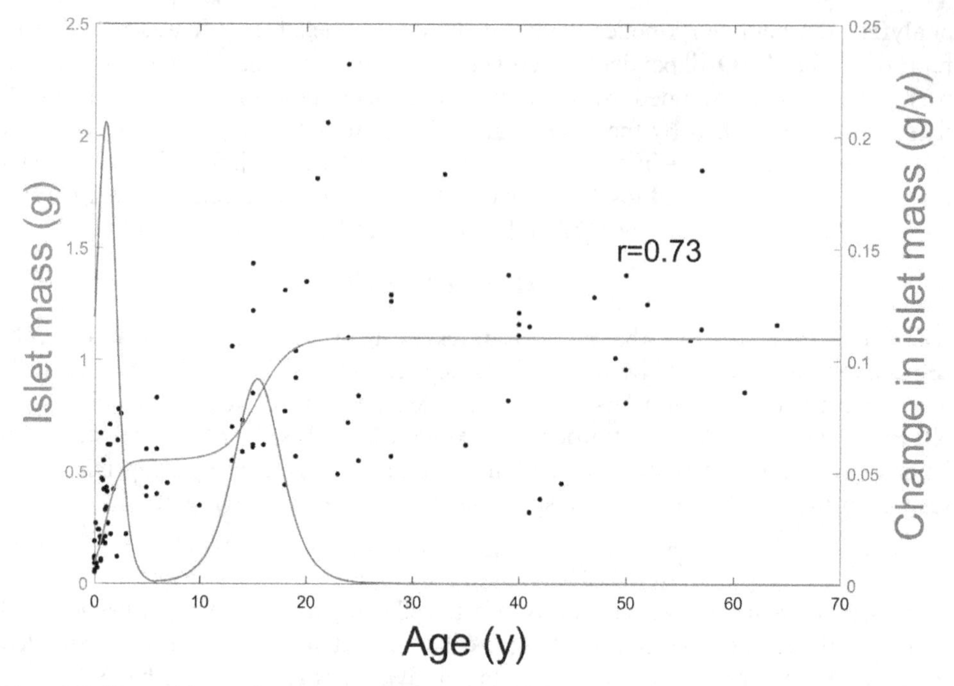

Figure 8.3. Elevated beta cell mass expansion rate during infancy and adolescence. The islet mass of 100 individuals given in [59] are plotted against their age. The blue line represents model fit, and the orange line the derived rate of change.

with $M = \frac{m}{c}$, $M_A = 0$, $T = 0$, and the beta cell number determined by its natural turnover $\dot{\beta} = \dot{\beta}_{\text{turnover}} = r\beta - e\beta + N$. This describes a healthy normal individual, where there is no infiltrating immune cells (i.e. no insulitis) and no autoimmune-mediated beta cell destruction.

When there is beta cell autoantigen ($n \neq 0$ and $A \neq 0$), a general stability analysis of equations (8.4) is difficult due to the many non-linear terms. This can be resolved by solving the equations under specific conditions whereby they can be linearized using perturbation techniques [60]. Since we are interested in the AAO problem, we examined the physiological differences between young- and adult-onset T1D. One fact that should catch readers' attention is that beta-cell mass continues to grow after birth to compensate the demand from a growing body [61]. Although the dynamics of beta cells is generally slow relative to immune cell dynamics (see chapter 9), it is considered fast during the first few years of life and during puberty compared to during adulthood, due to rapid organ growth and remodeling. Ogilvie studied pancreata from 100 normally nourished, non-diabetic individuals ranging from 0–64 years old [59]. These measurements were extracted and summarized in figure 8.3 (dots). Visual inspection suggests high islet mass growth rates during infancy (0–3 years) and adolescence (13–21 years). By doing a localized regression (figure 8.3, blue line) to the data and determining the corresponding rate of change (green line), two peaks of growth during infancy and adolescence were observed.

Utilizing this knowledge, one can perform linear expansion of equations (8.4) under high (e/m, r/m, $N/m \gg 1$, in dimensionless units) and low (e/m, r/m, $N/m \ll 1$) beta cell turnover limits, respectively, and carry out stability analysis of the linearized equations separately [34]. When taking the high turnover rate limit (corresponding to young age group), equations (8.4) yielded no steady state, indicating that if beta cell killing is initiated under this condition, no stable sub-healthy state exists, macrophages will be activated ($M_A \neq 0$), T cells will infiltrate the islets ($T \neq 0$), and beta cell destruction will proceed rapidly. This is consistent with the fact that young-onset T1D is a rapidly progressing disease (table 8.1).

On the other hand, under slow beta cell turnover limit, a steady state exists with:

$$\begin{aligned} M_A &= \frac{eN}{rl}\frac{\Delta}{\Delta + kb} \qquad M = \frac{a}{c} + \frac{ab}{mc}\frac{eN}{rl}\frac{\Delta}{\Delta + kb} \\ T &= 0 \qquad\qquad\qquad\qquad\qquad\qquad\qquad\qquad , \\ \beta &= \frac{N}{(e-r)}\left(1 - \frac{e}{r}\frac{\Delta}{\Delta + kb}\right) \qquad A = \frac{ne}{e-r}\frac{a}{m}N\frac{kb}{\Delta + kb} \end{aligned} \tag{8.5}$$

where $\Delta = \frac{ngabl}{cm}\frac{r}{e-r}$. Note that this is a state with no T-cell insulitis (infiltration of the islets). To preserve the positivity of state variable and maintain stability, the following condition must be satisfied

$$f_c = \frac{ngal}{kmc} < 1, \tag{8.6}$$

where f_c is a dimensionless parameter called the Copenhagen constant, in recognition of the pioneering work of Freiesleben De Blasio *et al* [53]. The stability of this state further required the amount of activated macrophages to satisfy $M_A < \frac{tm}{sa}$ [34]. The existence of this state suggests that when autoimmunity is initiated during adulthood (corresponding to low beta cells turnover), a slow disease procession is obtained, which agrees with clinical observations of adult-onset T1D (table 8.1).

When compared to clinical/laboratory findings in more detail, a general agreement was found to model predictions. First, the model predicted that a system with elevated beta cell turnover rates is less robust against autoimmunity. Epidemiological studies revealed that adolescents with more rapid organ development rate [62], such as pregnant women with more active beta-cell turnover [63] have higher risk for T1D. In a transgenic mice study, it was found that by turning on or off a beta cell turnover gene Tag (which causes abnormal beta cell turnover), the autoimmune destruction of beta cells can be initiated or reversed [64]. In animal models of T1D, it has been proposed that the neonatal wave of beta cell apoptosis could be the trigger of autoimmunity, backed by a number of supporting evidences [65]. Secondly, the model suggested that compromised phagocytosis increases the risk for perpetuating autoimmunity. It is known that defective phagocytosis (lower m value) of apoptotic bodies contribute to inflammatory and autoimmune diseases [66]; apoptotic cells if not phagocytosed in time, will undergo secondary necrosis and acquire enhanced immunogenicity [67]. Impaired phagocytosis has long been observed in T1D patients, and in at-risk (first degree relatives of probands) adult *healthy* individuals [68]. Interestingly, the phagocytosis rate m is also significantly lower *prior to* disease in the diabetic-prone strains than in the resistant strains in both rat [69] and mouse [70, 71] models of T1D. This impaired phagocytosis in macrophages will be studied in detail in the next section. Thirdly, absence of T-cell insulitis is a necessary condition for a slow-onset process in adult patients. A literature review of all published reports on insulitis (totaling 559 patients, table 8.2 of [34]) found that adult patients are much less likely to have insulitis than young patients (18.3% versus 63.6%, $p < 0.001$). This is the first mathematical model that was able to explain the AAO-dependent heterogeneity in T1D. These agreements indicate that the extended Copenhagen model (figure 8.2 and equations (8.2)) has captured key factors contributing to T1D initiation.

8.2.3 Critical quantitative traits, pathways and candidate disease genes

Complex disease involves multiple interacting genes and environmental factors that do not contribute independently. A mathematical framework allows one to understand how interplay of the factors leads to disease, and hence to dissect disease pathways in a systematic manner. The extended Copenhagen model (figure 8.2 and equations (8.4)) suggested that for both young and adult subjects, the autoantigen presentation was essential for disease initiation (see chapter 9), which presumably

Table 8.2. Top GO pathways identified from genome linkage scan data mining. The HLA locus was excluded. The overlap with the QTs predicted by the extended Copenhagen model is indicated in the last column.

FDR	Pathways	QTS it regulates
<0.01	Cell-matrix adhesion	m
<0.01	DNA-dependent regulation of transcription	
<0.01	Caspase activity	l, m
<0.01	Epigenetic regulation of gene expression	
<0.01	Ligase activity	
<0.01	Oxidoreductase activity	m, l
<0.01	Frizzled signaling pathway	
<0.01	Zinc finger, C2H2 type, domain	l
0.01	Amine biosynthesis	l
0.01	Cell adhesion receptor activity	l
0.01	Exocytosis	Beta cell function
0.01	Endocytosis	m
0.01	tRNA processing/metabolism/modification	Beta cell function
0.01	DNA methylation	
0.01	Voltage-gated sodium channel activity	Beta cell function
0.01	Oxidative phosphorylation	Beta cell stress
0.01	RNA modification	
0.01	Viral nucleocapsid	

corresponded to the shared HLA locus. For adult-onset T1D, the model predicted additional risk contributing factors from the condition of $f_c = \frac{ngal}{kmc} < 1$. Each individual parameter is a rate constant of a specific physiological process. Together they present a comprehensive picture of driving factors of the AAO heterogeneity. Many of these physiological processes have been studied extensively and their genetic regulating pathways well annotated. They therefore can be utilized to pinpoint the disease pathways and candidate disease genes, as depicted in figure 8.4 [34, 72]. For example, phagocytosis is a well-characterized gene ontology (GO) term, and the proteins involved are listed in GO database; substantial literature exists for signal transduction pathways in macrophage activation (factor g, k) and phagocytosis [55, 73, 74]. Transcription regulatory networks that significantly affect beta cell sensitivity to cytokine-induced death (l) in T1D context has been comprehensively profiled [55]. One can collect the scattered information and knowledge from public databases and in literature, and utilize them to build our candidate disease list.

8.2.4 Bioinformatics of disease pathway identification

The mathematical modeling of disease pathways can be further assisted by bioinformatics. Over the past three decades, extensive studies utilizing the advancement in high-throughput technologies have generated unprecedented large amounts of data that contain rich information about disease mechanisms and pathways [30]. In fact, pathway mining of genetic and genomic data has been an active research field in bioinformatics. The basic idea is to evaluate the statistical significance of enriched presence of a pathway among genes showing differential genetic or expression variation between affected and control cohorts [75–78].

Presently, a large number of tools exist for such analysis. The early ones treat genes in a pathway as a set of unrelated, equally important genes, ignoring their interactions. It is now appreciated that genetic variations which result in alterations to the topological structure of interaction networks are more meaningful than the mere change in individual genes. Genes which influence the expression of many other genes (i.e. hub genes) are more critical in determining network function [79].

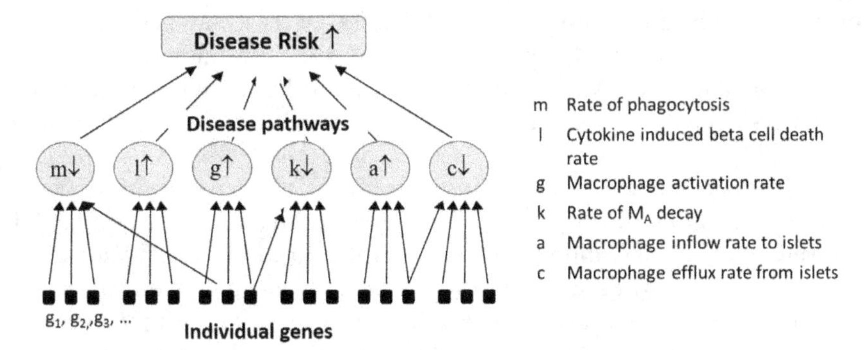

Figure 8.4. Differential driving factors of AAO heterogeneity predicted by the extended Copenhagen model. They together offer a framework to identify disease pathways and candidate disease genes.

Indeed, a central dogma in systems biology is (interaction) structure defines function in a complex system; functions, on the other hand, arise from interactions and crosstalk, and would not exist otherwise.

In the 'third generation' pathway analysis tools, the consideration of network structure are incorporated [80, 81], one example being TAPPA (Topological Analysis of Pathway Phenotype Association). Given a pathway where the interactions (of any type, protein, regulatory, genetic, etc) among its genes can be described by an adjacency matrix $A = (a_{i,j})$, with $a_{ij} = 1$ if gene i and j interact and 0 otherwise; assuming that the significance of each gene (in differential association or expression, for example) is given by its Z-score (z_i), we define the network connectivity index (NCI) [81]

$$NCI = \frac{1}{\sqrt{k}} \sum_{i \in A} \sum_{j \in A} |z_i|^{0.5} * a_{ij} * |z_j|^{0.5} * \mathrm{sgn}(z_i + z_j), \tag{8.7}$$

where sgn $(z_i + z_j)$ is the sign of the value of $(z_i + z_j)$. Having a quantitative measure of a pathway's overall significance, its relevance to a trait of interest can then be evaluated through the correlation between pathway NCI and the trait variations. For binary traits, the Mann–Whitney test can be used. For quantitative traits, the Spearman correlation can be adopted. Further, FDR can be derived through a permutation test. The resulting algorithm has been implemented in a publicly available Java plugin to Cytoscape named jActiveModuleTopo, and in TopASeq by Ihnatova *et al* (https://www.bioconductor.org/packages/release/bioc/html/ToPASeq.html) [81–84]. NCI captures the topological property of the pathway, and contributions from both the individual genes, and the interactions among them. Subtle but consistent gene expression change would lead to significant changes in NCI. Further, it naturally over-weighs hub genes connected to a larger number of other genes. This tool was applied to mine genome wide linkage, association and functional genomics data [82–85].

Many T1D genome linkage scan studies have been carried out prior to the GWAS and deep-sequencing era [86–90]. Data from the Type 1 Diabetes Genetics Consortium (T1DGC) [91] of 1,439 multiplex families with 1643 affected sib pairs (ASP), with its large sample size, quality of data and relative cohort demographic homogeneity, can be used. Results revealed that 982 of these families had all affected sibs onset <19 years, whilst 244 families had all affected sibs onset during adulthood ⩾19 years. Up till now, no AAO analysis of the cohort has been reported. In total 10 linkage regions were identified with $p < 0.01$: *2q31-33, 3p13-14, 6p21.31 (the HLA loci), 9q33-q34, 10p14-q11, 11p15, 12q14-q12, 16p12-q11.1, 16q22-q24, and 19p13.3-p13.2*. The physical maps were extracted from stsInfo2 and stsMap files from UCSC's genomic database (http://genome.ucsc.edu), and all genes within the 1-LOD regions were obtained and annotated, totaling 1301 positional candidate genes. The linkage analysis result by itself is noisy, where each linkage region may be harboring zero to a few disease genes among many that bear no relationship to the disease. Some disease genes may reside outside the linkage regions. The majority of the 1301 genes are not expected to be T1D disease genes, and many disease genes are likely missed by them. Examination of the common biological theme shared among genes in linkage loci at the pathway level can potentially amplify the signal-to-noise ratio. The top pathways from mining of the linkage data are listed in table 8.2. These pathways together point to two major categories of pathways: beta cell stress/death, and the immune system's capacity to handle it. Interestingly, the major pathway predictions overlap largely with those predicted by the extended Copenhagen model (compare table 8.2 and figure 8.4). Examples include phagocytosis and beta cell function, stress/death genes (voltage gated ion channel, exocytosis, caspase activities).

In the past two decades, linkage analysis was largely superseded by the GWASs. After a period of disillusionment and cynicism in the past two decades [92, 93], a renewed interest in linkage analysis in the era of whole genome sequencing (WGS) was initiated, due to its potential in the identification of rare variants associated with complex traits with high penetrance [29, 31]. Linkage analysis coupled with WGS has led to the successful identification of a number of novel disease susceptibility genes recently [31].

A number of T1D GWAS datasets are also available now; an incomplete list includes: (1) From dbGAP: 3000 T1DGC samples with age of onset, profiled with Ilumina 550K. (2) From WTCCC: 2000 control samples profiled with Ilumina 550K chip; 2000 control samples with Affymetrix 500K chip; 1504 from the 1958 British Birth Cohort Control, Affymetrix 500K chip; 1999 samples of T2D patients, Affymetrix 500K; 2000 T1D patients, Affymetrix 500K. (3) From T1DGC: scan of 3072 SNP in the HLA region. 4981 samples with age of onset and pedigree information. The latest T1DGC meta-analysis of these datasets produced over 40 loci that are putatively associated with T1D [10, 22, 23]. Genes under these loci can be collected and pathway mining can be performed.

Another major source for pathway information is functional genomics data, obtained using the microarray or sequencing technologies. They have demonstrated great potential in providing insights to the complex pathways and interacting networks that drives complex human diseases [94–96]. Network consideration is particularly important in pathway mining of functional genomics data, since disease driving genes are not necessarily those that exhibit the most significant expression changes. For instance, it is known that disease genes are more likely to be regulatory genes [97–99]. Mutations in a regulatory gene can cause significant changes in its downstream genes, leading to little changes in its own expression. A single-gene level statistical evaluation would often miss such regulatory genes.

In T1D, the primary tissues of interest, those at the site of beta cell destruction, the pancreatic islets and the pancreatic lymph nodes, are not readily available. Until recently, most functional genomics studies of human T1D either compared gene expression differences in blood cells between affected and control cohorts [100–102], or the differences in induced expression profiles in a control blood cell sample by sera from affected and control cohorts [103–107]. Due to the nature of the samples being profiled (immune cells, not beta cells), and that most of the studies focused on young-onset cases (where the HLA locus dominate the genetic risk), typically proinflammatory, IL-1 biased signature were observed with many genes involved in innate immunity and adaptive immunity [108].

This is beginning to change. With networks like the Network for Pancreatic Organ Donors with Diabetes (https://www.jdrfnpod.org/) and the Human Islet Research Network (https://hirnetwork.org/), T1D donor pancreas are being systematically collected and distributed for research. New technology advancements including those in the areas of single cell transcriptomics, epigenetics and epigenomics, proteomics are actively being exploited to study pancreatic tissues from T1D donors [109–112]. More data will be available that enable a more comprehensive investigation of how genetic variants may disrupt gene regulatory networks and signaling pathways in the pancreas and other cell types that contribute to the development of T1D. The bottleneck will be the development of more sophisticated statistical and computational methods for dissection of the complex phenotype–genotype relationship.

8.3 Candidate disease genes prioritization and validation for T1D AOO

8.3.1 Compile candidate disease genes of T1D AOO

The extended Copenhagen model described in the previous subsection offered a framework to identify disease pathways that are critical in determining the T1D risk in different age groups. Utilizing model predictions it was possible to collect candidate disease genes through the following:

1. From literature where signaling pathways regulating the Copenhagen traits (figure 8.4) were studied comprehensively. This includes macrophage activation (factor *g, k*) and phagocytosis (*m*) [73, 74], beta cell development and cell function (*e,r N*) [113], and beta cell death regulation during T1D [55].
2. From pathway databases, including KEGG, Biocarta, and GO, where those that are relevant to Copenhagen traits can be identified.
3. Genes in the top pathways from genomic data mining.
4. Genes within the 40 loci in the latest T1DGC meta-analysis [10, 22, 23].
5. Known functional candidate genes. At least 105 candidate genes have been putatively linked with T1D through association studies [18].

The complete list is available at https://github.com/xjw7736/T1D-candidate-genes.

8.3.2 Candidate gene prioritization

With candidate disease genes, ideally one can test them all at once, with currently available technology, in an independent sample. However, there are several major challenges: the sample size needs to be large to adequately address the multi-testing problem for thousands of genes, which is usually not feasible for individual research groups; new computational methods are needed to model the contribution of their interactions to disease, as there is not enough statistical power to enumerate through all possible interactions. To tackle these challenges, a platform for prioritization of candidate genes was developed to select the most critical few to be tested with the limited available resources. Types of information can be utilized generally include: (1) strength in evidence of disease association; (2) interaction (protein, regulatory, etc) with confirmed disease genes; (3) similarity in properties with confirmed disease genes; (4) functional importance (which can be inferred from topological importance) in disease pathways. The first type of information is self-evident; genes with higher statistical significance measures in disease association are more likely to be true disease genes. The rest are explained in more detail below.

Protein–protein interaction (PPI) between two genes is one of the strongest manifestations of a functional relation between them. It is generally accepted that direct PPI with known disease genes is indicative of involvement in the same disease [83, 98, 114, 115]. For instance, this idea was examined in T1D genetics using the 266 known disease genes and the 983 positional (linkage) candidates compiled by T1Dbase (http://t1dbase.org) prior to the deep sequencing era [83]. The positional candidate genes (located in the known linkage regions) were found to be 17.1 fold more likely than random to be first degree PPI neighbors of known T1D genes. For those that were first degree PPI neighbors of known T1D genes, they share ontological similarity with the latter ($p < 1.3e - 10$), were significantly ($p < 1e - 7$) more likely to appear in T1D-related publications (PubMed https://www.ncbi.nlm.nih.gov/pubmed/) even after excluding co-citation with the known T1D genes, and their proteins were more likely to contain T1D-relevant domains ($p < 0.0004$) [82, 83]. Additionally, it was found that their PPI network has distinct topological features from random with significantly higher number of interactions among themselves, even after adjusting for their high degrees ($p < 1e - 5$) [83].

Disease genes share some general common properties, such as more likely to have regulatory roles compared to other genes [97–99], tend to have longer nucleotide sequence length [98], share functional protein sequence domains [82], and share similar functional annotations etc. For example, it was found that the length distribution of known T1D genes is significantly biased toward the long end ($p < 2e - 11$, KS-test) [82].

The last type of useful information mentioned earlier is the functional importance in disease pathways. One important indicator of functional importance is the 'topological position' of a gene in the pathway, or in the PPI or regulatory network of the pathway. It is believed that 'hub' genes in a PPI network with a high degree of connections with others (i.e. of topologically importance) are the most critical ones to function [79]. In [98], topological measures including number of interaction to others genes were compared, and difference between disease genes and non-disease-non-essential genes was found. It has been proposed that disease genes are more likely to be hubs [116], and this idea was explicitly validated in studies of diseases of the brain and central nervous system [117, 118]. In T1D, it was found that interaction hubs in PPI networks are more likely to be disease genes [82, 83]. Many known disease genes of complex disorders are transcription factors (TF). Examples include the important diabetes genes HNF1a, HNF4a, and HNF6, which regulate the transcription for thousands of genes in liver and pancreas [119].

Recently, investigators began to utilize this property to identify candidate disease genes [97–99].

Presently, there are a number of interactome databases for humans and several model systems that provide network information. However, annotations in these databases are typically curated from a large amount of data obtained from different studies of different tissue types under a wide range of different conditions, and only represent expected likely interactions on average. When studying a specific tissue in the context of a certain condition (disease), it was shown that unique interactions or alterations to the norm are often more relevant. For this reason, context specific network inference from gene expression data, especially time course gene expression data, has been an active research field [120]. Commonly used approaches to infer interaction between gene pairs often rely on the correlation in their gene expression variations [121].

Several algorithms have been developed for network inference from time series gene expression data [122, 123], including the dynamic Bayesian network (DBYN) approach [124–126]. In a BYN nodes represent genes and edges represent conditional dependencies between gene pairs. BYN has a few distinct advantages, making it ideal for network modeling of expression data. It employs probabilistic semantics to describe the relationships between nodes, and is capable of handling the stochastic aspect and inherent noise in expression regulation [125], and the missing data and incomplete knowledge issues that plagued many gene expression data [127], most notably in single cell functional genomics data (the 0-inflation problem) [128, 129]. More importantly, its framework allows easy incorporation of other types of data as prior knowledge to improve the performance; some examples of useful other data types include exiting annotations in interactome databases, ChIP-chip/ChIP-seq [119, 130], and eQTL [131, 132] mapping data repositories. This is important given the high-noise nature of gene expression data and the often-small sample size (tens to dozens) compared to the number of genes (tens of thousands). Given enough training data, BYN has demonstrated its tremendous power in the recent deep learning algorithm applications [133].

DBYN is a scale up of BYN to investigate the dynamics of regulation from time series data by encompassing feedback loops [124–126, 134]. Existing knowledge from PubMed co-citation and GO semantic similarities was utilized to construct a prejudiced sampling set which reflect the likelihood of the node pair being linked. To address the high false positive and high computational demand issues commonly faced in network reconstruction, a Markov Chain Monte Carlo (MCMC) approach was adopted, which outputs average candidate models above a threshold rather than attempting to return the best model along with fuzzy theory-based rules in simulation to avoid being trapped at a local minimum. When applied to pancreatic development and growth data, results revealed that including prior knowledge significantly improved the performance, with true positive interactions increased twofold (from 6 to 12) while the false positive rate remained unchanged [135].

Another algorithm catered toward network modeling of time course data, termed the phase locking analysis, was also developed [136]. It is a technique to infer interaction from the coordination in timing of changes of two time series [137, 138], as compared to the coordination of the amplitude of changes evaluated in correlation approaches. The idea is rooted from the field of nonlinear dynamics, where it is known that synchrony through oscillations is a common, and maybe the most efficient way in a complex system to coordinate responses to external signals. Synchronization and frequency modulating is increasingly observed and appreciated in transcription regulation networks, as well as in other biological systems [139]. Some quick examples include NF-κB (nuclear factor K-light-chain-enhancer of activated B cells) [140–144], and the tumor suppressor p53 [140, 142–146]) in mammals and the SOS response to DNA damage in bacteria [147]. Using the Stanford yeast cell cycle data [148, 149], it was found that a significant phase locking exists among cell cycle genes, between transcription factors and their targets, and between gene pairs with prior evidence of physical or genetic interactions. Interaction inference utilizing phase locking is efficient at identifying network modules of focused biological themes. Compared to the commonly used correlation-based approaches, this metric has several advantages. It can automatically address issues of arbitrary time lags (most gene pairs are not exactly in phase, locking can occur at any amount of phase lag) or different dynamic time scales (~10% of the phase-locked gene pairs exhibit higher order than 1:1 locking). Additionally, temporal domain information is often more robust against noise than the amplitude domain information, and can be combined with the latter to further improve the performance of network modeling.

Presently, there is a large amount of data of functional genomic studies of T1D that are available to model T1D specific regulatory networks. A good place to start with is DKnet (https://dknet.org/); it provides information on T1D-relevant data and links to other databases such as GEO (Gene Expression Omnibus http://www.ncbi.nlm.nih.gov/geo/), where one can download the data. Additionally, a number of auxiliary data types will be useful in regulatory network modeling, most notably Encode (Encyclopedia of DNA Elements, https://www.encodeproject.org/) and Genotype-Tissue Expression (GTEx, http://www.gtexportal.org/home/). A major gap in T1D has been the lack of studies that profile the cell populations in human pancreatic islets and pancreatic lymph nodes, the important sites where the cellular interactions leading to disease initiation. The Human Pancreas Analysis Program (HPAP https://hirnetwork.org/consortium/hpap) of the Human Islet Research Network (https://hirnetwork.org/) supported by NIDDK, NIH is poised to fill this gap. A large amount of data is expected to come out of HPAP in the next few years that can be utilized to model candidate disease genes for T1D.

So far in this subsection we discussed types of data that can be exploited in candidate disease gene prioritization. Efficient integration of these diverse data types is also a challenge. Given the noisy nature of high-throughput data and our incomplete understanding of disease biology, in a recent work, a Naïve Bayesian Network (NBN) classifier was developed to integrate attributes of candidate genes and rank them [82]. The following features for each gene were included: similarity of ontological annotations of molecular function and biological process to known disease genes, being in known linkage region, gene sequence length, protein motif according to Interpro, topological positions in protein–protein interaction networks.

The NBN determines a probability score (PS) for all human genes in terms of their likelihood of involvement in disease. Training datasets were obtained from T1Dbase (http://t1dbase.org), which includes a complete list of 266 functional candidate genes and 983 positional candidates. Cross-validation using the receiver operating characteristic (ROC) curve suggests good performance with AUC ~0.82. It was found that genes with higher PS are more likely to be cited in T1D related publications (PubMed https://www.ncbi.nlm.nih.gov/**pubmed**/), with significantly shifted p-values toward the lower end in GWAS data [82].

8.3.3 Genotyping of predicted candidate genes

The ultimate goal of a systems biology study is to translate to, and to accelerate the progress in, solving biomedical challenges. Here, the integrative genomics pipeline described in the previous subsections was tested by genotyping and validating some of the top predicted candidate genes in pathways regulating APC activation and phagocytosis (parameters m and g, figure 8.4). They were selected for possessing the following features: (i) within the shared disease pathways between mathematical modeling and genomic data mining; (ii) being a regulator with more than five target genes in the disease pathway according to the preliminary DBYN simulation; and (iii) being a transcriptional factor. This selection yielded several known T1D genes, including PTPN22 and TNFa, as well as a number of genes whose T1D association has not been reported before, including STAT1, GLP1R, ATF2, and MAPK8, which were typed.

All four genes have extended gene regions with multiple linkage disequilibrium (LD) blocks and tag SNPs. To select markers for genotyping, the gene regions were first flanked on each side by 1K, SNP data from the HapMAP's CEPH population (a Caucasian cohort) were analyzed to define the haplotype blocks using a previously developed tool [150, 151], and tag SNPs within the gene regions with minor allele frequency (MAF) > 10% were retrieved. For each gene, one tag SNP was then selected using two programs: SNPselector [152] and QuickSNP [153], and were listed in table 8.3. Typing of the SNP markers was carried out in two cohorts: (1) a world's-largest collection of both young-(<19 years) and adult-onset (>19 years) T1D cohorts that were obtained from Finland, entitled FINN (table 8.3), totaling 1220 singleton cases, 1549 controls, and 140 multiplex families. The Finnish population is well-isolated with little stratification, ideal for genetic study of human disease [154]. In FINN, 64% singleton cases were adult onset and 36% are young-onset. The onset ages were 28.8 ± 6.6 years and 14.4 ± 4.4 years in each group, while the participants' ages at the time of blood sample collection were 44.8+/−9.3 years and 39.3+/−11.3 years, respectively. (2) Families with T1D that were collected through the Children's Hospital of Wisconsin, entitled WINN (table 8.3). WINN comprised of 212 proband families that are primarily Caucasians, and the cases were mostly young-onset (210 out of 212 with onset age <19 years).

With the genotyping results, one can perform case-control (CC) association analysis, as well as family-based analysis using the FBAT software [155, 156], and Cox regression to examine the AAO effect. The results are summarized in table 8.3.

Table 8.3. T1D candidate gene typing results in Finn and Winn.

Priority score	Gene	Cytoband	SNP marker	SNP	MAF in FINN control	FINN CC *p*-value (1220/1549)[a]	FINN FBAT *p*-value (140)	FINN TDT *p*-value (323)	FINN Cox regression AAO (1220)	WINN CC *p*-value (212)	WINN TDT *p*-value (21)
3	ATF2	2q32	rs212350	C/T	0.05	**0.0105**[b]	*	**0.0520**	*	0.937	0.788
3	STAT1	2q32.2	rs10199181	A/T	0.34	**0.021**	0.76	0.858	n.s.	0.354	**0.0412**
2	GLP1R	6p21	rs2268644	C/G	0.44	0.19	**0.033**	0.174	n.s.	0.984	**0.0401**
2	MAPK8	10q11.23	rs10508903	A/C	0.47	0.27	0.18	0.557	$P \sim 0.005$	0.449	0.517

* typing failed

CC: Case control, n.s.: Not significant.

[a] In parenthesis are sample size.

[b] Only 323 cases typed for this marker.

Overall, three of the four SNPs showed significant association with T1D, in at least one of the cohorts. GLP1R is marginally significant at $p = 0.033$ in FBAT analysis, and MAPK8 showed significant association with the AAO. Specifically, ATF2 shows suggestive p values in both CC and the transmission disequilibrium test (TDT), in FINN, which consists of mostly adult-onset cases. A close examination finds that this resulted from the complete absence of the T/T genotype in the FINN cases with the confidence interval of odds ratio ~(0.02–0.73). In STAT1 though, only a suggestive p value is observed in CC analysis, the odds ratios are significant being (1.21–3.37) with A/T genotype. The fact that they are only associated with T1D in FINN agrees with the model prediction of these genes contributing to adult-onset T1D. It is important to point out that although the p-values are moderate compared to what is typically seen from GWAS, these genes came with prior biological knowledge of their T1D involvement; in addition, there is no need of multi-test adjustment on their p-values since only four markers were typed.

8.4 Summary and discussion

In this section, an example of how systems biology can be incorporated into the genetic study of a complex disease like T1D was described. It employs an integrative genomics approach to identify and narrow down candidate disease genes by integrating mathematical models of disease dynamics, genomic and genetic data mining of disease pathways, and network modeling of genes in the disease pathways in order to prioritize them. This approach relies heavily on mathematical and computational techniques, and represents a highly cost-effective approach that complements the laboratory genetic mapping efforts. Compared to the many reported integrative genomics efforts [96, 157–163], what makes the one presented here unique is the component of mathematical modeling of cellular interactions leading to the initiation of a disease process (perpetuating beta cell destruction in this case, see figure 8.1) [51]. Additionally, the design of this approach is modular, such that each module can be updated, improved, and further developed as our knowledge of the disease improves, technologies advance, and more data become available. For instance, (figure 8.1) the mathematical model of disease initiation only considered events in pancreatic islets, but it can be extended to include a compartment that describes the interactions of immune cell populations in the pancreatic lymph nodes; in step 2 the network modeling of disease pathways can be applied to newer data, as they become available, that profile cell populations in the islet and lymph nodes at single cell level (from the Human Islet Research Network https://hirnetwork.org/, for example); and in step 3, more sophisticated statistical and data science models (such as deep learning algorithms) can be incorporated to evaluate the contribution from the gene interactions. The role of environment, gene-environment interaction, epigenetics, and microbiomes, to name a few, are also important aspects to be included.

While this section concentrated on T1D, the methodology presented is eminently suitable for other complex diseases. There are two take-home messages from this section. First, understanding disease biology is important for dissecting the genetics of complex disease, unavoidable even with the advancement of 'next generation' technologies. Secondly, when the general problem is challenging, focus on more specific ones first, such as what was done here, by focusing on the AAO problem of T1D, rather than attempting to solve the genetics of T1D.

Dissecting genetics of complex traits has been a major challenge [26]. The conventional approach of enumerating through disease association of single individual genetic variants can utilize some systems-level thinking and disease biology modeling [32, 34]. Disease results from failure of certain physiological functions, which are regulated by intricate networks of molecules and signaling pathways. In a system of multiple interacting components, the function depends on the properties of individual components, the structure of their interactions, and the interplay between the two. As stated by Linus Pauling: 'Life is a relationship among molecules and not a property of any molecule'.

A systems approach is not only important to dissecting the genetic architecture of complex traits, but also relevant to the translation of genetic knowledge to medicine. A series of studies have demonstrated the value of genetic knowledge in developing disease risk prediction scores, intervention strategies and therapy for T1D [164, 165].

Moving forward, there are also a number of conceptual and theoretical challenges. The need of a systems approach to study biology is long appreciated, and the basic idea perhaps can be traced all the way back to the Waddington landscape [166]. However, it remains a major challenge how to implement this highly abstract, fundamental idea in concrete examples, develop mechanistic and predictive models, and demonstrate its real-world values. One technical hurdle is, perhaps, the lack of a quantitative metric to characterize the 'landscape' in one or a set of measures, such that its dependence on genetic variants, and its role in shaping biological functions and traits, can be mathematically modeled.

Exercises

1. Derive the steady states of the original Copenhagen model, are there any stable steady states?
2. Derive the nondimensionalized form of the extended Copenhagen model given in equations (8.4), and identify the degrees of freedom of the model.
3. Using equations (8.4) derive the stable state solutions when $A = 0$.
4. Digitize the data in figure 8.3 and redo the fitting. Did you find anything new that is not discussed in this section?
5. Name at least three major barriers to unraveling the genetic architecture of Type 1 diabetes' progression.
6. What role do the following genes play in the human body? (Look up journal articles and read the abstract or conclusion):
   (a) CTLA-4, PTPN22, and IL2RA.
   (b) How could their roles contribute to Type 1 diabetes' progression?
7. Name three major dynamic differences between young and adult onsets of Type 1 diabetes?

8. Take equations (8.3) and (8.4), choose parameters in the range that make beta cell turnover slow compared to the kinetics of other processes, simulate population dynamics for all cell populations in the presence of an initial beta cell insult and beta cell autoantigen. Gradually increase beta cell turnover rates, and examine how the population dynamics vary.
9. Discuss if and why interactions between proteins may be more meaningful than correlations in expression variation between genes.
10. What assumptions and analyses presented in this section can be applied to other complex diseases? Why?

References

[1] Gale E A 2001 The discovery of type 1 diabetes *Diabetes* **50** 217–26

[2] Redondo M J, Yu L, Hawa M, Mackenzie T, Pyke D A, Eisenbarth G S and Leslie R D 2001 Heterogeneity of type i diabetes: analysis of monozygotic twins in Great Britain and the United States *Diabetologia* **44** 354–62

[3] Caillat-Zucman S, Garchon H J, Timsit J, Assan R, Boitard C, Djilali-Saiah I, Bougneres P and Bach J F 1992 Age-dependent HLA genetic heterogeneity of type 1 insulin-dependent diabetes mellitus *J. Clin. Invest.* **90** 2242–50

[4] Eringsmark Regnell S and Lernmark A 2013 The environment and the origins of islet autoimmunity and type 1 diabetes *Diabet. Med.* **30** 155–60

[5] Atkinson M A and Eisenbarth G S 2001 Type 1 diabetes: new perspectives on disease pathogenesis and treatment *Lancet* **358** 221–9

[6] Redondo M J, Jeffrey J, Fain P R, Eisenbarth G S and Orban T 2008 Concordance for islet autoimmunity among monozygotic twins *N. Engl. J. Med.* **359** 2849–50

[7] Steck A K, Barriga K J, Emery L M, Fiallo-Scharer R V, Gottlieb P A and Rewers M J 2005 Secondary attack rate of type 1 diabetes in colorado families *Diabetes Care* **28** 296–300

[8] Hyttinen V, Kaprio J, Kinnunen L, Koskenvuo M and Tuomilehto J 2003 Genetic liability of type 1 diabetes and the onset age among 22,650 young Finnish twin pairs: a nationwide follow-up study *Diabetes* **52** 1052–5

[9] Groop L and Pociot F 2014 Genetics of diabetes–are we missing the genes or the disease *Mol. Cell Endocrinol.* **382** 726–39

[10] Polychronakos C and Li Q 2011 Understanding type 1 diabetes through genetics: advances and prospects *Nat. Rev. Genet.* **12** 781–92

[11] Redondo M J, Steck A K and Pugliese A 2018 Genetics of type 1 diabetes *Pediatr. Diabetes* **19** 346–53

[12] Nyaga D M, Vickers M H, Jefferies C, Perry J K and O'Sullivan J M 2018 The genetic architecture of type 1 diabetes mellitus *Mol. Cell Endocrinol.* **477** 70–80

[13] Robertson C C and Rich S S 2018 Genetics of type 1 diabetes *Curr. Opin. Genet. Dev.* **50** 7–16

[14] Terwilliger J D and Ott J 1994 *Handbook of Human Genetic Linkage* (Baltimore, MD: Johns Hopkins University Press)

[15] Concannon P *et al* 2009 Genome-wide scan for linkage to type 1 diabetes in 2,496 multiplex families from the Type 1 Diabetes Genetics Consortium *Diabetes* **58** 1018–22

[16] Paterson A D and Petronis A 2000 Age and sex based genetic locus heterogeneity in type 1 diabetes *J. Med. Genet.* **37** 186–91

[17] Hessner M J, Wang X and Ghosh S 2003 *Genetics of Type 1 Diabetes* 3rd edn (Philadelphia, PA: Lippincott Williams & Wilkins)
[18] Atkinson M A 2005 Thirty years of investigating the autoimmune basis for type 1 diabetes: Why can't we prevent or reverse this disease? *Diabetes* **54** 1253–63
[19] Gaudieri S, Dawkins R L, Habara K, Kulski J K and Gojobori T 2000 SNP profile within the human major histocompatibility complex reveals an extreme and interrupted level of nucleotide diversity *Genome Res.* **10** 1579–86
[20] Cudworth A G and Woodrow J C 1974 Letter: HL-A antigens and diabetes mellitus *Lancet* **2** 1153
[21] Tabor H K, Risch N J and Myers R M 2002 Candidate-gene approaches for studying complex genetic traits: practical considerations *Nat. Rev. Genet.* **3** 391–7
[22] Barrett J C *et al* 2009 Genome-wide association study and meta-analysis find that over 40 loci affect risk of type 1 diabetes *Nat. Genet.* **41** 703–7
[23] Onengut-Gumuscu S *et al* 2015 Fine mapping of type 1 diabetes susceptibility loci and evidence for colocalization of causal variants with lymphoid gene enhancers *Nat. Genet.* **47** 381–6
[24] Floyel T, Kaur S and Pociot F 2015 Genes affecting beta-cell function in type 1 diabetes *Curr. Diab. Rep.* **15** 97
[25] Pociot F and Lernmark A 2016 Genetic risk factors for type 1 diabetes *Lancet* **387** 2331–9
[26] Rich S S 2016 Diabetes: still a geneticist's nightmare *Nature* **536** 37–8
[27] Clayton D G 2009 Prediction and interaction in complex disease genetics: experience in type 1 diabetes *PLoS Genet.* **5** e1000540
[28] Rao D C 2008 An overview of the genetic dissection of complex traits *Adv. Genet.* **60** 3–34
[29] Borecki I B and Province M A 2008 Linkage and association: basic concepts *Adv. Genet.* **60** 51–74
[30] Altshuler D, Daly M J and Lander E S 2008 Genetic mapping in human disease *Science* **322** 881–8
[31] Ott J, Wang J and Leal S M 2015 Genetic linkage analysis in the age of whole-genome sequencing *Nat. Rev. Genet.* **16** 275–84
[32] Hartwell L H, Hopfield J J, Leibler S and Murray A W 1999 From molecular to modular cell biology *Nature* **402** C47–52
[33] Imagawa A, Hanafusa T, Miyagawa J and Matsuzawa Y 2000 A proposal of three distinct subtypes of type 1 diabetes mellitus based on clinical and pathological evidence *Ann. Med.* **32** 539–43
[34] Wang X, He Z and Ghosh S 2006 Investigation of the age-at-onset heterogeneity in type 1 diabetes through mathematical modeling *Math. Biosci.* **203** 79–99
[35] Paterson A D and Petronis A 2000 Age of diagnosis-based linkage analysis in type 1 diabetes *Eur. J. Hum. Genet.* **8** 145–8
[36] Hao W, Gitelman S, DiMeglio L A, Boulware D and Greenbaum C J 2016 Fall in C-peptide during first 4 years from diagnosis of type 1 diabetes: variable relation to age, HbA_{1c}, and insulin dose *Diabetes Care* **39** 1664–70
[37] Barker A *et al* 2014 Age-dependent decline of beta-cell function in type 1 diabetes after diagnosis: a multi-centre longitudinal study *Diabetes Obes. Metab.* **16** 262–7
[38] Ludvigsson J, Carlsson A, Deli A, Forsander G, Ivarsson S A, Kockum I, Lindblad B, Marcus C, Lernmark A and Samuelsson U 2013 Decline of C-peptide during the first year

after diagnosis of type 1 diabetes in children and adolescents *Diabetes Res. Clin. Pract.* **100** 203–9

[39] DIAMOND Project Group 2006 Incidence and trends of childhood type 1 diabetes worldwide 1990–1999 *Diabet. Med.* **23** 857–66

[40] LaPorte R, Matsushima M and Chang Y 1995 *Prevalence and Incidence of Insulin-Dependent Diabetes* (Bethesda, MD: NIH)

[41] Pipeleers D and Ling Z 1992 Pancreatic beta cells in insulin-dependent diabetes *Diabetes Metab. Rev.* **8** 209–27

[42] Madsbad S, Faber O K, Binder C, McNair P, Christiansen C and Transbol I 1978 Prevalence of residual beta-cell function in insulin-dependent diabetics in relation to age at onset and duration of diabetes *Diabetes* **27** 262–4

[43] Lernmark A *et al* 1995 Heterogeneity of islet pathology in two infants with recent onset diabetes mellitus *Virchows Arch.* **425** 631–40

[44] Kitano H 2002 Systems biology: a brief overview *Science* **295** 1662–4

[45] (TM) OMIM *Online Mendelian Inheretance in Man* www.ncbi.nlm.nih.gov/omim/

[46] Valdes A M, Thomson G, Erlich H A and Noble J A 1999 Association between type 1 diabetes age of onset and hla among sibling pairs *Diabetes* **48** 1658–61

[47] Valdes A M, Erlich H A and Noble J A 2005 Human leukocyte antigen class I B and C loci contribute to type 1 diabetes (T1D) susceptibility and age at T1D onset *Hum. Immunol.* **66** 301–13

[48] Quinn M, Ficociello L H and Rosner B 2003 Change in glycemic control predicts change in weight in adolescent boys with type 1 diabetes *Pediatr. Diabetes* **4** 162–7

[49] Moreland E C, Tovar A, Zuehlke J B, Butler D A, Milaszewski K and Laffel L M 2004 The impact of physiological, therapeutic and psychosocial variables on glycemic control in youth with type 1 diabetes mellitus *J. Pediatr. Endocrinol. Metab.* **17** 1533–44

[50] Orchard T J, Chang Y F, Ferrell R E, Petro N and Ellis D E 2002 Nephropathy in type 1 diabetes: a manifestation of insulin resistance and multiple genetic susceptibilities? further evidence from the Pittsburgh epidemiology of diabetes complication study *Kidney Int.* **62** 963–70

[51] Mathis D, Vence L and Benoist C 2001 beta-cell death during progression to diabetes *Nature* **414** 792–8

[52] Gartner L P and Hiatt J L 1997 *Color Textbook of Histology* (Philadelphia, PA: W.B. Saunders)

[53] De Blasio B F, Bak P, Pociot F, Karlsen A E and Nerup J 1999 Onset of type 1 diabetes: a dynamical instability *Diabetes* **48** 1677–85

[54] Matzinger P 2002 The danger model: a renewed sense of self *Science* **296** 301–5

[55] Eizirik D L and Mandrup-Poulsen T 2001 A choice of death–the signal-transduction of immune-mediated beta-cell apoptosis *Diabetologia* **44** 2115–33

[56] Eizirik D L and Darville M I 2001 beta-cell apoptosis and defense mechanisms: lessons from type 1 diabetes *Diabetes* **50** S64–9

[57] Hellerstrom C and Swenne I 1991 Functional maturation and proliferation of fetal pancreatic beta-cells *Diabetes* **40** 89–93

[58] Fowden A L and Hill D J 2001 Intra-uterine programming of the endocrine pancreas *Br. Med. Bull.* **60** 123–42

[59] Ogilvie R F 1937 A quantitative estimation of the pancreatic islet tissue *Q. J. Med.* **6** 287–300

[60] Holmes M H 1995 *Introduction to Perturbation Methods* (Berlin: Springer)

[61] Bouwens L and Rooman I 2005 Regulation of pancreatic beta-cell mass *Physiol. Rev.* **85** 1255–70

[62] Buschard K, Buch I, Molsted-Pedersen L, Hougaard P and Kuhl C 1987 Increased incidence of true type i diabetes acquired during pregnancy *Br. Med. J. (Clin. Res. Ed.)* **294** 275–9

[63] Laybutt D R, Weir G C, Kaneto H, Lebet J, Palmiter R D, Sharma A and Bonner-Weir S 2002 Overexpression of c-Myc in β-cells of transgenic mice causes proliferation and apoptosis, downregulation of insulin gene expression, and diabetes *Diabetes* **51** 1793–804

[64] Berkovich I and Efrat S 2001 Inducible and reversible beta-cell autoimmunity and hyperplasia in transgenic mice expressing a conditional oncogene *Diabetes* **50** 2260–7

[65] Trudeau J D, Dutz J P, Arany E, Hill D J, Fieldus W E and Finegood D T 2000 Neonatal beta-cell apoptosis: a trigger for autoimmune diabetes? *Diabetes* **49** 1–7

[66] Maderna P and Godson C 2003 Phagocytosis of apoptotic cells and the resolution of inflammation *Biochim. Biophys. Acta* **1639** 141–51

[67] Wu X, Molinaro C, Johnson N and Casiano C A 2001 Secondary necrosis is a source of proteolytically modified forms of specific intracellular autoantigens: implications for systemic autoimmunity *Arthritis Rheum.* **44** 2642–52

[68] Kohler E, Bock U, Knospe S, Michaelis D and Rjasanowski I 1988 Phagocytic activity of blood cells in diabetic risk probands and newly diagnosed type 1 diabetics *Exp. Clin. Endocrinol.* **91** 259–64

[69] O'Brien B A, Fieldus W E, Field C J and Finegood D T 2002 Clearance of apoptotic beta-cells is reduced in neonatal autoimmune diabetes-prone rats *Cell Death Differ.* **9** 457–64

[70] O'Brien B A, Huang Y, Geng X, Dutz J P and Finegood D T 2002 Phagocytosis of apoptotic cells by macrophages from NOD mice is reduced *Diabetes* **51** 2481–8

[71] Maree A F, Komba M, Finegood D T and Edelstein-Keshet L 2008 A quantitative comparison of rates of phagocytosis and digestion of apoptotic cells by macrophages from normal BALB/c and diabetes-prone NOD mice *J. Appl. Physiol.* **104** 157–69

[72] Wang X, He Z and Ghosh S 2004 Mathematical modeling reveals that the recent developmental secular trend may contribute to the epidemiological changes of type 1 diabetes *Proc. of the Int. Conf. on Mathematics and Engineering Techniques in Medicine and Biological Sciences*

[73] Kwiatkowska K and Sobota A 1999 Signaling pathways in phagocytosis *Bioessays* **21** 422–31

[74] Ma J, Chen T, Mandelin J, Ceponis A, Miller N E, Hukkanen M, Ma G F and Konttinen Y T 2003 Regulation of macrophage activation *Cell Mol. Life Sci.* **60** 2334–46

[75] Hosack D A Jr, Dennis G, Sherman B T, Lane H C and Lempicki R A 2003 Identifying biological themes within lists of genes with ease *Genome Biol.* **4** R70

[76] Draghici S, Khatri P, Bhavsar P, Shah A, Krawetz S A and Tainsky M A 2003 Onto-tools, the toolkit of the modern biologist: Onto-express, onto-compare, onto-design and onto-translate *Nucleic Acids Res.* **31** 3775–81

[77] Khatri P, Draghici S, Ostermeier G C and Krawetz S A 2002 Profiling gene expression using onto-express *Genomics* **79** 266–70

[78] Khatri P, Bhavsar P, Bawa G and Draghici S 2004 Onto-tools: an ensemble of web-accessible, ontology-based tools for the functional design and interpretation of high-throughput gene expression experiments *Nucleic Acids Res.* **32** W449–56

[79] Carter S L, Brechbuhler C M, Griffin M and Bond A T 2004 Gene co-expression network topology provides a framework for molecular characterization of cellular state *Bioinformatics* **20** 2242–50
[80] Mitrea C, Taghavi Z, Bokanizad B, Hanoudi S, Tagett R, Donato M, Voichita C and Draghici S 2013 Methods and approaches in the topology-based analysis of biological pathways *Front. Physiol.* **4** 278
[81] Gao S and Wang X 2007 Tappa: topological analysis of pathway phenotype association *Bioinformatics* **23** 3100–2
[82] Gao S, Jia S, Hessner M J and Wang X 2012 Predicting disease related subnetworks for type 1 diabetes using a new network activity score *Omics* **16** 566–78
[83] Gao S and Wang X 2009 Predicting type 1 diabetes candidate genes using human protein-protein interaction networks *J. Comput. Sci. Syst. Biol.* **2** 133–46
[84] Gao S and Wang X 2012 Construction of tissue specific trait-pathway network in a mouse model of obesity-induced diabetes *PLoS One* **7** e44544
[85] Gao S, Wolanyk N, Chen Y, Jia S, Hessner M J and Wang X 2017 Investigation of coordination and order in transcription regulation of innate and adaptive immunity genes in type 1 diabetes *BMC Med. Genomics* **10** 7
[86] Davies J L *et al* 1994 A genome-wide search for human type 1 diabetes susceptibility genes *Nature* **371** 130–6
[87] Hashimoto L *et al* 1994 Genetic mapping of a susceptibility locus for insulin-dependent diabetes mellitus on chromosome 11q *Nature* **371** 161–4
[88] Concannon P *et al* 1998 A second-generation screen of the human genome for susceptibility to insulin-dependent diabetes mellitus *Nat. Genet.* **19** 292–6
[89] Mein C A *et al* 1998 A search for type 1 diabetes susceptibility genes in families from the United Kingdom *Nat. Genet.* **19** 297–300
[90] Cox N J, Wapelhorst B, Morrison V A, Johnson L, Pinchuk L, Spielman R S, Todd J A and Concannon P 2001 Seven regions of the genome show evidence of linkage to type 1 diabetes in a consensus analysis of 767 multiplex families *Am. J. Hum. Genet.* **69** 820–30
[91] Concannon P, Erlich H A, Julier C, Morahan G, Nerup J, Pociot F, Todd J A and Rich S S 2005 Type 1 diabetes: evidence for susceptibility loci from four genome-wide scans in 1435 multiplex families *Diabetes* **54** 2995–3001
[92] Lander E and Kruglyak L 1995 Genetic dissection of complex traits: guidelines for interpreting and reporting linkage results *Nat. Genet.* **11** 241–7
[93] Risch N and Merikangas K 1996 The future of genetic studies of complex human diseases *Science* **273** 1516–7
[94] Hughes T R *et al* 2000 Functional discovery via a compendium of expression profiles *Cell* **102** 109–26
[95] Karp C L *et al* 2000 Identification of complement factor 5 as a susceptibility locus for experimental allergic asthma *Nat. Immunol.* **1** 221–6
[96] Schadt E E *et al* 2005 An integrative genomics approach to infer causal associations between gene expression and disease *Nat. Genet.* **37** 710–7
[97] Franke L, Bakel H, Fokkens L, de Jong E D, Egmont-Petersen M and Wijmenga C 2006 Reconstruction of a functional human gene network, with an application for prioritizing positional candidate genes *Am. J. Hum. Genet.* **78** 1011–25
[98] Xu J and Li Y 2006 Discovering disease-genes by topological features in human protein-protein interaction network *Bioinformatics* **22** 2800–5

[99] Oti M, Snel B, Huynen M A and Brunner H G 2006 Predicting disease genes using protein–protein interactions *J. Med. Genet.* **43** 691–8

[100] Kaizer E C, Glaser C L, Chaussabel D, Banchereau J, Pascual V and White P C 2007 Gene expression in peripheral blood mononuclear cells from children with diabetes *J. Clin. Endocrinol. Metab.* **92** 3705–11

[101] Stechova K *et al* 2012 Healthy first-degree relatives of patients with type 1 diabetes exhibit significant differences in basal gene expression pattern of immunocompetent cells compared to controls: expression pattern as predeterminant of autoimmune diabetes *Scand. J. Immunol.* **75** 210–9

[102] Reynier F *et al* 2010 Specific gene expression signature associated with development of autoimmune type-I diabetes using whole-blood microarray analysis *Genes Immun.* **11** 269–78

[103] Wang X, Jia S, Geoffrey R, Alemzadeh R, Ghosh S and Hessner M J 2008 Identification of a molecular signature in human type 1 diabetes mellitus using serum and functional genomics *J. Immunol.* **180** 1929–37

[104] Kaldunski M, Jia S, Geoffrey R, Basken J, Prosser S, Kansra S, Mordes J P, Lernmark A, Wang X and Hessner M J 2010 Identification of a serum-induced transcriptional signature associated with type 1 diabetes in the biobreeding rat *Diabetes* **59** 2375–85

[105] Jia S, Kaldunski M, Jailwala P, Geoffrey R, Kramer J, Wang X and Hessner M J 2011 Use of transcriptional signatures induced in lymphoid and myeloid cell lines as an inflammatory biomarker in type 1 diabetes *Physiol. Genomics* **43** 697–709

[106] Levy H *et al* 2012 Transcriptional signatures as a disease-specific and predictive inflammatory biomarker for type 1 diabetes *Genes Immun.* **13** 593–604

[107] Chen Y G, Mordes J P, Blankenhorn E P, Kashmiri H, Kaldunski M L, Jia S, Geoffrey R, Wang X and Hessner M J 2013 Temporal induction of immunoregulatory processes coincides with age-dependent resistance to viral-induced type 1 diabetes *Genes Immun.* **14** 387–400

[108] Cabrera S M, Chen Y G, Hagopian W A and Hessner M J 2016 Blood-based signatures in type 1 diabetes *Diabetologia* **59** 414–25

[109] Carrano A C, Mulas F, Zeng C and Sander M 2017 Interrogating islets in health and disease with single-cell technologies *Mol. Metab.* **6** 991–1001

[110] Aylward A, Chiou J, Okino M L, Kadakia N and Gaulton K J 2018 Shared genetic risk contributes to type 1 and type 2 diabetes etiology *Hum. Mol. Genet.* ddy314

[111] Wang Y J *et al* 2019 Multiplexed *in situ* imaging mass cytometry analysis of the human endocrine pancreas and immune system in type 1 diabetes *Cell Metab.* **29** 769–83

[112] Rosen E D, Kaestner K H, Natarajan R, Patti M E, Sallari R, Sander M and Susztak K 2018 Epigenetics and epigenomics: Implications for diabetes and obesity *Diabetes* **67** 1923–31

[113] Servitja J M and Ferrer J 2004 Transcriptional networks controlling pancreatic development and beta cell function *Diabetologia* **47** 597–613

[114] George R A, Liu J Y, Feng L L, Bryson-Richardson R J, Fatkin D and Wouters M A 2006 Analysis of protein sequence and interaction data for candidate disease gene prediction *Nucleic Acids Res.* **34** e130

[115] Kann M G 2007 Protein interactions and disease: computational approaches to uncover the etiology of diseases *Brief Bioinform.* **8** 333–46

[116] Batada N N, Hurst L D and Tyers M 2006 Evolutionary and physiological importance of hub proteins *PLoS Comput. Biol.* **2** e88

[117] Samson K 2006 Gene networks unique to human brains shed light on evolutionary divergence from chimpanzees *Neurol. Today* **6** 6–7

[118] Oldham M C, Horvath S and Geschwind D H 2006 Conservation and evolution of gene coexpression networks in human and chimpanzee brains *Proc. Natl Acad. Sci. USA* **103** 17973–8

[119] Odom D T *et al* 2004 Control of pancreas and liver gene expression by HNF transcription factors *Science* **303** 1378–81

[120] Przytycka T M, Singh M and Slonim D K 2010 Toward the dynamic interactome: it's about time *Brief Bioinform.* **11** 15–29

[121] Langfelder P and Horvath S 2008 WGCNA: an R package for weighted correlation network analysis *BMC Bioinform.* **9** 559

[122] Sima C, Hua J P and Jung S W 2009 Inference of gene regulatory networks using time-series data: a survey *Curr. Genomics* **10** 416–29

[123] Bianco-Martinez E, Rubido N, Antonopoulos C G and Baptista M S 2016 Successful network inference from time-series data using mutual information rate *Chaos* **26** 043102

[124] Zou M and Conzen S D 2005 A new dynamic Bayesian network (DBN) approach for identifying gene regulatory networks from time course microarray data *Bioinformatics* **21** 71–9

[125] Friedman N, Linial M, Nachman I and Pe'er D 2000 Using bayesian networks to analyze expression data *J. Comput. Biol.* **7** 601–20

[126] Ong I M, Glasner J D and Page D 2002 Modelling regulatory pathways in E. Coli from time series expression profiles *Bioinformatics* **18** S241–8

[127] Troyanskaya O, Cantor M, Sherlock G, Brown P, Hastie T, Tibshirani R, Botstein D and Altman R B 2001 Missing value estimation methods for DNA microarrays *Bioinformatics* **17** 520–5

[128] Pierson E and Yau C 2015 Zifa: Dimensionality reduction for zero-inflated single-cell gene expression analysis *Genome Biol.* **16** 241

[129] Kharchenko P V, Silberstein L and Scadden D T 2014 Bayesian approach to single-cell differential expression analysis *Nat. Methods* **11** 740–2

[130] Buck M J and Lieb J D 2004 Chip-chip: considerations for the design, analysis, and application of genome-wide chromatin immunoprecipitation experiments *Genomics* **83** 349–60

[131] Morley M, Molony C M, Weber T M, Devlin J L, Ewens K G, Spielman R S and Cheung V G 2004 Genetic analysis of genome-wide variation in human gene expression *Nature* **430** 743–7

[132] Rockman M V and Kruglyak L 2006 Genetics of global gene expression *Nat. Rev. Genet.* **7** 862–72

[133] Silver D *et al* 2016 Mastering the game of go with deep neural networks and tree search *Nature* **529** 484–9

[134] Pe'er D 2005 Bayesian network analysis of signaling networks: a primer *Sci. STKE* **2005** pl4

[135] Gao S and Wang X 2011 Quantitative utilization of prior biological knowledge in the bayesian network modeling of gene expression data *BMC Bioinform.* **12** 359

[136] Gao S, Hartman J, Carter J L, Hessner M J and Wang X 2010 Global analysis of phase locking in gene expression during cell cycle: the potential in network modeling *BMC Syst. Biol.* **4** 167

[137] Schafer C, Rosenblum M G, Abel H H and Kurths J 1999 Synchronization in the human cardiorespiratory system *Phys. Rev. E Stat. Phys. Plasmas Fluids Relat. Interdiscip. Topics* **60** 857–70

[138] Rosenblum M, Pikovsky A, Kurths J, Schafer C and Tass P A 2001 *Phase Syncrhonization: From Theory to Data Analysis, Neuro-informatics and Neural Modeling* vol 4 (Amsterdam: Elsevier) pp 279–321

[139] Haus E 2007 Chronobiology in the endocrine system *Adv. Drug Deliv. Rev.* **59** 985–1014

[140] Ashall L *et al* 2009 Pulsatile stimulation determines timing and specificity of NF-κB-dependent transcription *Science* **324** 242–6

[141] Covert M W, Leung T H, Gaston J E and Baltimore D 2005 Achieving stability of lipopolysaccharide-induced NF-κB activation *Science* **309** 1854–7

[142] Friedrichsen S, Harper C V, Semprini S, Wilding M, Adamson A D, Spiller D G, Nelson G, Mullins J J, White M R and Davis J R 2006 Tumor necrosis factor-alpha activates the human prolactin gene promoter via nuclear factor-κB signaling *Endocrinology* **147** 773–81

[143] Nelson D E *et al* 2004 Oscillations in NF-κB signaling control the dynamics of gene expression *Science* **306** 704–8

[144] Tay S, Hughey J J, Lee T K, Lipniacki T, Quake S R and Covert M W 2010 Single-cell NF-κB dynamics reveal digital activation and analogue information processing *Nature* **466** 267–71

[145] Geva-Zatorsky N *et al* 2006 Oscillations and variability in the p53 system *Mol. Syst. Biol.* **2** 2006.0033

[146] Loewer A, Batchelor E, Gaglia G and Lahav G 2010 Basal dynamics of p53 reveal transcriptionally attenuated pulses in cycling cells *Cell* **142** 89–100

[147] Friedman N, Vardi S, Ronen M, Alon U and Stavans J 2005 Precise temporal modulation in the response of the sos dna repair network in individual bacteria *PLoS Biol.* **3** e238

[148] Spellman P T, Sherlock G, Zhang M Q, Iyer V R, Anders K, Eisen M B, Brown P O, Botstein D and Futcher B 1998 Comprehensive identification of cell cycle-regulated genes of the yeast Saccharomyces cerevisiae by microarray hybridization *Mol. Biol. Cell* **9** 3273–97

[149] Cho R J *et al* 1998 A genome-wide transcriptional analysis of the mitotic cell cycle *Mol. Cell* **2** 65–73

[150] Olivier M, Wang X, Cole R, Gau B, Kim J, Rubin E M and Pennacchio L A 2004 Haplotype analysis of the apolipoprotein gene cluster on human chromosome 11 *Genomics* **83** 912–23

[151] Smith E M, Wang X, Littrell J, Eckert J, Cole R, Kissebah A H and Olivier M 2006 Comparison of linkage disequilibrium patterns between the HapMap CEPH samples and a family-based cohort of northern European descent *Genomics* **88** 407–14

[152] Xu H, Gregory S G, Hauser E R, Stenger J E, Pericak-Vance M A, Vance J M, Zuchner S and Hauser M A 2005 SNPselector: a web tool for selecting SNPs for genetic association studies *Bioinformatics* **21** 4181–6

[153] Grover D, Woodfield A S, Verma R, Zandi P P, Levinson D F and Potash J B 2007 Quicksnp: an automated web server for selection of tagsnps *Nucl. Acids Res.* **35** W115–20

[154] Lander E S and Schork N J 1994 Genetic dissection of complex traits *Science* **265** 2037–48

[155] Horvath S, Wei E, Xu X, Palmer L J and Baur M 2001 Family-based association test method: age of onset traits and covariates *Genet. Epidemiol.* **21** S403–8

[156] Horvath S, Xu X and Laird N M 2001 The family based association test method: strategies for studying general genotype–phenotype associations *Eur. J. Hum. Genet.* **9** 301–6

[157] Dahia P L *et al* 2005 Novel pheochromocytoma susceptibility loci identified by integrative genomics *Cancer Res.* **65** 9651–8

[158] Wu Q *et al* 2005 Integrative genomics revealed *RAI3* is a cell growth-promoting gene and a novel p53 transcriptional target *J. Biol. Chem.* **280** 12935–43

[159] Glazier A M, Nadeau J H and Aitman T J 2002 Finding genes that underlie complex traits *Science* **298** 2345–9

[160] Calvo S, Jain M, Xie X, Sheth S A, Chang B, Goldberger O A, Spinazzola A, Zeviani M, Carr S A and Mootha V K 2006 Systematic identification of human mitochondrial disease genes through integrative genomics *Nat. Genet.* **38** 576–82

[161] Antinozzi P A, Garcia-Diaz A, Hu C and Rothman J E 2006 Functional mapping of disease susceptibility loci using cell biology *Proc. Natl Acad. Sci. USA* **103** 3698–703

[162] Giallourakis C, Henson C, Reich M, Xie X and Mootha V K 2005 Disease gene discovery through integrative genomics *Annu. Rev. Genomics Hum. Genet.* **6** 381–406

[163] Tiffin N *et al* 2006 Computational disease gene identification: a concert of methods prioritizes type 2 diabetes and obesity candidate genes *Nucl. Acids Res.* **34** 3067–81

[164] Groop L, Storm P and Rosengren A 2014 Can genetics improve precision of therapy in diabetes? *Trends Endocrinol. Metab.* **25** 440–3

[165] Cerolsaletti K, Hao W and Greenbaum C J 2019 Genetics coming of age in type 1 diabetes *Diabetes Care* **42** 189–91

[166] Waddington C H 1957 *The Strategy of the Genes; A Discussion of Some Aspects of Theoretical Biology* (New York: Macmillan)

IOP Publishing

Diabetes Systems Biology
Quantitative methods for understanding beta-cell dynamics and function
Anmar Khadra

Chapter 9

Immune-cell dynamics in type 1 diabetes

Hassan Jamaleddine and Anmar Khadra

9.1 Introduction

Type 1 diabetes (T1D) is an organ-specific autoimmune disease that results from the destruction of pancreatic beta cells by cytotoxic T lymphocytes (CTLs) [1–3]. This includes both $CD4^+$ and $CD8^+$ T cells [4–7] that are activated by cross priming antigen presenting cells (APCs), such as mature dendritic cells [5, 8] and macrophages [9]. The destruction of beta cells eventually leads to the abolition of insulin secretion crucial for regulating glucose homeostasis. The activation of T cells and their recognition of beta cells are dictated by the binding of T-cell receptors (TCRs) with surface molecules, called peptide-major histocompatibility complexes (pMHCs) class I and II, on APCs and beta cells [7, 10]. Dead beta cells are phagocytosed, at the start and during the autoimmune response, by antigen presenting cells (APCs) such as dendritic cells and macrophages [11, 12], leading to T cells recognizing a growing list of autoantigens [13] during disease progression. Earlier reports have shown that genetic susceptibility to T1D is linked to the human leukocyte antigen (HLA) region [14].

Animal models, such as non-obese diabetic (NOD) mice that spontaneously develop a form of T1D closely resembling human T1D, are typically used to study this disease. Experimental results reveal that the autoimmune nature of T1D is a manifestation of multiple types of immune cells being involved directly and/or indirectly in its onset and progression. The dynamics of these cells appear to be altered in affected individuals when compared to healthy ones. For example, it has been suggested that the dynamic properties of macrophages, a class of immune cells that play a role in antigen presentation and the activation of CTLs, differ between animal models that are prone to T1D and those that are disease-resistant [15, 16], whereas effector T cells appear to play a significant role in beta-cell destruction and in the induction of insulin-dependence [7, 17]. Moreover, it has been shown that mature B cells release a series of beta-cell-specific autoantibodies in high-risk subjects (HRS) several years prior to the onset of diabetes-related symptoms [18, 19].

doi:10.1088/978-0-7503-3739-7ch9

In this chapter, we will present a series of mathematical models that focus on a number of physiological aspects of T1D and present various methods, adopted from the field of nonlinear dynamics and bifurcation theory, to analyze the models. These methods include steady-state analysis, stability theory and numerical techniques. They will illustrate how such models can be used to unravel five important characteristics of the disease; namely, (i) immune-dominance of beta-cell specific autoantigens, (ii) abnormalities in macrophage dynamics, (iii) autoreactive T-cell population growth, (iv) remission-relapse in T1D, and (v) autoantibody release by B cells.

9.2 Protein processing in T1D

During beta-cell growth and metabolism, proteins are synthesized for incorporation into cellular domains, but they are also degraded, processed into peptides by proteasomes, and presented on the beta-cell surface as pMHC complexes. The number of pMHC complexes displayed per protein molecule degraded (typically in the range of 10^{-2}–10^{-3}) is called pMHC-processing efficiency [20]. Our goal here is to investigate how aspects of protein allocation and efficiency affect pathogenicity of beta-cell specific autoantigens.

As stated earlier, $CD4^+$ and $CD8^+$ T-cell activation and recruitment as pathogenic autoreactive CTLs during T1D development require that (i) self-pMHC complexes be presented to T cells by cross-priming APCs, such as dendritic cells (DCs), capable of shuttling beta-cell specific autoantigens from the pancreas to the pancreatic lymph nodes, and (ii) the same pMHC complexes be displayed on the beta-cell surface. The former elicits T-cell activation and differentiation into beta-cell specific effector pathogenic CTLs, while the latter marks beta cells for cognate recognition and destruction. Evidence suggests that the disease is initiated by insulin-reactive $CD4^+$ T cells [21], and that $CD8^+$ T cells contribute to T1D by killing beta cells [7].

In NOD mice, a fraction of diabetogenic $CD8^+$ T cells target a peptide from islet-specific glucose-6-phosphatase catalytic subunit-related protein (IGRP) [22]. IGRP-reactive T cells are unusually frequent in circulation [23] and are not subject to thymic negative selection because it is only expressed in beta cells (unlike other diabetogenic proteins that are expressed in the thymus as tissue restricted antigens [24]). Interestingly, IGRP autoreactivity accounts for ~40% of the intra-islet $CD8^+$ T-cell pool, indicating that IGRP is a 'dominant' autoantigen in murine T1D. That brings up an important question of why certain beta-cell proteins, such as IGRP in NOD mice, take on a more dominant autoantigenic role when compared to other diabetogenic proteins that are also exclusively expressed in beta cells. It is also unclear whether autoantigenicity (the ability of a given protein to elicit T cell responses) correlates with pathogenicity (the ability of these T cell responses to trigger target-cell death).

We address these important questions by developing and analyzing a mathematical model to explore protein dominance in the context of T1D and test the hypothesis that, unlike autoantigenicity, pathogenicity is inhibited at high levels of protein stability.

9.2.1 The model

For a given diabetogenic protein (such as IGRP), the model assumes that protein stability (see figure 9.1) is determined by an allocation parameter representing the fraction f of this protein directed to a pool (denoted by R) that is rapidly degraded by proteasomes versus a fraction $(1 - f)$ that remains resistant to degradation and resided in a functionally stable pool (labeled Q). An important aspect of both pools is their relative efficiency $\eta = \eta_q/\eta_r$ in processing beta-cell specific proteins into pMHC. Here, η_q (η_r) is the efficiency of processing proteins from the stable (unstable) pool into pMHC, which need not be identical for the two pools, even if the protein is itself the same (due to cellular localization that could influence efficiency). In fact, we expect η_q to be less than η_r, and so $\eta < 1$.

Based on recent experimental evidence, we can assume that (i) the unstable, rapidly degraded protein pool R contributes predominantly to pathogenicity through the pool of pMHC complexes displayed on the surface of beta cells, whereas (ii) the stable protein pool Q contributes mostly to autoantigenicity through cross-presentation of pMHC on APC. It is important to note that both the stable (Q) and unstable (R) protein pools will contribute to pMHC presentation on beta cells but generally at different rates and efficiencies. Moreover, several experimental observations suggest that DCs (one type of APCs) preferentially cross-present pMHC complexes derived from intact cellular proteins that they take up [25–27]. Therefore, the stable protein pool Q is more likely to survive beta-cell death and thus

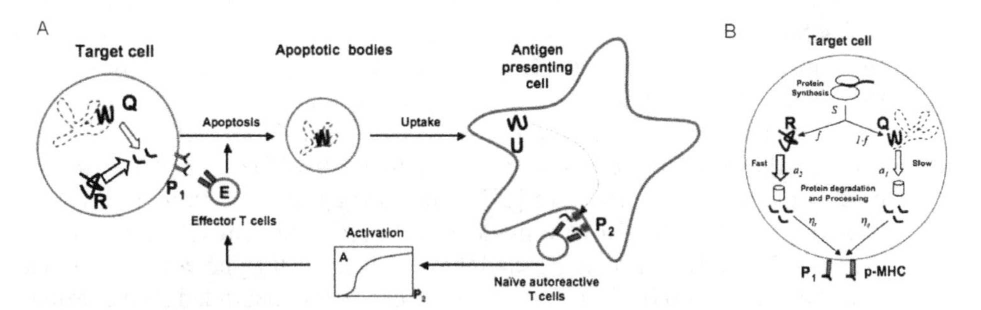

Figure 9.1. A scheme displaying the key pathways involved in antigen processing in T1D. (A) Beta-cell specific proteins are allocated to two pools for processing, a stable pool Q and an unstable pool R, each contributing differently to the production of pMHC (P_1) on the surface of beta cells. Proteins in pool Q survive apoptosis intact, get taken up by APCs upon T-cell induced beta-cell destruction, and finally get cross-presented as pMHC (P_2) that can trigger the activation of naive autoreactive T cells into effector T cells. Effector T cells then recognize and kill beta cells displaying similar pMHC, generating more apoptotic bodies and amplifying the cycle of autoimmune response. (B) Inside beta cells, proteins are synthesized at a rate S, a fraction of which (f) is unstable and allocated to a rapidly degraded pool (R), while the rest is funneled to a stable pool (Q). The rates of degradation $a_1 \ll a_2$ and the efficiencies $\eta_r > \eta_q$ differ, but both pools contribute to pMHC (P_1) on beta-cell surface. Reprinted from [65] by permission of The Japanese Society for Immunology.

access the cross-presentation pathway in APCs. This produces a positive feedback loop that drives the system autocatalytically (as shown in figure 9.1).

Based on the above discussion and the mechanism presented in figure 9.1, we can define the following time-dependent variables:

- $Q(t)$ = Level of a given beta-cell specific protein in the stable pool in the cell (e.g. number of copies of IGRP per beta cell);
- $R(t)$ = Level of a given beta-cell specific protein in the unstable pool targeted for rapid processing per cell (e.g. ubiquitinated and/or defective/misfolded IGRP);
- $E(t)$ = The population size of $CD8^+$ autoreactive T lymphocytes per target cell (e.g. IGRP-reactive effector T cells);
- $P_1(t)$ = The expression level of pMHC per beta cell displayed on the surface of the target cell (e.g. the average expression level of IGRP-MHC on the surface of a beta cell);
- $P_2(t)$ = The expression level of pMHC per APC displayed on the surface of an APC (e.g. the average expression level of IGRP-MHC on the surface of a DC).

We can now make the following assumptions about the model as suggested by figure 9.1.

1. All proteins and complexes have some mean rates of turnover. The parameters μ_i will be used to denote the turnover rate of pMHC corresponding to peptide i (the mean lifetime of pMHC is then $1/\mu_i$). The parameters a_1, a_2 represent the degradation rates of a protein in pools Q, R, respectively.
2. Protein is synthesized in the cell at a constant rate S. A fraction f is unstable and shunted to the pool R that will undergo rapid degradation. After processing, some of this becomes pMHC on the beta-cell surface (P_1). A fraction $1 - f$ of synthesized protein, on the other hand, is channeled to the stable pool Q and protected from rapid decay. The parameter f represents the balance between allocation to the two pathways.
3. Some fraction of degraded protein, of whatever kind, will be funneled through the protein processing pathway to form pMHC displayed on the beta-cell surface. According to [20], it takes roughly 2000 rapidly degraded protein molecules to form a single pMHC complex (in [20] the fraction $\eta_r \approx 1/2000$ pMHC/protein is called the 'efficiency'). It is, however, suggested that the efficiency of pMHC production from slowly degraded stable protein molecules (denoted by η_q) is much lower [28]. Thus, as stated before, we expect the relative efficiency $\eta = \eta_q/\eta_r$ to be less than 1.
4. Only the normal protein is ever found in a stable form in apoptotic bodies created when cells die, and only that stable form survives uptake by APCs (such as DCs) and presentation on APC surface as pMHC (at a level P_2), with an average efficiency of $\eta_d \approx 1/1406$ pMHC/protein [20].
5. Autoreactive $CD8^+$ T cells are activated to proliferate at a rate that depends on the expression level of pMHC (P_2) displayed on APCs. We will explore

several forms of said rate in this section. In the simplest scenario, we consider the rate of T-cell activation and proliferation to be simply proportional to P_2. In other scenarios, we instead include nonlinear dependencies.

6. We assume that the number of APCs (particularly DCs), A, the number of beta cells, β, and the number of circulating naive $CD8^+$ T cells are constant over the entire time span of interest.
7. Effector T cells kill beta cells at a rate that depends on their encounter of pMHC on the target-cell surface (i.e. the expression level of pMHC on beta cells, P_1), and have a natural turnover rate of γ.

Based on the above assumptions, we are now ready to describe the model equations as suggested by figure 9.2 (note that the equation for P_2 will be derived in detail in the following section).

$$\frac{dQ}{dt} = S(1 - f) - a_1 Q, \tag{9.1a}$$

$$\frac{dR}{dt} = Sf - a_2 R, \tag{9.1b}$$

$$\frac{dP_1}{dt} = a_3 R + a_4 Q - \mu_1 P_1, \tag{9.1c}$$

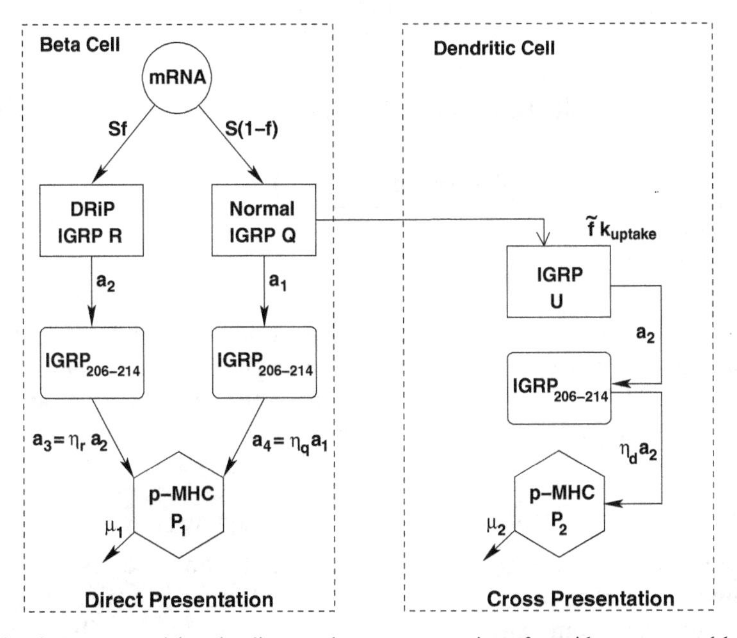

Figure 9.2. A scheme summarizing the direct and cross presentation of peptides on targeted beta cells and APCs. The rates involved in each pMHC class-I pathway are shown. The scheme illustrates in detail the model described by equations (9.1) and the mini-model described by equations (9.6).

$$\frac{dP_2}{dt} = a_5QE\,\mathcal{K}(P_1) - \mu_2 P_2, \tag{9.1d}$$

$$\frac{dE}{dt} = A(P_2) - \gamma E. \tag{9.1e}$$

There are two functions that remain to be determined in this model; namely, $\mathcal{K}(P_1)$ and $A(P_2)$. For the former, we assume that the rate of beta-cell killing is proportional to their surface display of autoantigen pMHC, P_1, i.e. we let

$$\mathcal{K}(P_1) = \kappa P_1, \tag{9.2}$$

where κ has units of per effector cell per pMHC complex displayed per unit time. As for the pMHC-(P_2)-dependent T-cell activation term, $A(P_2)$, there are three possible formalisms (labeled cases I–III) that can be considered.

Case I: Activation of naive T cells is simply proportional to the expression level of pMHC on APCs, i.e.

$$A(P_2) = \alpha\left(\frac{P_2}{k_p}\right). \tag{9.3}$$

Case II: There is a saturation in the pMHC-dependent activation rate, so that

$$A(P_2) = \alpha\left(\frac{P_2}{k_p + P_2}\right). \tag{9.4}$$

Case III: The peptide-dependent activation rate is a Hill-function, given by

$$A(P_2) = \alpha\left(\frac{P_2^n}{k_p^n + P_2^n}\right), \tag{9.5}$$

with $n > 1$.

In each case, k_p is a typical peptide level governing the response, and α is the number of T cells activated per unit time. In cases II–III, α is the maximal rate and k_p is the level of pMHC display on APCs at which activation is at 50% of its maximal level. Note that $A(0) = 0$ for all cases.

9.2.2 Derivation of the equation for P_2

The equation for P_2 as described in equation (9.1d) can be derived from a more detailed model that considers the killing of beta cells and their uptake by APCs, as shown in figure 9.3 (along with figure 9.2). Here we briefly describe such a model, which helps identify the assumptions made and the various parameters appearing in original simplified model.

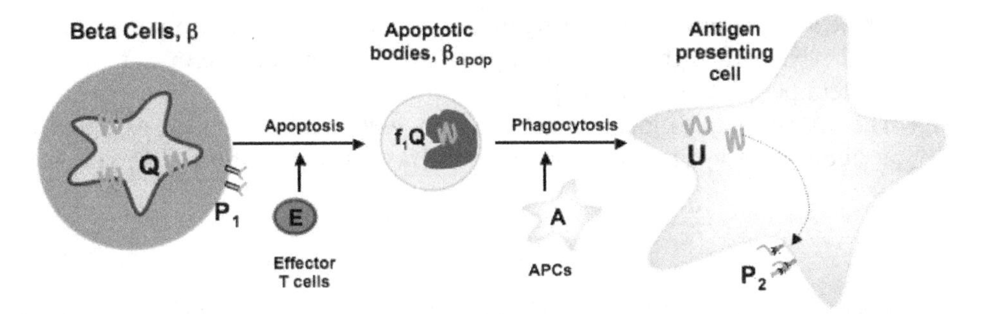

Figure 9.3. A schematic diagram showing the mechanism used to derive an equation for P_2, the pMHC expression level on the surface of an APC. Q is the number of normal protein molecules in a beta cell. Of these, a fraction f_1 survives in apoptotic bodies. When these are taken up by APCs, they form U protein molecules per APC, and are then processed to form pMHC on the APC surface, at level P_2.

As suggested earlier, we will assume that the number of beta cells, β, and the number of APCs, A, are constants. We now consider how the level of stable protein, Q, inside a targeted beta cell affects the various steps leading to pMHC presentation at the APC surface. We do so by assembling a three dimensional model that describes beta-cell death, uptake of protein by APCs, and cross-presentation of pMHC. This mini-model incorporates the following assumptions.

- Target cells being killed produce apoptotic bodies (β_{apop}). This happens at a rate that depends on the number of effector cells, E, and the pMHC displayed on beta cells, P_1.

$$\frac{d\beta_{\text{apop}}}{dt} = \underbrace{(\kappa P_1 \boldsymbol{E})}_{\text{[killing rate]}}\ \beta - \underbrace{k_{\text{uptake}} A}_{\text{[uptake by DCs]}} \cdot \beta_{\text{apop}},$$

  where κ is the rate of killing of beta cells per effector cell per pMHC display per minute, and k_{uptake} is the rate of uptake of apoptotic bodies per DC per minute (see also figure 9.2). Here the terms with underbraces have units of [per minute].
- Apoptotic bodies are assumed to contain some level of the original stable form of the protein of interest (e.g. IGRP). Generally, each apoptotic body may have only some fraction $f_1 < 1$ (dimensionless) of the amount of normal protein in a beta cell, so $f_1 Q$ is the amount of such protein assumed to be carried over by each of those apoptotic bodies.
- By taking up apoptotic material, a given APC will obtain some copies of the protein molecules of interest. This number of protein molecules per APC, U, carries similar units as do Q and R. Then the equation that governs the dynamics of U is given by

$$\frac{dU}{dt} = \underbrace{k_{\text{uptake}}}_{\text{[uptake rate per DC]}} \cdot \underbrace{\beta_{\text{apop}} f_1 Q}_{\text{[amount of protein]}} - a_2 U,$$

where a_2 is the rate of processing of the protein pool U into peptides inside an APC[1]. Note that the units are appropriate, as the first term has units of [protein per APC per minute].

- The pMHC expression level, P_2, on an APC surface is then assumed to accumulate at a rate proportional to the rate of degradation of U, i.e.

$$\frac{dP_2}{dt} = \underbrace{\eta_d}_{\text{[fraction into pMHC]}} \cdot a_2 U - \mu_2 P_2,$$

where η_d is the efficiency of pMHC production in an APC, i.e. the number of pMHC complexes made per protein molecule degraded inside an APC. Note that the constants a_2, μ_2 have units of [per minute].

Based in the discussion above, the detailed model for pMHC processing in an APC thus consists of the following set of equations:

$$\frac{d\beta_{\text{apop}}}{dt} = \kappa P_1 E\beta - k_{\text{uptake}} A\beta_{\text{apop}}, \tag{9.6a}$$

$$\frac{dU}{dt} = k_{\text{uptake}}\beta_{\text{apop}} f_1 Q - a_2 U, \tag{9.6b}$$

$$\frac{dP_2}{dt} = \eta_d a_2 U - \mu_2 P_2. \tag{9.6c}$$

To arrive at equation (9.1d), we apply quasi-steady state (QSS) approximation on β_{apop} and U. This is done by setting $dB_{\text{apop}}/dt = dU/dt = 0$ in equations (9.6a) and (9.6b), to obtain

$$k_{\text{uptake}}\beta_{\text{apop}} \approx (\kappa P_1 E)\frac{\beta}{A},$$

and

$$a_2 U \approx f_1 Q(k_{\text{uptake}}\beta_{\text{apop}}) \approx f_1 Q(\kappa P_1 E)\frac{\beta}{A}.$$

Substituting this quantity into equation (9.6c), we get

$$\frac{dP_2}{dt} = \eta_d\left[f_1 Q(\kappa P_1 E)\frac{\beta}{A}\right] - \mu_2 P_2.$$

Restated in the format of the original equation (9.1d), for P_2, we find

[1] The parameter a_2 was previously used to denote the rapid degradation rate of the unstable protein pool, R, within beta cells. In a related way, we expect APCs to be highly efficient in processing protein taken up from apoptotic bodies, such that the degradation rate of protein pool U would be of similar magnitude to that of R. For simplicity, we set the two rates to be equal to each other.

Table 9.1. Parameter values. See text in section 9.2.8 for discussion on how these estimates were made. The symbol '~' refers to fitted values. DC: dendritic cells, the primary APCs.

Symbol	Meaning	Value	Range
S	Rate of protein synthesis	5.33×10^4 proteins/min	
f	Fraction of rapidly degraded	0.25	[0–1]
a_1, a_2	Decay rates of Q and R	0.0002, 0.07 min^{-1}	
μ_1, μ_2	pMHC decay rates (β, DC)	0.01 min^{-1}	[0.001–0.01]
κ	Killing rate per T cell	3×10^{-12} (T cells pMHC min)$^{-1}$	$5 \times [10^{-13} - 10^{-9}]$
α	T-cell expansion rate	4.02×10^{-3} cell/ min	$[3.47–6.94] \times 10^{-3}$
k_p	pMHC level for $\frac{1}{2}$-max activation	6 pMHC	
k	$= \mu_2 k_p/(\eta_r S)$	3.16×10^{-5}	$[10^{-8} - 10^{-2}]$
γ	Turnover rate of E	2.083×10^{-4} min^{-1}	
η_r, η_q	Net pMHC efficiencies (R, Q)	1/2000, -pMHC/protein	[1/3122 – 1/994]
η_d	Net pMHC efficiency in DC	1/1406 pMHC/protein	[1/1780 – 1/1211]
$\eta = \frac{a_2 a_4}{a_1 a_3}$	Efficiency ratio $\eta = \eta_q/\eta_r$	~ 0.1	[0 – 1]
a_5	$= \eta_d f_1 B/A$	$\sim 4.5 \times 10^{-4}$	

$$\frac{dP_2}{dt} = \left[\eta_d f_1 \frac{\beta}{A}\right](\kappa P_1 E Q) - \mu_2 P_2.$$

Thus the parameter in the square braces is what we have called a_5 in equation (9.1d), where

$$a_5 = \eta_d f_1 \frac{\beta}{A},$$

is dimensionless. It is the product of three fractions: the number of targeted beta cells per APC (β/A), the fraction of normal protein that survives apoptosis and is carried by apoptotic bodies, f_1, and the number of pMHC complexes produced per protein molecule degraded, η_d. A summary of the model parameters with their definitions can be found in table 9.1.

9.2.3 Protein steady states

The variables Q, R, P_1 in equations (9.1) do not depend on E or P_2, so the system of equations can be decoupled. In a normally functioning cell, internal pools of protein and surface pMHC presentation would be maintained at some homeostatic level, so it is reasonable to assume that Q, R, and P_1 are at quasi-steady state. Setting $dQ/dt = 0$, $dR/dt = 0$, $dP_1/dt = 0$, we can solve for Q, R, P_1, to obtain

$$Q_s = \frac{S}{a_1}(1 - f), \tag{9.7a}$$

$$R_s = \frac{S}{a_2}f, \tag{9.7b}$$

$$\begin{aligned} P_{1s} &= \frac{a_3R_s + a_4Q_s}{\mu_1} \\ &= S\frac{(a_3/a_2)f + (a_4/a_1)(1-f)}{\mu_1} = \frac{S}{\mu_1}\left(\eta_r f + \eta_q(1-f)\right). \end{aligned} \tag{9.7c}$$

9.2.4 Rescaling the model

We scale equations (9.1) to reduce the number of parameters and to simplify our analysis. First note that units of parameters in table 9.1 are as follows: $[S]$ = protein/min, $[a_1] = [a_2] = [\mu_1] = [\mu_2] = [\gamma] = \text{min}^{-1}$, $[a_3] = [a_4]$ = pMHC/(protein min). As noted before, $a_3 = \eta_r a_2$ and $a_4 = \eta_q a_1$, where $\eta_r \approx 1/2000$ pMHC/proteins and $\eta_q < \eta_r$ are the pMHC efficiencies in beta cells [20, 28]. This means that the efficiency ratio $\eta \equiv (a_2a_4)/(a_1a_3) = \eta_q/\eta_r$. a_5 is dimensionless (see subsection 9.2.2), while $[\kappa]$ = (T cells pMHC min)$^{-1}$.

Notice that, in the extreme case when $f = 0$, the steady-state value of Q becomes

$$\bar{Q} = \frac{S}{a_1}.$$

At the other extreme where $f = 1$, with $a_4 \approx 0$, the steady states $\bar{R}$ and $\bar{P}_1$ may be written as

$$\bar{R} = \frac{S}{a_2} \quad \text{and} \quad \bar{P}_1 = \frac{a_3S}{a_2\mu_1},$$

respectively. Also, if we take $A(P_2) \approx \alpha$, then from equation (9.1e) we have that

$$\bar{E} = \frac{\alpha}{\gamma}.$$

These values give insights into how to rescale the model and generate dimensionless variables.

In order to do this, we now recast the model in terms of relative variables, as suggested by the parameter combinations above. We define new variables that are ratios of the actual levels to these representative levels ($q = Q/\bar{Q}$, $r = R/\bar{R}$, etc). That is, we normalize all variables on a convenient relative scale. This corresponds to making the substitutions

$$Q = \frac{S}{a_1}q, \quad R = \frac{S}{a_2}r, \quad P_1 = \frac{a_3S}{a_2\mu_1}p_1, \quad P_2 = \frac{a_3S}{a_2\mu_2}p_2, \quad E = \frac{\alpha}{\gamma}e. \tag{9.8}$$

After simplification, the resulting scaled model becomes

$$\frac{dq}{dt} = a_1[1 - f - q], \tag{9.9a}$$

$$\frac{dr}{dt} = a_2[f - r], \tag{9.9b}$$

$$\frac{dp_1}{dt} = \mu_1[r + \eta\, q - p_1], \tag{9.9c}$$

$$\frac{dp_2}{dt} = \mu_2[\chi\, q\, p_1\, e - p_2], \tag{9.9d}$$

$$\frac{de}{dt} = \gamma[\tilde{A}(p_2) - e], \tag{9.9e}$$

where

$$\tilde{A}(p_2) = \begin{cases} p_2/k & \text{(Case I)} \quad (9.10a) \\ p_2/(k + p_2) & \text{(Case II)} \quad (9.10b) \\ p_2^n/(k^n + p_2^n), \quad n \geqslant 2 & \text{(Case III)}, \quad (9.10c) \end{cases}$$

and

$$k = \frac{a_2\mu_2 k_p}{a_3 S} = \frac{\mu_2 k_p}{\eta_r S}, \quad \eta = \frac{a_2 a_4}{a_1 a_3} = \frac{\eta_q}{\eta_r}, \quad \chi = \frac{a_5 \alpha \kappa S}{a_1 \mu_1 \gamma}.$$

All parameters appearing in front of the square brackets have units of per minute. In this mode, the time variable was not nondimensionalized (i.e. rescaled) in order to display the results in a form that is easier to interpret. Analysing normalized level of physiological quantities varying over time in conventional units is more convenient.

9.2.5 Interpretations of dimensionless parameters

As previously described, $\eta < 1$ is the ratio of the pMHC production efficiencies of the stable and rapidly processed protein. The parameter k, on the other hand, is a scaled version of the peptide level for 50%-maximal T-cell activation. It is the ratio of the peptide level to the 'typical peptide level', $\bar{P}_1$, defined earlier.

As for the parameter χ, it can be written as

$$\chi = \left(\beta\kappa\frac{\alpha}{\gamma}\right) \cdot \left(\frac{Sf_1}{a_1}\right) \cdot \left(\frac{\eta_d}{A}\right) \cdot \left(\frac{\mu_2}{\mu_1}\right)\frac{1}{\mu_2}.$$

The product of the first three terms is the rate of pMHC production per APC (by a maximal T-cell population when a maximal level of stable protein is taken up from apoptotic bodies). The final term is the time constant of pMHC degradation on APCs, whereas the term $\mu_2/\mu_1 \approx 1$ is the ratio of time constants for pMHC degradation on beta cells and on APCs.

Finally, it is important to point out that the constants multiplying the terms in brackets in equations (9.9a)–(9.9c) will only influence transient behaviour, and not the steady states of the system.

9.2.6 Model reduction

As in the original model, the variables q, r, and p_1 in the scaled model described by equations (9.9a)–(9.9c) do not depend on e or p_2, so the system of equations can be decoupled. Quasi-steady state levels of q, r, and p_1 (satisfying $dq/dt = 0$, $dr/dt = 0$, $dp_1/dt = 0$) can again be used to reduce the dimensionality of the model. Thus, by QSS approximation, we obtain

$$0 \approx a_1[1 - f - q], \tag{9.11a}$$

$$0 \approx a_2[f - r], \tag{9.11b}$$

$$0 \approx \mu_1[r + \eta\, q - p_1], \tag{9.11c}$$

$$\frac{dp_2}{dt} = \mu_2[\, \chi\, q\, p_1\, e - p_2], \tag{9.11d}$$

$$\frac{de}{dt} = \gamma[\tilde{A}(p_2) - e]. \tag{9.11e}$$

Then equations (9.11d)–(9.11e), with substitutions based on equations (9.11a)–(9.11c), reduce to the system of two equations in the scaled variables p_2, e, given by

$$\frac{dp_2}{dt} = \mu_2[\chi\,(1 - f)\,(f + \eta(1 - f))\, e - p_2], \tag{9.12a}$$

$$\frac{de}{dt} = \gamma[\tilde{A}(p_2) - e]. \tag{9.12b}$$

The steady state solution $(p_2, e) = (0, 0)$ represents the healthy state of the scaled model, where there is no cross-presentation and no accumulation in T-cell population. We need to explore the stability properties of that steady state and the existence of other states characteristic of the disease.

9.2.7 Steady states and stability analysis

Aside from the healthy state, other steady state solutions corresponding to autoimmunity can exist, depending on the assumed form of T-cell activation function, $\tilde{A}(p_2)$ (as per Cases I–III). Moreover, whether the healthy state is stable depends both on the case considered and on the parameter values assumed.

Stability of healthy state

Regardless of the values of q, r, p_1, which are generally nonzero, the healthy steady state is $p_2 = 0$, $e = 0$, which corresponds to no T-cell accumulation and no pMHC presentation on APCs. We examine the stability of this steady state by noting the signs of the eigenvalues of the Jacobian matrix for the model.

$$J = \begin{bmatrix} -\mu_2 & \mu_2\chi\,(1-f)\,(f+\eta(1-f)) \\ \gamma\tilde{A}'(0) & -\gamma \end{bmatrix},$$

where

$$\tilde{A}'(0) = \begin{cases} 1/k, & n = 1 \text{ (Cases I \& II)} \\ 0, & n = 2 \text{ (Case III)} \end{cases}$$

Notice that the trace of J, $\mathrm{Tr}\,(J) = -(\mu_2 + \gamma)$, is always negative, so the sign of the real part of eigenvalues depends on the determinant of J. In case III, $\mathrm{Det}\,(J) = \mu_2\gamma > 0$, implying that the healthy state is locally stable. In cases I and II,

$$\mathrm{Det}(J) = \mu_2\gamma\left(1 - \frac{\chi}{k}(1-f)\,(f+\eta(1-f))\right).$$

If this determinant is negative ($\mathrm{Det}\,(J) < 0$), then the two eigenvalues are of opposite signs, with the positive eigenvalue signifying that the healthy state is a saddle. This means that small disturbances in the values of $p_2(t)$, $e(t)$ will be amplified, implying susceptibility to autoimmune disease. In other words, in cases I and II, the healthy state will become unstable, and even small random events could initiate autoimmunity whenever

$$(1-f)\,(f+\eta(1-f)) > \frac{k}{\chi} =: L. \tag{9.13}$$

We will show below that there are still some differences in dynamics between cases I and II.

It remains to analyze case III, where the healthy state is always locally stable, i.e. stable to sufficiently small disturbances. In case III, we have up to three possible steady states. All of these steady states must satisfy $dp_2/dt = 0$, $de/dt = 0$. Thus, for $n = 2$ (an assumption we make to simplify the analysis), we have

$$0 = \left[\chi\,(1-f)\,(f+\eta(1-f))\,e - p_2\right], \tag{9.14}$$

$$0 = \left[\frac{p_2^2}{k^2 + p_2^2} - e\right]. \tag{9.15}$$

This means that either $p_2 = 0$, $e = 0$, or else

$$\frac{p_2}{k^2 + p_2^2} = \frac{1}{\chi\,(1-f)\,(f+\eta(1-f))} =: \frac{1}{\phi},$$

where $\phi = \chi(1-f)\,(f+\eta(1-f))$. Thus the nontrivial steady states are the roots of

$$p_2^2 - \phi p_2 + k^2 = 0, \quad \Rightarrow \quad p_2 = \frac{\phi \pm \sqrt{\phi^2 - 4k^2}}{2}. \tag{9.16}$$

These states exist provided that $\phi > 2k$, i.e.

$$\phi = \chi(1 - f)(f + \eta(1 - f)) > 2k,$$

which implies that

$$(1 - f)(f + \eta(1 - f)) > 2\left(\frac{k}{\chi}\right) = 2L. \tag{9.17}$$

When this condition is satisfied, the larger of the two roots of equation (9.16) becomes the autoimmune state with elevated T-cell population. The smaller of the two is a saddle whose stable manifold forms a border between the basin of attraction of the healthy state and that of the autoimmune state. Thus in case III, even though the healthy state remains stable, the appearance of the autoimmune state hinges on an inequality similar to that satisfied in cases I and II.

To summarize, the nondimensionalized model has the following three possible dynamic behaviours, depending on the assumed T-cell activation function:

Case I: $\tilde{A}(p_2)$ is given by equation (9.10a), and $\tilde{A}'(0) = 1/k > 0$. We find that autoimmunity could be triggered if the fraction f satisfies inequality (9.13). The quantity $L = k/\chi$ encapsulates a dimensionless ratio of parameters that governs multiple effects other than protein allocation. Figure 9.4(a) shows the phase-plane structure of the model defined by equation (9.12) as determined by its nullclines. This case is unrealistic in that it predicts unbounded growth of T cells and pMHC once the healthy state is destabilized. However, it gives

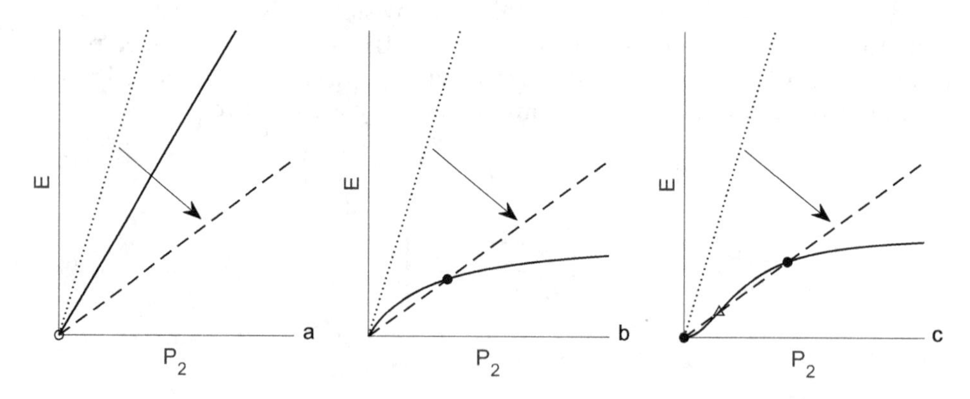

Figure 9.4. The nullclines of the nondimensionalized model given by the two-dimensional model described by equation (9.12) for cases I–III: The assumed pMHC-dependent T-cell activation function $\tilde{A}(p_2)$ is assumed to be (a) linear (case I); (b) Michaelis–Menten (saturated) (case II); and (c) Hill function with a Hill coefficient $n = 2$ (case III). Solid lines are the e-nullclines ($e = \tilde{A}(p_2)$). Dashed and dotted lines are the p_2-nullclines, where dashed (dotted) lines correspond to cases where one of the two inequalities (9.13) and (9.17) is satisfied (not satisfied) and autoimmunity is (is not) triggered.

a worst-case scenario of autoimmune initiation from a healthy state when activation of T cells is proportional to pMHC displayed.

Case II: $\tilde{A}(p_2)$ is given by equation (9.10b). The stability of the healthy state follows exactly the same pattern as in case I. Conditions for autoimmunity are precisely the same as before. A more realistic feature of this case is the presence of bounded growth. As shown in figure 9.4(b), loss of stability of the steady state at (0, 0) is accompanied by the appearance of a second (stable) steady state representing the autoimmune state (black dot at the intersection of the nullclines). When this happens, all initial conditions will eventually lead to autoimmunity in this case.

Case III: $\tilde{A}(p_2)$ is given by equation (9.10c) and $\tilde{A}'(0) = 0$. This case differs from the above two in that the healthy state is always stable to sufficiently small perturbations. However, as shown in figure 9.4(c), other steady states, including a stable/elevated (autoimmune) state can coexist with the healthy state provided that inequality (9.17) is satisfied. When this happens, autoimmunity could occur given a sufficiently large initial insult. In this case, even if the system is initially at the healthy state, a large enough disturbance can place the system in the basin of attraction of the autoimmune state, leading to pathogenesis.

In all three cases, a balance exists between the expression $(1 - f)(f + \eta(1 - f))$ and parameters k and χ on the right hand side of inequalities (9.13) and (9.17). Autoimmunity can occur when $(1 - f)(f + \eta(1 - f)) > L$ for cases I and II. In case III, the condition is more stringent, i.e. $(1 - f)(f + \eta(1 - f)) > 2L$, so initiation is less likely, all else being equal.

Stability analysis of the full model

Analysis of stability of the healthy state in the full model consisting of equations (9.9) leads to the same results. We consider here the case II type activation function, for which the healthy state can be destabilized. Note that case I is essentially similar (with simpler expressions), while in case III, the healthy state remains locally stable. The Jacobian matrix of the system of equations (9.9) in case II is

$$J = \begin{bmatrix} -a_1 & 0 & 0 & 0 & 0 \\ 0 & -a_2 & 0 & 0 & 0 \\ \mu_1 \eta & \mu_1 & -\mu_1 & 0 & 0 \\ \mu_2 \chi p_1 e & 0 & \mu_2 \chi qe & -\mu_2 & \mu_2 \chi qp_1 \\ 0 & 0 & 0 & \gamma \dfrac{k}{(k + p_2)^2} & -\gamma \end{bmatrix}.$$

We substitute the healthy state, $q = 1 - f$, $r = f$, $p_2 = 0$, $e = 0$ into this Jacobian matrix and find that the eigenvalues are

$$\lambda_1 = -a_1, \tag{9.18}$$

$$\lambda_2 = -a_2, \tag{9.19}$$

$$\lambda_3 = -\mu_1, \tag{9.20}$$

$$\lambda_{4,5} = -\frac{1}{2}\left((\gamma + \mu_2) \pm \sqrt{(\gamma - \mu_2)^2 + 4\frac{\mu_2\chi}{k}(1 - f)p_1\gamma}\right), \tag{9.21}$$

where $p_1 = f + \eta(1 - f)$ corresponds to the value of p_1 in the healthy state. Since $f \leqslant 1$, $(\gamma - \mu_2)^2 > 0$, it is clear that all eigenvalues (including $\lambda_{4,5}$) are real. Moreover, λ_1, λ_2, $\lambda_3 < 0$. Thus stability hinges on whether one of $\lambda_{4,5}$ is positive. To get one positive eigenvalue (which destabilizes the healthy state), we must have

$$(\gamma + \mu_2)^2 < (\gamma - \mu_2)^2 + 4\frac{\mu_2\chi}{k}(1 - f)p_1\gamma.$$

Algebraic simplification reduces this to

$$(1 - f)p_1 > \frac{k}{\chi}.$$

Substituting the value of $p_1 = f + \eta(1 - f)$ produces the main inequality (9.13) required for the instability of the healthy state.

9.2.8 Details of parameter estimates

One of the most difficult aspects of developing mathematical models of physiological systems is generating ballpark estimates of the parameters associated with the model. The difficulty arises from the limitation of experimental data available to generate such estimates. For the model presented here, we will provide a detailed explanation as to how 'data fitting' and steady state analysis can be used to generate estimates for parameter values. Data from various experimental set ups and cell lines can be employed to achieved both of these tasks.

According to [20, 29, 30], stable proteins have a turnover rate of 1.1% per hour, which is approximately 0.0002 min^{-1}. This implies that $a_1 \approx 0.0002\ \text{min}^{-1}$. It was also found in [20] that unstable proteins have a half-life of about 10 minutes, which would mean that $a_2 \approx \ln(2)/10 \approx 0.07\ \text{min}^{-1}$ (notice that $a_1 < a_2$, as expected). As for the efficiency ratio $\eta = (a_2a_4)/(a_1a_3)$, we choose the default value of 0.1, but one can still use bifurcation analysis to examine the dynamics of the model over the entire range of η given by [0 – 1], as demonstrated below.

To estimate pMHC turnover rate, we note that cells express around $[10^4 - 10^5]$ pMHC on their surface and they produce 10^2 class-I molecules per min [20, 29]. It follows that $\mu_1 \approx [0.001 - 0.01]\ \text{min}^{-1}$. We also expect $\mu_2 \approx \mu_1$.

The two parameters α and γ have been previously estimated to be 5.79 cell/day $\approx 4.02 \times 10^{-3}$ cell/min and 0.3 $\text{day}^{-1} \approx 2.08 \times 10^{-4}\ \text{min}^{-1}$, respectively [31, 32].

In healthy conditions in NOD mice, there are 5×10^5 beta cells. These cells are killed at an average rate of 4300 cell/day by a population of 10^6 T cells, with an initial killing rate of 6×10^4 cells/day during the first 5 days of disease onset [33].

Assuming that CTLs require 2–10 pMHC on beta cells to induce apoptosis, we conclude that the killing rate of beta cells by T cells, κ, is given by $\kappa \approx 5 \times [10^{-13} - 10^{-9}]$ (T cells pMHC min)$^{-1}$.

The two remaining parameters a_5 and k_p will be determined in the context of T1D. We could estimate a_5 by computing its value obtained in subsection 9.2.2. Substituting the steady state values of the quantities β (5×10^5 cells) and A (4×10^5 cells) into the equation $a_5 = \eta_d f_1 B/A$ [32] and using the average value of η_d (~ 1/1406 pMHC/protein) listed in table 3 of [20], we obtain $a_5 \approx 4.5 \times 10^{-4}$, for $f_1 \approx 0.5$. As for k_p, it has been suggested in [31] that there are a total of 2×10^5 possible binding sites on DCs (the primary APCs) in pancreatic lymph nodes, 10% of which are assumed to express pMHC at a sufficient level. It was also stated that the required level of expression of pMHC in each site for the activation of half of T cells is 120 pMHC/site. This implies that $k_p \approx 6$ pMHC per DC. It follows that $k \approx [0.1 - 3.5] \times 10^{-3}$. When numerically investigating the model in the next sections, we take a larger range for k, namely $k = [10^{-8} - 10^{-2}]$.

9.2.9 Bifurcation analysis

Here we study how varying parameters affects the behaviour of the model using bifurcation diagrams. These diagrams illustrate the long term behaviour of the system as a function of one model parameter. We consider the influence of T-cell avidity (reciprocal of k), efficiency ratio (η) and protein fraction (f) on disease onset. We also compare saturated and sigmoidal T-cell activation (i.e. cases II and III, respectively) on the dynamics of the model. Following common convention, stable steady states in these bifurcation diagrams are depicted with solid lines, and unstable steady states with dashed lines.

Reciprocal of T-cell avidity, k

We first begin by analyzing how the steady state behaviour of the scaled model, given by equations (9.9), changes with respect to variations in k. Figure 9.5 shows the bifurcation diagrams of e (T cells on a relative scale of $0 \leqslant e \leqslant 1$, on the vertical axis) as a function of k (shown on a logarithmic scale on the horizontal axis). On each panel, moving from right to left (decreasing k and $\log(k)$) corresponds to increasing T-cell avidity.

Panels (a) and (b) contrast the behaviour of the model with the saturated ($n = 1$, case II) and the sigmoidal ($n = 2$ case III) T cell activation functions, respectively. In both cases, when T-cell avidity is too low (or k is too large), the healthy state is the only steady state (solid line at $e = 0$) and it is stable.

When k is decreased, signifying higher T-cell avidity, a stable state representing autoimmunity comes into existence (solid curve at higher level of e). This transition (or bifurcation) takes place close to $\log(k) \approx -4$, i.e. $k \approx 10^{-4}$. Figure 9.5 shows the contrast between cases II and III. In the former (panel A), the disease state appears to undergo a transcritical bifurcation by exchanging stability with the healthy state. In fact, for values of $\log(k) < -4$, i.e. $k < 10^{-4}$, only the disease state is stable in that case). In case III (panel B), however, there is a saddle-node bifurcation and two

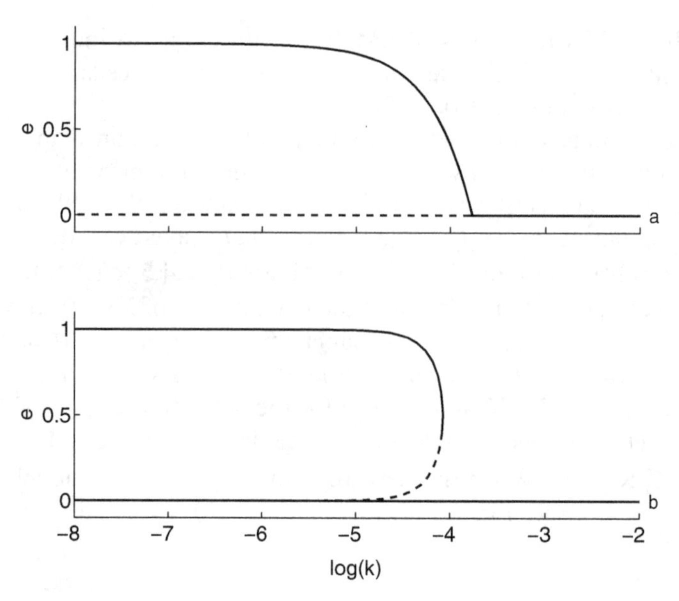

Figure 9.5. Bifurcation diagrams of e with respect to k (in logarithmic scale), using the scaled model described by equation (9.9), for two different values of the Hill coefficient appearing in equation (9.9e), while keeping all other parameters fixed. (a) $n = 1$ (case II); (b) $n = 2$ (case III). Steady states are plotted in solid (stable) and dashed (unstable) lines. Although bistability is the feature that distinguishes case (b) from (a), both panels show similar responses to varying T-cell avidity. Three distinct regimes in the range of k are identifiable and are of interest in both cases.

steady states simultaneously coming into existence. The healthy state and the disease state are both locally stable, while a saddle (dashed line) separates the regions attracted to either one. In figure 9.5(b), the saddle is so close to the healthy state it can barely be resolved from it. This means that the basin of attraction of the healthy state is very small, and most initial conditions will evolve towards the disease state.

Another important aspect of the model associated with case III is that it exhibits a jump discontinuity at the saddle node, unlike case II (see figure 9.5). This implies that solution trajectories starting from close proximity of the autoimmune state will exhibit a sharp decline in effector T-cell population (towards the 'healthy' state) when crossing the saddle node as k increases. Such behaviour will not be observed in case II, because the transition is smoother at the transcritical bifurcation. One can thus use this criterion to determine which one of these two cases is more likely when compared to experimental observations.

Effects of avidity and protein allocation on dynamics
The full model predicts that increasing T-cell avidity, or lowering k (which decreases the value of L in equation (9.13)) promotes autoimmunity in two ways: (i) by making the autoimmune state possible, where it was previously absent, and (ii) by increasing the basin of attraction of the autoimmune state, hence making it more likely to be attained, given a sufficient amount of perturbation away from the healthy state. In figure 9.6(A), we show the effect of the protein allocation fraction f

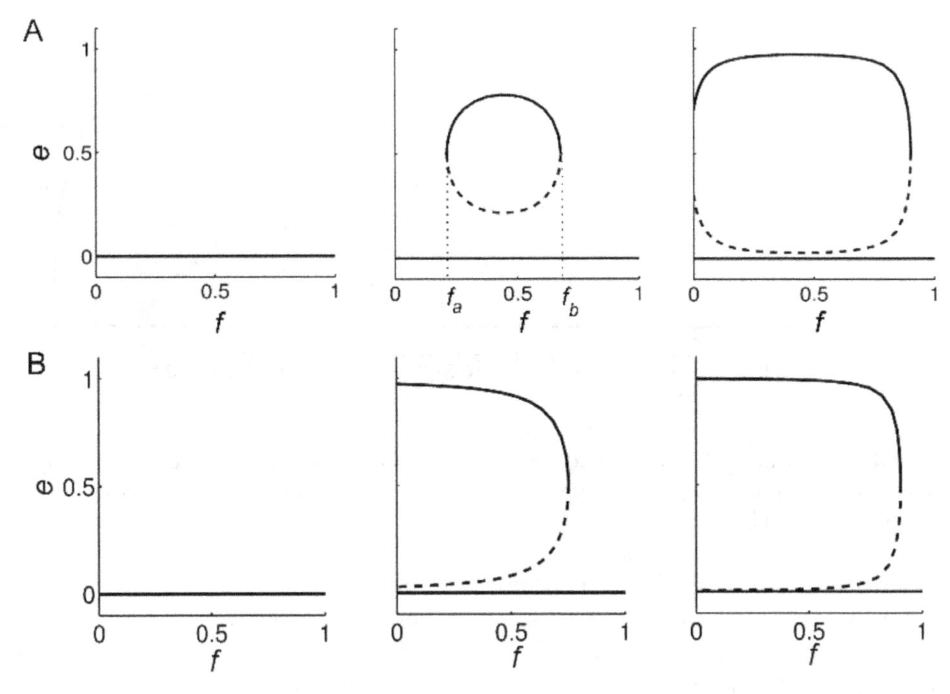

Figure 9.6. Bifurcation diagrams of the scaled effector T cells (e) with respect to protein allocation fraction (f) for increasing values of avidity ($\sim k^{-1}$), as determined by the model in case III (sigmoidal T-cell activation). The steady-state levels of the scaled T-cell population (vertical axis) versus f (horizontal axis) at (A) $\eta = 0.1$, and (B) $\eta = 0.7$ for low: $k = 10^{-3}$ (left), intermediate: $k = 8 \times 10^{-5}$ (middle) and high: $k = 3 \times 10^{-5}$ (right) T-cell avidity. Stable steady states are plotted as solid lines, while saddle points are plotted as dashed lines. Increasing the avidity of T cells (decreasing k) increases the range of f over which an elevated T-cell population exists with a maximum pathogenic potential occurring at $f \sim 0.5$ in (A), and at $f = 0$ in (B). Reprinted from [65] by permission of The Japanese Society for Immunology.

(horizontal axes) on the presence of the autoimmune state when T-cell avidity (k^{-1}) is successively increased and the value of relative efficiency is kept constant at $\eta = 0.1$. At low avidity, (left panel), only the healthy steady state is present (solid horizontal line at $e = 0$; p_2 is not shown), and all initial states evolve towards health (i.e. no elevation in the effector T-cell population and total survival of beta cells). For higher avidity (middle panel), a pair of new steady states with elevated levels of T cells appears over some range $f_a \leqslant f \leqslant f_b$ centered near $f = 0.5$. One of them is stable (solid line) representing the autoimmune state while the other is a saddle (dashed line). One can see that when $f \in [f_a, f_b]$, the model exhibits bistability, where some initial states evolve to the healthy state and others to the autoimmune state. Within this interval, the most effectively pathogenic values of f (i.e. values of f corresponding to autoimmune states with the highest level of T cells) are those close to 0.5. Increasing T-cell avidity even more, as in the right panel, broadens the dangerous range ($[f_a, f_b]$) and widens the regime of bistability, so that it occupies most of the interval [0, 1]. This makes the autoimmune state more persistent and also increases its basin of attraction.

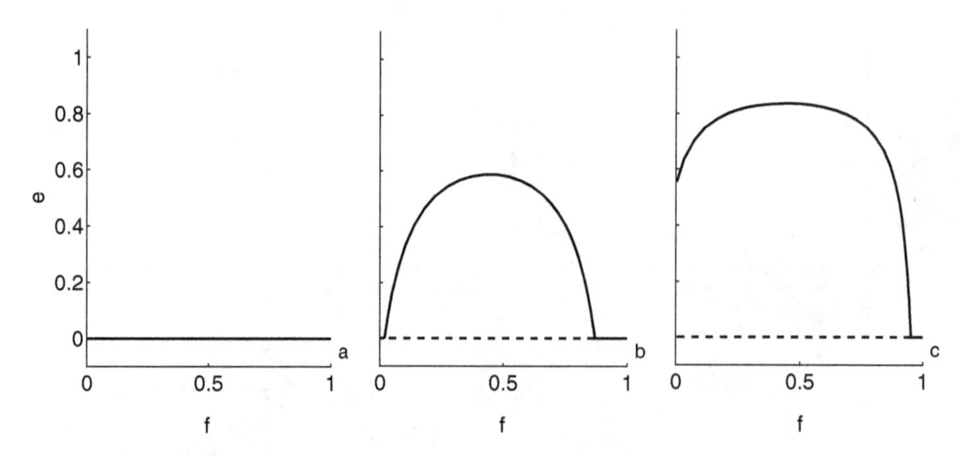

Figure 9.7. As in figure 9.6(A), but with saturated T-cell activation model (case II). Notice that increasing the avidity of T cells (by decreasing k) here decreases the parameter range of f within which the healthy state is stable, as was previously the case.

Increasing the value of η to 0.7 and carrying out the same simulations reveal similar results: the range of bistability expands with increasing T cell avidity (decreasing k), as shown in the left (low avidity), middle (intermediate avidity) and right (high avidity) panels of figure 9.6(B). The only noticeable difference, however, between the panels in (A) and (B) is the observed shift of the most pathogenic value of f to the left in the middle and right panels of (B) relative to (A). In these two cases, the panels in (B) show that $f = 0$ carries the most pathogenic potential, a result already predicted by our previous analysis.

In the above simulations, T-cell activation was assumed to depend sigmoidally on the expression level of pMHC (p_2) as with case III. For comparison, we show in figure 9.7 the same diagram for saturated T-cell activation, i.e. when $n = 1$ (case II). As before, the higher the avidity of T cells, the larger the range of f within which the stable autoimmune state is observed. This is consistent with the analysis discussed in figure 9.4. The only difference between the dynamics observe in this scenario in comparison to case III, where T-cell activation is sigmoidal ($n = 2$), is that the autoimmune state in the former disappears at a saddle-node bifurcation point, whereas in the latter, it disappears at a transcritical bifurcation point. In fact, bistability is not observed within the pathogenic range of f in case II.

We will show in section 9.4 that such dynamics are left intact even after the inclusion of other processes associated with T-cell dynamics. The increases in complexity within the model will illustrate the contribution of each component of the autoimmune response to T1D onset and progression, but also demonstrate that the underlying dynamics of the model are maintained.

Taken together, the results from the model presented in this section suggest that the development of autoimmunity depends upon a balance of protein allocation to either the stable or unstable pools, processing of these pools into pMHC, and T cell avidity. Non-trivially, for η sufficiently large (i.e. higher efficiency of pMHC production from the stable protein pool relative to that from the unstable pool), having a smaller allocation to the rapid degrading pool (i.e. making f closer to 0) makes the protein more immunogenic.

9.3 Phagocytosis in macrophages during T1D

It is well known that macrophages play the role of scavenger cells responsible for clearing pathogens and apoptotic cells by phagocytosis. During development in rodents and in primates, beta cells undergo programmed cell death, a naturally occurring phenomenon that is commonly called the apoptotic wave [34]. In the context of T1D, it has been hypothesized that clearance of apoptotic bodies following the apoptotic wave is defective, leading to the induction of autoimmune responses against surviving beta cells [35, 36]. As we have shown in the previous section, clearance of apoptotic bodies can lead to T-cell activation through cross presentation. Experimental observations of *in vitro* assays reveal that fewer apoptotic cells are seen inside macrophages from diabetes prone NOD mice compared with other strains of mice that are not prone to T1D [36]. This could be due to:

- upregulated degradation;
- impaired engulfment;
- defective macrophage activation.

The goal in this section is to test each one of these hypotheses by developing Markov chain models that account for the different classes of macrophages with a certain number of internalized apoptotic bodies inside them, as shown in figure 9.8. These models can help us examine the mechanisms underlying phagocytosis and derive shortcuts for calculating ballpark estimates of the kinetic parameters associated with the engulfment of apoptotic bodies, their internalization and digestion. Several models will be tested to achieve these goals.

9.3.1 Markov chain models

The following is a list of assumptions that will be considered when developing a mathematical model to describe the dynamics of this system during T1D:

1. Engulfment of apoptotic cells follows mass-action kinetics with a constant rate k_e.
2. Binding and engulfment of apoptotic bodies is fast and irreversible.

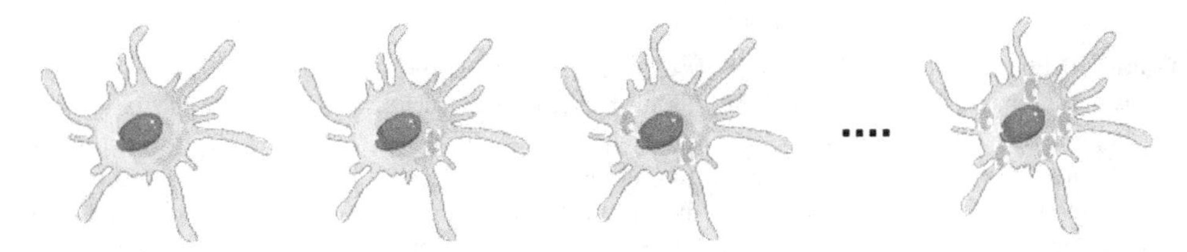

Figure 9.8. Macrophages classified based on the number of apoptotic bodies (yellow circles) inside them. This could range from no apoptotic bodies to seven in total.

3. Digestion of apoptotic bodies can be saturated, i.e. sequential (one at a time). This condition can be relaxed as we shall see in the model.
4. Macrophage death is neglected because it occurs at a slow time scale (in weeks) compared to phagocytosis that occurs at a faster time scale (in hours). Based on this assumption, the rate constants considered in the model will all be in hours.

Let M_i represent the class of macrophages (i.e. the number of cells per volume) with i undigested apoptotic bodies inside them ($i = 1, 2, \ldots, N$, where N, called macrophage capacity, was determined experimentally to be at most 7). Based on assumptions 1–4 listed above, one can present four different Markov models of phagocytosis in microphages as shown in figure 9.9. In all of these models, the transitions between the different states are governed by rate constants that can be estimated using experimental data. Model 1 assumes that the engulfment (digestion) of apoptotic bodies occur at a rate k_e (k_d) and that the activation of macrophages is equivalent to the engulfment of the first apoptotic body, model 2 assumes that macrophage activation is reversible and occurs at a rate k_a upon the engulfment of the first apoptotic body (a rate that is different from the engulfment of subsequent apoptotic bodies), while model 3 assumes that macrophage activation is irreversible (occurring at a rate k_a) and is independent from the engulfment of apoptotic bodies. In all of these three Markov models, the digestion of apoptotic bodies is assumed to be saturated. To relax this condition, model 4 is presented as an alternative to model

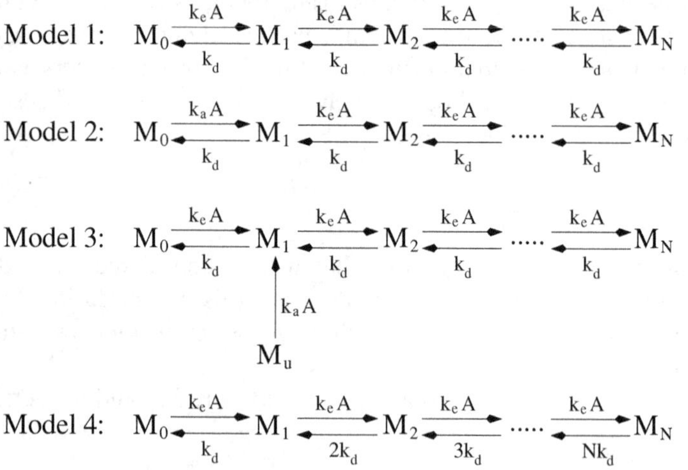

Model 1: $$M_0 \underset{k_d}{\overset{k_e A}{\rightleftharpoons}} M_1 \underset{k_d}{\overset{k_e A}{\rightleftharpoons}} M_2 \underset{k_d}{\overset{k_e A}{\rightleftharpoons}} \cdots\cdots \underset{k_d}{\overset{k_e A}{\rightleftharpoons}} M_N$$

Model 2: $$M_0 \underset{k_d}{\overset{k_a A}{\rightleftharpoons}} M_1 \underset{k_d}{\overset{k_e A}{\rightleftharpoons}} M_2 \underset{k_d}{\overset{k_e A}{\rightleftharpoons}} \cdots\cdots \underset{k_d}{\overset{k_e A}{\rightleftharpoons}} M_N$$

Model 3: $$M_0 \underset{k_d}{\overset{k_e A}{\rightleftharpoons}} M_1 \underset{k_d}{\overset{k_e A}{\rightleftharpoons}} M_2 \underset{k_d}{\overset{k_e A}{\rightleftharpoons}} \cdots\cdots \underset{k_d}{\overset{k_e A}{\rightleftharpoons}} M_N, \qquad M_u \xrightarrow{k_a A} M_1$$

Model 4: $$M_0 \underset{k_d}{\overset{k_e A}{\rightleftharpoons}} M_1 \underset{2k_d}{\overset{k_e A}{\rightleftharpoons}} M_2 \underset{3k_d}{\overset{k_e A}{\rightleftharpoons}} \cdots\cdots \underset{Nk_d}{\overset{k_e A}{\rightleftharpoons}} M_N$$

Figure 9.9. Four different Markov models describing activation and phagocytosis in macrophages. Models 1–3 assume saturated digestion (assumption 3), while model 4 ignores such an assumption (and so the digestion rate is proportional to the number of apoptotic bodies). M_i, $i = 1, 2, \ldots, N$, represent the class of macrophages with i apoptotic bodies inside them, M_u (in model 3) the class of unactivated macrophages, A the number of apoptotic bodies per ml, k_e (k_d) the engulfment (digestion) rate of apoptotic bodies, and k_a (in models 2 and 3) the activation rate of macrophages. Notice that model 2 assumes macrophage activation is reversible and is induced by engulfment of apoptotic bodies, and that model 4 can be modified to incorperate various aspects of models 2 and 3.

1, in which digestion is assumed to be unsaturated and dependent on the number of apoptotic bodies present inside each class of macrophages. Certainly, one can modify model 4 to incorporate other aspects of macrophage activation as was done in models 2 and 3.

We will show later on that comparison methods, based on numerical techniques, can be used to determine the likelihood of each scheme. Examples of such methods include the Markov Chain Monte Carlo techniques. For these techniques to produce reasonable results, one must first provide educated initial guesses (i.e. proper ballpark estimates) of the kinetic parameters involved in these models. In the next section, we will show how to use steady state analysis to provide such estimates.

9.3.2 Model 1

Mathematical description

The dynamics of model 1, shown in figure 9.9, can be described by the following set of equations

$$\frac{dM_0}{dt} = -k_e M_0 A + k_d M_1, \tag{9.22a}$$

$$\frac{dM_1}{dt} = k_e M_0 A - (k_d + k_e A)M_1 + k_d M_2, \tag{9.22b}$$

$$\frac{dM_2}{dt} = k_e M_1 A - (k_d + k_e A)M_2 + k_d M_3, \tag{9.22c}$$

$$\frac{dM_i}{dt} = k_e M_{i-1} A - (k_d + k_e A)M_i + k_d M_{i+1}, \quad i = 1, 2, \dots, N-1, \tag{9.22d}$$

$$\frac{dM_N}{dt} = k_e M_{N-1} A - k_d M_N, \tag{9.22e}$$

where the dynamics of apoptotic bodies A is governed by the equation

$$\frac{dA}{dt} = -k_e A \sum_{i=0}^{N-1} M_i.$$

If we let $M = \sum_{i=0}^{N} M_i$ be the total number of macrophages in a phagocytic culture and assume that N is very large (i.e. $N \to \infty$), we can conclude that

$$\frac{dA}{dt} = -k_e AM \quad \Longrightarrow \quad A = A_0 \exp(-k_e Mt). \tag{9.23}$$

The initial conditions associated with system (9.22a)–(9.22e) are given by

$$M_0(0) = M, \qquad M_i(0) = 0 \ (i \geqslant 1), \qquad A(0) = A_0. \tag{9.24}$$

The kinetic parameters of the model described by equations (9.22a)–(9.24) can be determined using two experimentally measurable quantities that are typically used to study phagocytic culture; namely,

1. Percent phagocytosis (ϕ), defined as the percent of macrophages that have visible engulfed apoptotic bodies. It is described mathematically by the following expression

$$\phi = \frac{100}{M} \sum_{i=1}^{N} M_i. \tag{9.25}$$

2. Phagocytic index (I_ϕ), defined as the average number of engulfed apoptotic bodies per 100 macrophages. It is described mathematically by the following expression

$$I_\phi = \frac{100}{M} \sum_{i=1}^{N} iM_i. \tag{9.26}$$

We will show how these quantities can be used to obtain ballpark estimates of the parameters of the model.

Model analysis

At the start of activation of this system, i.e. close to time $t = 0$, we can assume that $M_i \approx 0$, for all $i \geqslant 2$. Furthermore, during this very short period of time, we would expect to observe a decay in M_0 and a rise in M_1. We can thus conclude, based in equations (9.22a) and (9.22b), that

$$\begin{cases} \dfrac{dM_0}{dt} &= -k_e M_0 A \\ \dfrac{dM_1}{dt} &= k_e M_0 A. \end{cases}$$

Since $M_0A = MA_0$ at $t = 0$ as per equation (9.24), it follows that

$$k_e \approx \begin{cases} -\left(\left. \dfrac{\Delta M_0}{\Delta T} \right|_{t=0} \right) \Big/ (MA_0) \\ \left(\left. \dfrac{\Delta M_1}{\Delta T} \right|_{t=0} \right) \Big/ (MA_0), \end{cases} \tag{9.27}$$

where ΔM_0 and ΔM_1 are the changes in M_0 and M_1 that occur during the time duration ΔT, respectively. These two quantities can be measured experimentally over the period ΔT to estimate k_e. This can be done as follows. Suppose that in Balb/c mice, M_0 drops by 12% within 5 min when 200 000 macrophages (per ml) are co-cultured with 10^6 apoptotic cells (per ml). To derive an estimate for k_e in hours, we convert 5 min into hours (5 min = 1/12 h) and use equation (9.27) to obtain

$$\begin{aligned} k_e &\approx -\frac{-0.12 \times 200\,000\,(\text{cell/mL})}{1/12\,(\text{h})} \\ &\quad \times \frac{1}{200\,000\,(\text{cell/mL}) \times 10^6\,(\text{apoptotic cells/mL})} \\ &\approx 14.4 \times 10^{-7}\,\text{mL/(cell h)}. \end{aligned}$$

If the drop in the value of M_0 in NOD mice is 3%, we obtain the estimate $k_e \approx 3.6 \times 10^{-7}$ mL(cell h)$^{-1}$. Comparing the two estimates for k_e in Balb/c and NOD mice suggests that engulfment of apoptotic bodies by macrophages in NOD mice is impaired (as it occurs at a lower rate), which could explain the reduced number of internalized apoptotic bodies in NOD macrophages.

The rest of the parameters of model 1 can be estimated using QSS approximation in which we assume that all classes of macrophages M_i, $i = 1, 2, \ldots, N$, equilibrate after 1 hour. This is done by setting equations (9.22a)–(9.22e) to zeros as follows

$$\left.\begin{aligned} 0 &\approx -k_eM_0A + k_dM_1 \\ 0 &\approx \cancel{k_eM_0A} - (\cancel{k_d} + k_eA)M_1 + k_dM_2 \\ 0 &\approx \cancel{k_eM_1A} - (\cancel{k_d} + k_eA)M_2 + k_dM_3 \\ &\vdots \end{aligned}\right\} \Longrightarrow \left\{\begin{aligned} M_1 &= \left(\frac{k_eA}{k_d}\right)M_0 \\ M_2 &= \left(\frac{k_eA}{k_d}\right)M_1 \\ M_3 &= \left(\frac{k_eA}{k_d}\right)M_2. \\ &\vdots \end{aligned}\right.$$

Thus, for all $i \geqslant 0$, we have

$$M_{i+1} = \lambda M_i, \tag{9.28}$$

where

$$\lambda = \frac{k_eA}{k_d}. \tag{9.29}$$

By quantifying the ratio M_{i+1}/M_i experimentally, one can determine the parameter combination λ based on equation (9.28), and then use it, along with equation (9.29), to generate an estimate for k_d, as follows

$$k_d = \frac{k_eA}{\lambda} = \frac{k_eAM_i}{M_{i+1}}. \tag{9.30}$$

From equation (9.28), it can be shown that $M_i = \lambda^i M_0$, which implies that

$$M = M_0(1 + \lambda + \lambda^2 + \ldots + \lambda^N) = M_0 \sum_{i=0}^{N} \lambda^i.$$

The latter represents a geometric series in λ, provided that $\lambda < 1$. By assuming that $\lambda < 1$, we conclude that

$$M = M_0 \frac{1 - \lambda^{N+1}}{1 - \lambda}.$$

For large N (i.e. at the limit when $N \to \infty$), we have

$$M \approx \frac{M_0}{1 - \lambda}. \tag{9.31}$$

It follows that

$$\lambda \approx \frac{M - M_0}{M},$$

and

$$k_d \approx \frac{k_e M A}{M - M_0}.$$

It was indicated earlier that the two quantities 'percent phagocytosis' (ϕ) and 'phagocytic index' (I_ϕ), defined by equations (9.25) and (9.26), respectively, can be measured experimentally. Based on the various estimates obtained above and the assumption that $\lambda < 1$, one can determine ϕ and I_ϕ as follows. According to equations (9.25), (9.30) and (9.31), we have

$$\begin{aligned}
\phi &= \frac{100}{M} \sum_{i=1}^{N} M_i = \frac{100}{M}(M_1 + M_2 + \cdots + M_N) \\
&= \frac{100}{M}(\lambda M_0 + \lambda^2 M_0 + \cdots + \lambda^N M_0) \\
&= \frac{100\lambda M_0}{M} \sum_{i=0}^{N-1} \lambda^i \\
&\approx \frac{100\lambda M_0}{M(1 - \lambda)} = 100\lambda \quad (\text{as } N \to \infty).
\end{aligned}$$

Similarly, by equations (9.26) and (9.31), we have

$$
\begin{aligned}
I_\phi &= \frac{100}{M}\sum_{i=1}^{N} iM_i = \frac{100}{M}\sum_{i=1}^{N} i\lambda^i M_0 = \frac{100}{M}M_0\lambda\sum_{i=1}^{N} i\lambda^{i-1}\\
&= \frac{100}{M}M_0\lambda\frac{d}{d\lambda}\left(\sum_{i=1}^{N}\lambda^i\right) = \frac{100}{M}M_0\lambda\frac{d}{d\lambda}\left(\sum_{i=0}^{N}\lambda^i\right)\\
&\approx \frac{100}{M}M_0\frac{\lambda}{(1-\lambda)^2} \quad (\text{as } N \to \infty)\\
&\approx 100\frac{\lambda}{1-\lambda}.
\end{aligned}
$$

It follows that

$$
k_d \approx \left(\frac{100}{I_\phi} + 1\right)k_e A.
$$

Similar approaches can be used to generate estimates for the key parameters of models 2 and 3. In problem 3, the reader is asked to generate ballpark estimates of k_a, k_e and k_d.

Upon the determination of the percent phagocytosis and phagocytic index experimentally, ballpark estimates of k_d can be generated. These estimates, along with those that were made for k_e, can then be used as initial educated guesses for various numerical techniques designed to fit the general model described by equations (9.22a)–(9.22e) to the experimental data associated with various classes of macrophages. Some of these techniques could be iterative methods similar to those Monte-Carlo (MCMC) techniques mentioned in chapter 4, or they could be based on evolutionary algorithms such as the genetic algorithm [37]. In the next section, we present another approach based on 'Akaike Information Criterion'.

9.3.3 Model comparisons

In section 9.3.1, several models were presented that could potentially explain the dynamics of macrophage phagocytosis and allow for comparisons between normal (Balb/c) macrophages and abnormal (NOD) macrophages. In [38], the Akaike Information Criterion (AIC) [39] was used to achieve this. AIC allows for model comparisons by determining which mathematical model minimizes the Kullback–Leibler divergence between data and model (i.e. the discrepancy between the two). It maximizes the probability that a candidate model generates the observed data by rewarding descriptive accuracy, while penalizing for increases in the number of free parameters. In [38], a second-order bias correction was proposed; it was given by

$$
\mathcal{L} = n\ln\left(\frac{\mathcal{X}}{n}\right) + 2K + \frac{2K(K+1)}{n-K-1},
$$

where

$$\mathcal{X} = \|\text{data} - \text{model}\|_2^2,$$

is the sum of squared error between the data and the fitted model ($\|\cdot\|_2$ is the L_2-norm), n is the number of data points and K is the number of estimated parameters. The use of this second-order bias correction AIC is recommended when the ratio of sample size to the number of parameters is smaller than 40 [40]. The likelihood of each model presented in section 9.3.1 can then be determined by evaluating

$$p(\text{Model } i|\text{data}) = \frac{\exp(-0.5\Delta_i(\mathcal{L}))}{\sum_{j=1}^{M_{\max}} \exp(-0.5\Delta_j(\mathcal{L}))}, \tag{9.32}$$

where $p(\text{Model } i|\text{data})$ is the probability of obtaining model i ($i = 1, 2, \ldots, M_{\max}$, and $M_{\max} = 4$ is the total number of models), given the observed data, and

$$\Delta_i(\mathcal{L}) = \mathcal{L}_i - \min_{j=1,2,\ldots,M_{\max}} (\mathcal{L}_j).$$

The denominator in equation (9.32) ensures that probability of all models add up to 1. One may apply this criterion to discriminate between the different models under consideration. For example, when comparing two models, labeled A and B, the likelihood of model A being correct is

$$\begin{aligned} p(A|\text{data}) &= \frac{\exp(-0.5\Delta_A(\mathcal{L}))}{\exp(-0.5\Delta_A(\mathcal{L})) + \exp(-0.5\Delta_B(\mathcal{L}))} \\ &= \frac{\exp(-0.5\Delta_A(\mathcal{L}))}{\exp(-0.5\Delta_A(\mathcal{L})) + \exp(-0.5\Delta_B(\mathcal{L}))} \times \frac{\exp(0.5\Delta_B(\mathcal{L}))}{\exp(0.5\Delta_B(\mathcal{L}))} \\ &= \frac{\exp(-0.5\Delta(\mathcal{L}))}{1 + \exp(-0.5\Delta(\mathcal{L}))}, \end{aligned}$$

where $\Delta(\mathcal{L}) = |\, \mathcal{L}_B - \mathcal{L}_A \,|$. A similar approach can be used to compare the four models listed in section 9.3.1.

9.4 Autoimmune dynamics and transient stability

As indicated previously, the complications resulting from the initiation and progression of T1D depend critically on the level of autoreactive and cytotoxic T-lymphocytes in circulation. These activated effector T cells play a key role in destroying pancreatic islets and inducing insulin dependency. Recent evidence from the NOD mouse model, however, has suggested that antibody-secreting mature B-lymphocytes (or plasma cells) may also contribute to pathogenesis [41]. Interestingly, helper T cells activate B cells into effector plasma cells capable of secreting soluble forms of islet-specific immunoglobulin (or autoantibodies) that bind to beta-cell specific autoantigens [42].

It has been experimentally shown that high-risk human subjects (HRS) exhibit multiple islet autoantibodies and high-avidity T cells. These autoantibodies, along with other biomarkers and genetic factors, can be used as predictors of clinical T1D development among HRS [43–49]. The presence of these biomarkers indicates that autoimmunity leading to beta-cell destruction has been initiated. Previous studies have shown that conventional autoantibodies, although useful, seem insufficient to predict T1D [43–50]. More recent observations, on the other hand, suggest that novel autoantibodies (such as those that are reactive to the extracellular domain of the neuroendocrine antigen IA-2) in combination with conventional biomarkers can identify rapid progressors of T1D onset when compared to the use of conventional ones alone [43, 51, 52]. This is confirmed by studies showing that a subgroup of HRS who are positive for certain conventional and novel autoantibodies confers a cumulative risk of 100% by 10 year follow-up, whereas the presence of two conventional autoantibody markers confers a cumulative risk of 74% at 15 year follow-up in another subgroup [43, 46, 50]. The former group was termed rapid progressors while the latter was called slow progressors.

It has been hypothesized that the pace of disease progression in both groups is governed by T-cell avidity, which in turn is correlated with the avidity of the autoantibodies reactive to the same islet-specific autoantigens [19, 49, 53, 54]. In other words, each beta-cell specific epitope will dictate the type of autoantibodies and autoreactive T cells that will generate the autoimmune response, and how fast disease progression will be. The correlation between the avidity of autoantibodies and T cells and the interconnections between T and plasma cells have been investigated mathematically. Here we show how this was done and elucidate how T-cell killing efficacy plays a role in shaping up the dynamics. We will illustrate the methodology used to examine all possible responses associated with different levels of T-cell avidity and killing efficacy in generating an autoimmune response, including the B-cell response. A dynamic one-clone model, consisting of one antigenic specificity [19, 53], will be used for this purpose and for investigating the discrepancy in predicting T1D disease onset between rapid and slow progressors based on the notion of T-cell avidity. Because of limited evidence, the role of B cells, or plasma cells, in pathogenicity will be ignored.

9.4.1 One-clone model

Model assumptions

Based on the scheme in figure 9.10, we consider a homogeneous model comprised of islet-specific autoreactive T cells ($T = T(t)$ representing the population size of T cells at a given time t), islet specific autoreactive B cells ($B = B(t)$ representing the population size of B cells), autoantibody-secreting plasma cells ($P_c = P_c(t)$ representing the population size of plasma cells), soluble forms of immunoglobulin or autoantibodies ($I_g = I_g(t)$ representing the concentration of autoantibodies that can be used as genetic biomarkers to predict T1D), pancreatic beta cells ($\beta = \beta(t)$ representing the population size of surviving beta cells), and autoantigens expressed as pMHC complexes ($P = P(t)$ representing the expression level of pMHC on APCs). In other words, the components of this model are assumed to have one autoantigenic specificity. We further assume that:

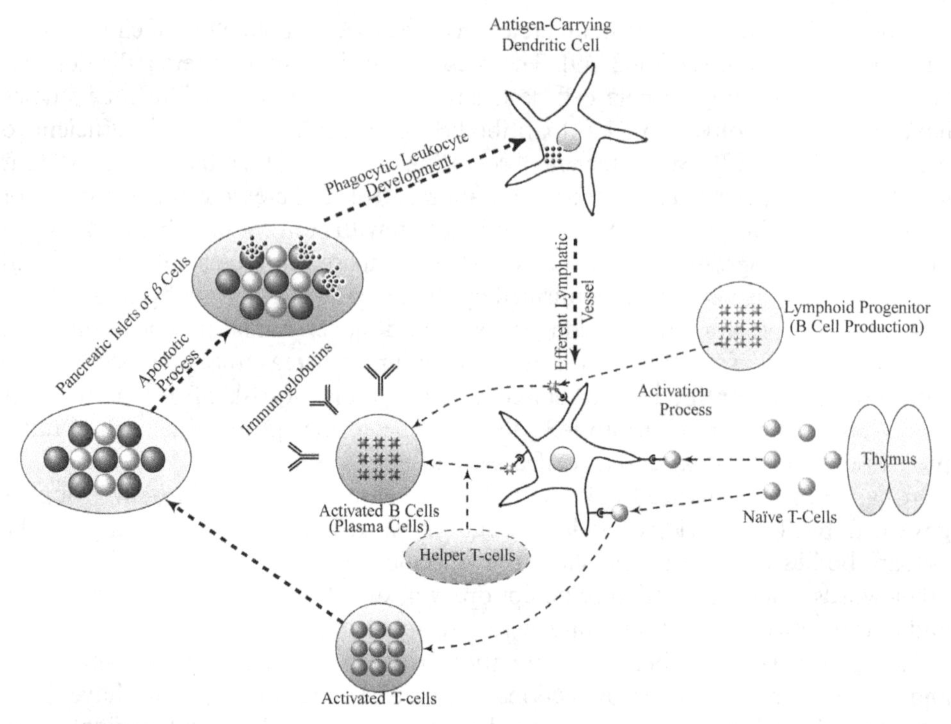

Figure 9.10. Model diagram of T1D progression showing the subpopulations of immune cells implicated in the disease. Reduced expression of autoantigens in the thymus may predispose to T1D and autoimmunity by inducing T-cell positive selection [52]. Naïve T cells leaving the thymus are activated into effector T cells in the lymph nodes by APCs in a manner similar to figure 9.1. Effector T cells then infiltrate the islets and cause beta-cell destruction at a relatively slow rate. Protein fragments from apoptotic bodies are subsequently taken up by APCs for processing, causing epitope spreading and further amplification of the autoreactive T-cell response. B-cell activation into autoantibody-secreting plasma cells is induced by helper T cells and APCs via a similar mechanism. Reproduced from [53]. Copyright 2014 Jaberi-Douraki *et al.* CC BY 4.0.

1. The T-cell pool is occupied mostly by terminally differentiated effector cells and a small compartment of memory cells responsible for self-renewal. The proliferation rate of the latter compartment is small and roughly proportional to the population size of APCs (assumed to be roughly constant). The ability of memory cells to self-replicate can keep, in certain cases, the level of T cells elevated for prolonged durations extending beyond the time when a critical number of beta cells is lost. Hence, we do not include in this model a separate non-vanishing pool of memory T cells to account for a continuously elevated level of T cells (as suggested in [55]), an assumption that simplifies the analysis significantly.
2. The failure of negative selection in the thymus can lead to the escape of pathogenic T cells into the periphery. This peptide-dependent thymus input of T cells can act as another source of newly activated T cells in the model. The elevated level of peripheral autoreactive T cells in HRS (i.e. the initial

level of T cells) that escape negative selection in the thymus, on the other hand, can be accounted for by the initial conditions of the model.

3. The activation of T cells is carried out by APCs averaged over three types of cells: macrophages, dendritic cells (DCs) and B cells. As suggested by figure 9.10, direct involvement of the B-cell pool in activating T cells is ignored in the model.
4. The decay in the population size of surviving beta cells is assumed to depend linearly on the level of effector T cells. We do not include a source term for beta-cell replication or neogenesis, because experimental evidence for such behaviour is lacking. The spatial distribution of beta cells in islets is also ignored in this formulation and the total number of beta cells within one pool is considered instead. With such formalism, the stochastic effects of having a small number of T cells infiltrating these islets become negligible.
5. Islet-specific B-cell activation and maturation into plasma cells is regulated by a class of $CD4^+$ T cells, called helper T cells (assumed to have constant level), and APCs (via B-cell receptor binding to pMHC, P), in the presence of various cytokines secreted by T cells. Using QSS approximation, we can assume that cytokine concentrations are proportional to the population size of T cells (at steady state). In other words, B-cell maturation is a T-cell- and peptide-dependent process.
6. The avidity of T and B cells are correlated, i.e. their dependence on pMHC expression level for activation is very similar (or even identical).
7. Beta-cell specific autoantibodies are released mainly by plasma cells with a small fraction released by B cells. The latter is responsible for maintaining 'basal level' of circulating antoantibodies in the absence of plasma cells.

Model equations

Based on the assumptions listed in the previous section, the differential equations governing the dynamics of this system are given by

$$\frac{dT}{dt} = \Sigma_T + \alpha\, T\, \mathcal{F}_1(P) - \epsilon\, T^2 - \delta_T\, T \tag{9.33a}$$

$$\frac{dB}{dt} = \gamma + (\eta_1\, \mathcal{F}_1(P) - \eta_2\, T\, \mathcal{F}_1(P) - \delta_B)\, B \tag{9.33b}$$

$$\frac{dP_c}{dt} = \eta_2\, T\, \mathcal{F}_1(P) - \delta_{P_c}\, P_c, \tag{9.33c}$$

$$\frac{dI_g}{dt} = a_1\, B + a_2\, P_c - \delta_{I_g}\, I_g \tag{9.33d}$$

$$\frac{d\beta}{dt} = \mathcal{G}(\beta) - \mathcal{H}_1(T, \beta) - \delta_\beta\, \beta \tag{9.33e}$$

$$\frac{dP}{dt} = \mathcal{H}_2(T, \beta) - \delta_P P, \tag{9.33f}$$

where Σ_T is the input of naïve T cells from the thymus, ϵ is the intra-clonal competition rate between T cells that are reactive to the same autoantigen (to maintain T-cell homeostasis due to limited physical space), δ_T, δ_B, δ_{P_c}, δ_β are the mortality (turnover) rates of T, B, plasma and beta cells, respectively, δ_{I_g}, δ_P are the autoantibody and pMHC degradation rates, respectively, γ is the basal level of B-cell production rate from bone marrow, and a_1, a_2 are the rates of autoantibody release by B cells and plasma cells, respectively, with $a_1 \ll a_2$.

In this model, we assume that the pool of T cells retains a subset of memory T cells capable of replicating in a pMHC-dependent manner according to the term $\alpha\, T\, \mathcal{F}_1(P)$ in equation (9.33a), where α is the replication rate and $\mathcal{F}_1(P)$ is the probability of T-cell activation. The two terms $\eta_1\, \mathcal{F}_1(P)\, B$ and $\eta_2\, T\, \mathcal{F}_1(P)\, B$ in equation (9.33b), on the other hand, represent pMHC-dependent B-cell self-renewal by APCs with a rate η_1, and B-cell maturation into plasma cells with a rate η_2 (contingent upon both pMHC-expression on APCs and cytokine secretion), respectively. In other words, the pool of B cells is also assumed to retain a subset of memory B cells capable of replicating. The function $\mathcal{H}_1(T, \beta)$ describes beta-cell killing by T cells and harmful cytokines, whereas the function $\mathcal{G}(\beta)$ describes beta-cell neogenesis and replication both lumped up in this term. Finally, the function $\mathcal{H}_2$ describes pMHC production per T-cell per beta-cell.

The dependence of B-cell maturation and self-renewal on the expression level of pMHC on APCs is determined by the probability function $\mathcal{F}_1(P)$, which happens to be identical to that used for T-cell activation in equation (9.33a). The same probability function was employed in both cell types to imply that the avidities T and B cells are correlated (see assumption 6). This makes the model more physiological [19, 53] and more capable of testing the hypothesis that avidity determines T1D disease outcomes (in terms of slow and fast progressors). Furthermore, it was assumed that $a_1 \ll a_2$ to imply that the vast majority of circulating autoantibodies is produced by plasma cells and minimally by B cells [19, 53] (see assumption 7). By taking a_1 to be nonzero, on the other hand, ensures that there is a basal level of circulating islet-specific autoantibodies even after the halt of the autoimmune response. Although there is evidence for B-cell involvement in the destruction of beta cells during T1D [56], such effect appears minimal compared to T-cell induced destruction. We have therefore ignored such an effect in our model.

Subsequent to beta-cell apoptosis and their internalization by APCs (see figure 9.10), beta-cell proteins are processed by proteasomes into autoantigenic peptides within APCs and expressed as pMHCs on their surface. Equation (9.33f) accounts for all of these antigenic processing mechanisms, but it does not account for the entire system of T1D-specific autoantigens. In other words, the model only focuses on the dynamics of one autoantigenic peptide, as an averaged representative of all antigenic specificities, for simplicity.

Model functions
Since T-cell activation is regulated by the level of pMHC expression level on APCs, we expect that

$$\frac{d\mathcal{F}_1}{dP} \geqslant 0.$$

Based on T-cell dose-response curves to growing level of pMHC on APCs shown in [57], one can typically describe the probability function $\mathcal{F}_1(P)$ as a Hill function with a Hill coefficient $n = 1$, given by

$$\mathcal{F}_1(P) = \frac{P}{P + k}, \tag{9.34}$$

where k represents the expression level of pMHC for 50% maximum activation of T cells. The larger the value of k, the higher the expression level of pMHC required for T-cell activation, which implies that k^{-1} can be used as a measure for T-cell avidity (as seen with the protein-processing model in section 9.2). The empirical determination of the Hill function that approximates T-cell activation (e.g. by measuring IFN-γ secretion from $CD4^+$ T cells in response to variations to GAD-expression level on APCs) was also done in [58]. As explained in [19], the value of k for GAD-reactive T cells was found to lie between [0.02,0.18] μM for high-avidity T cells [58], but this range was then extended to [0.1, 9] μM for low-avidity T cells. We take advantage of a broad spectrum for k ($\in$[0, 20] μM) in our analysis of the model (9.33a)–(9.33f) to study the effects of T-cell avidity on disease onset and progression.

The function $\mathcal{H}_1$ in equation (9.33e) determines beta-cell killing by ($CD8^+$ and $CD4^+$) T cells occurring at an effective rate κ (expected to remain approximately constant throughout disease progression within the range $[10^{-11}, 10^{-7}]$ $(\text{cell day})^{-1}$ [19, 32, 59]). On the other hand, the function $\mathcal{H}_2$ in equation (9.33f), which describes pMHC production from processing beta-cell specific proteins in APCs, is assumed to be proportional to $\mathcal{H}_1(T, \beta)$. In other words,

$$\mathcal{H}_2(T, \beta) = R\,\mathcal{H}_1(T, \beta), \tag{9.35}$$

where R is the production rate of pMHC per T-cell per beta-cell. By using mass-action kinetics to describe the autoimmune attack by $CD8^+$ and $CD4^+$ T cells and by assuming that beta-cell killing is saturated by beta-cell number, we arrive at the following formalism for $\mathcal{H}_1$:

$$\mathcal{H}_1(T, \beta) = \kappa\, T\, \frac{\beta}{1 + \mu\,\beta}, \tag{9.36}$$

where μ is the saturation parameter per beta-cell for beta-cell killing. In some special cases, we may assume that the saturation parameter is identically zero to make $\mathcal{H}_1$ a bilinear function of T-cell and beta-cell population sizes. Beta-cell renewal (via beta-cell replication or neogenesis) can be described by a Hill function (see [32]), given by

$$\mathcal{G}(\beta) = s\frac{\beta}{\beta + k_\beta}, \tag{9.37}$$

where s is the maximal rate of beta-cell renewal and k_β is the number of beta cells required for 50% maximal renewal.

In equation (9.33a), we used the source term Σ_T to represent the input of naïve cells from the thymus, which does not necessarily remain constant throughout disease progression. In fact, we would expect this term to be given by

$$\Sigma_T = \sigma\, \mathcal{F}_1(P). \tag{9.38}$$

By making use of equations (9.34)–(9.38), we can give a detailed and explicit mathematical description to the one-clone model (9.33a)–(9.33f) as follows

$$\frac{dT}{dt} = \sigma \frac{P}{P+k} + \alpha\, T \frac{P}{P+k} - \epsilon\, T^2 - \delta_T\, T \tag{9.39a}$$

$$\frac{dB}{dt} = \gamma + \left[(\eta_1 - \eta_2 T) \frac{P}{P+k} - \delta_B \right] B \tag{9.39b}$$

$$\frac{dP_c}{dt} = \eta_2 T \frac{P}{P+k} B - \delta_{P_c} P_c \tag{9.39c}$$

$$\frac{dI_g}{dt} = a_1\, B + a_2 P_c - \delta_{I_g}\, I_g \tag{9.39d}$$

$$\frac{d\beta}{dt} = s\frac{\beta}{\beta + k_\beta} - \kappa\, T \frac{\beta}{1 + \mu\, \beta} - \delta_\beta\, \beta \tag{9.39e}$$

$$\frac{dP}{dt} = R\, \kappa\, T \frac{\beta}{1 + \mu\, \beta} - \delta_P\, P. \tag{9.39f}$$

This new model possesses three important properties; namely, the presence of a term describing thymus input Σ_T, the dependence of pMHC production rate on κ (T-cell killing efficacy), and the dependence of T- and B-cell activation/maturation on the same pMHC-dependent Hill function $\mathcal{F}_1$ with a Hill coefficient $n = 1$. The latter assumption makes the model more physiological by linking T-cell avidity to the binding affinity of autoantibodies via the parameter k.

9.4.2 Model rescaling

In order to further simplify the model and to reduce the total number of parameters, we nondimensionalize equations (9.39a)–(9.39f). To do this, we make the following substitutions:

$$\left.\begin{aligned} \mathcal{T} &= \frac{T}{\hat{R}}, & \mathcal{B} &= \frac{\delta_B\, B}{\gamma}, & \mathcal{P}_c &= \frac{\delta_{P_c} P_c}{\gamma} \\ \mathcal{I}_g &= \frac{\delta_{I_g} \delta_{P_c} I_g}{a_2\, \gamma}, & \upbeta &= \frac{\beta}{\beta_0}, & \mathcal{P} &= \frac{\delta_P\, P}{R\, \kappa\, \hat{R}\, \beta_0} \end{aligned}\right\}, \tag{9.40}$$

where β_0 is the initial number of beta cells right before the autoimmune attack (with the ratio β/β_0 denoted by a non-italicized variable $\upbeta$) and

$$\hat{R} = \frac{(\sqrt{\alpha} - \sqrt{\delta_T})^2}{\epsilon}.$$

For simplicity, we assume that $s = \mu = \delta_\beta = 0$ given their experimental insignificance and/or the little effect they exert on the dynamics [32]. By applying the substitutions introduced in equation (9.40) to the model (9.39a)–(9.39f) and using non-italicized font for the newly generated parametric quantities hereafter, we obtain

$$\frac{d\mathcal{T}}{dt} = \upsigma \frac{\mathcal{P}}{\mathcal{P} + \mathrm{k}} + \alpha\, \mathcal{T} \frac{\mathcal{P}}{\mathcal{P} + \mathrm{k}} - (\sqrt{\alpha} - \sqrt{\delta_T})^2\, \mathcal{T}^2 - \delta_T\, \mathcal{T} \tag{9.41a}$$

$$\frac{d\mathcal{B}}{dt} = \delta_B + \left[(\eta_1 - \upeta_2 \mathcal{T}) \frac{\mathcal{P}}{\mathcal{P} + \mathrm{k}} - \delta_B \right] \mathcal{B} \tag{9.41b}$$

$$\frac{d\mathcal{P}_c}{dt} = \delta_{P_c} \left(\frac{\upeta_2 \mathcal{T}}{\delta_B} \frac{\mathcal{P}}{\mathcal{P} + \mathrm{k}} \mathcal{B} - \mathcal{P}_c \right) \tag{9.41c}$$

$$\frac{d\mathcal{I}_g}{dt} = \delta_{I_g} (\mathrm{a}\, \mathcal{B} + \mathcal{P}_c - \mathcal{I}_g) \tag{9.41d}$$

$$\frac{d\upbeta}{dt} = -\kappa\, \hat{R}\, \upbeta\, \mathcal{T} \tag{9.41e}$$

$$\frac{d\mathcal{P}}{dt} = \delta_P\, (\upbeta\, \mathcal{T} - \mathcal{P}), \tag{9.41f}$$

where the new parametric quantities are given by

$$\upsigma = \frac{\sigma}{\hat{R}}, \quad \mathrm{k} = \frac{\delta_P\, k}{R\, \kappa\, \hat{R} \beta_0}, \quad \upeta_2 = \eta_2 \hat{R}, \quad \mathrm{a} = \frac{a_1\, \delta_{P_c}}{a_2\, \delta_B}.$$

Thus the number of parameters that need to be estimated have been reduced by introducing these parametric quantities. Note that once again, we did not rescale the time variable t for clarity.

9.4.3 Phase-plane analysis

In this section, we will study the global dynamics of the one clone model (9.33a)–(9.33f) using steady state and stability analysis. We will show that the behaviour of this model is similar to that observed in the one clone model of [19]. In the following, we use a reductionist approach to determine these steady states and analyze their stability properties.

We begin first by noting that due to homeostatic mechanisms, beta-cell loss happens at a very slow time scale relative to T-cell dynamics and pMHC processing. Therefore, we may assume that the scaled variable β is roughly a constant. Furthermore, since $\mathcal{T}$ and $\mathcal{P}$ are decoupled from β, $\mathcal{P}_c$ and $\mathcal{I}_g$, one can ignore the differential equations of these latter variables and only focus on the two-variable model, given by

$$\frac{d\mathcal{T}}{dt} = F(\mathcal{T}, \mathcal{P}) := \sigma \frac{\mathcal{P}}{\mathcal{P} + \mathrm{k}} + \alpha\, \mathcal{T} \frac{\mathcal{P}}{\mathcal{P} + \mathrm{k}} - \left(\sqrt{\alpha} - \sqrt{\delta_T}\right)^2 \mathcal{T}^2 - \delta_T\, \mathcal{T} \tag{9.42a}$$

$$\frac{d\mathcal{P}}{dt} = G(\mathcal{T}, \mathcal{P}) := \delta_P\, (\beta \mathcal{T} - \mathcal{P}), \tag{9.42b}$$

where β is assumed to be a constant (≈ 1 at the start and during the initial phase of the autoimmune attack). The dynamic behaviour of this subsystem (9.42a)–(9.42b), referred to as the reduced one-clone model hereafter, can be investigated using the phase portrait of figure 9.11 showing the model's $\mathcal{T}$- (black solid line) and $\mathcal{P}$- (gray-solid line) nullclines (when $\beta = 1$). For any $\beta \leqslant 1$, these nullclines are given by

$$\mathcal{T}\text{-nullclines:}\quad \mathcal{T}' = 0 \implies \begin{cases} \mathcal{T} = 0 \;\&\; \mathcal{P} = \dfrac{\mathrm{k}\Psi(\mathcal{T}; 0)}{1 - \Psi(\mathcal{T}; 0)}, & \text{when } \sigma = 0 \\ \mathcal{P} = \dfrac{\mathrm{k}\Psi(\mathcal{T}; \sigma)}{1 - \Psi(\mathcal{T}; \sigma)}, & \text{when } \sigma \neq 0, \end{cases}$$

where

$$\Psi(\mathcal{T}; \sigma) = \frac{\delta_T \mathcal{T} + \left(\sqrt{\alpha} - \sqrt{\delta_T}\right)^2 \mathcal{T}^2}{\sigma + \alpha\, \mathcal{T}}, \quad \Psi(\mathcal{T}; 0) = \frac{1}{\alpha}\left[\delta_T + \left(\sqrt{\alpha} - \sqrt{\delta_T}\right)^2 \mathcal{T}\right],$$

and

$$\mathcal{P}\text{-nullcline:}\qquad \mathcal{P}' = 0 \qquad \implies \qquad \mathcal{P} = \beta \mathcal{T} \quad (\iff \mathcal{T} = \mathcal{P}/\beta),$$

respectively. They intersect at the steady states of the reduced model and subdivide the positive quadrant of the $\mathcal{P}$, $\mathcal{T}$-plane into regions possessing different signs for $\mathcal{P}'$ and $\mathcal{T}'$. Although the number of steady states depends on the value of $\sigma \geqslant 0$, the origin $\boldsymbol{E}_0 = (0, 0)$ always remains a steady state for the model, representing the disease-free (or healthy) state. The existence of other steady states (lying at the intersection of the sigmoidal $\mathcal{T}$-nullcline and $\mathcal{P}$-nullcline), on the other hand, depends on whether $\sigma = 0$ or > 0. They can be determined by solving the two equations

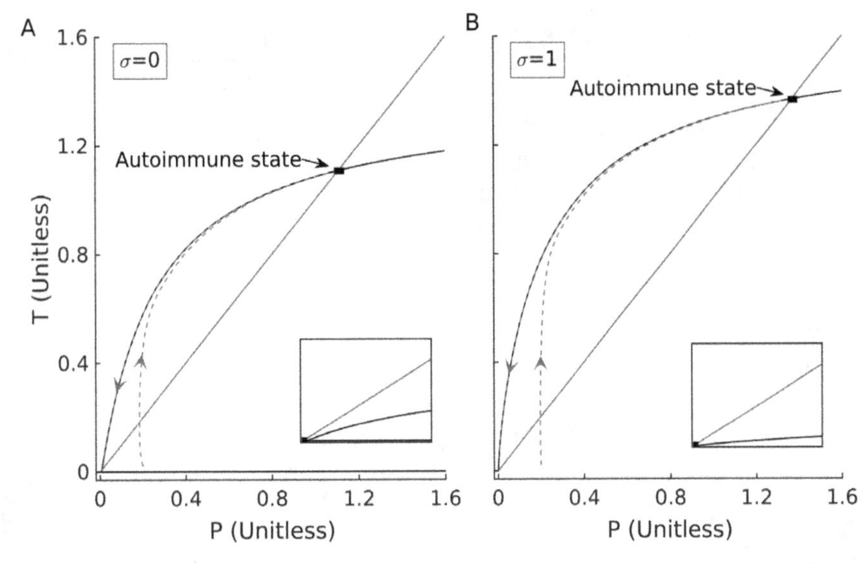

Figure 9.11. Phase portrait of the reduced (scaled) one-clone model (9.42a)–(9.42b) displaying the $\mathcal{T}$- (solid black) and $\mathcal{P}$- (solid gray) nullclines when (A) $\sigma = 0$ and (B) $\sigma = 1$ ($\ell > 0$). The $\mathcal{P}$-axis in (A) is also a $\mathcal{T}$-nullcline. Solution trajectory of the full model (9.41a)–(9.41f) (dashed line) transiently approaches the elevated autoimmune state (black rectangle) and follows it along the $\mathcal{T}$-nullcline, during beta-cell loss, until it disappears at either a saddle node bifurcation point (A) or at a transcritical bifurcation point (B) (see figure 9.12), at which it then converges to the 'healthy state' $\boldsymbol{E}_0 = (0, 0)$. Insets: increasing the value of k to $k = 2$ in (A) and $k = 20$ in (B) shifts the 'sigmoidal' $\mathcal{T}$-nullcline rightward, making it lie entirely below the $\mathcal{P}$-nullcline. In both cases, the healthy state $\boldsymbol{E}_0 = (0, 0)$ becomes the only steady state of the model.

$$\mathcal{P}^* = \beta\mathcal{T}^*, \tag{9.43}$$

and

$$\sigma\frac{1}{\mathcal{T}^* + \mathrm{k}/\beta} + \alpha\frac{\mathcal{T}^*}{\mathcal{T}^* + \mathrm{k}/\beta} = \delta_T + \left(\sqrt{\alpha} - \sqrt{\delta_T}\right)^2\mathcal{T}^*, \tag{9.44}$$

where $\mathcal{T}^* \neq 0$ and β is a constant (due to slow beta-cell loss). Figure 9.11(A) shows that when these two equations are satisfied at $\sigma = 0$, two additional interior equilibria $\boldsymbol{E}_1^{(A)} = (\mathcal{T}_1^{(A)}, \mathcal{P}_1^{(A)})$ and $\boldsymbol{E}_2^{(A)} = (\mathcal{T}_2^{(A)}, \mathcal{P}_2^{(A)})$ may also exist, with the latter representing the autoimmune state. For $\sigma = 1$, on the other hand, the two equations produce exactly one interior equilibria $\boldsymbol{E}^{(B)} = (\mathcal{T}^{(B)}, \mathcal{P}^{(B)})$ representing the autoimmune state (see figure 9.11(B)).

Generally speaking, increasing the value of k shifts the sigmoidal $\mathcal{T}$-nullcline rightward, causing the number of intersections between the nullclines to decrease and the autoimmune state to eventually disappear (as shown in the insets of figure 9.11). The slow decrease in the beta-cell number during autoimmunity can produce such a rightward shift through the ratio k/β. To analyze how this may happen, we need to first solve for $\mathcal{T}^*$ in equation (9.44), as follows

$$\mathcal{T}^* = \frac{(\alpha - \delta_T - \ell^2\mathrm{k}/\beta) \pm \sqrt{(\alpha - \delta_T - \ell^2\mathrm{k}/\beta)^2 + 4(\sigma - \delta_T\mathrm{k}/\beta)\ell^2}}{2\ell^2}, \tag{9.45}$$

where $\ell = \sqrt{\alpha} - \sqrt{\delta_T}$ is taken to be positive by assuming $\alpha > \delta_T$ (a physiologically reasonable assumption). Based on this and on the value of the discriminant, we conclude that there are two important cases to consider

(A) The following two inequalities

$$0 \leqslant \sigma < \frac{\delta_T \mathrm{k}}{\beta}, \tag{9.46}$$

and

$$\alpha - \delta_T - \frac{\ell^2\mathrm{k}}{\beta} > 0 \quad \Longleftrightarrow \mathrm{k} < \beta\frac{\sqrt{\alpha} + \sqrt{\delta_T}}{\ell}, \tag{9.47}$$

are satisfied.

(B) The following inequality

$$\sigma \geqslant \frac{\delta_T \mathrm{k}}{\beta},$$

is satisfied.

We will show below that in the former case, we obtain at most three steady states (see figure 9.11(A) when $\sigma = 0$), and in the latter, we obtain exactly two steady states (see figure 9.11(B) when $\sigma = 1$). In both cases, the healthy state is always present.

In case (A), the number of (nontrivial) interior steady states depends on the sign of the discriminant in equation (9.45). By solving for k in the quadratic equation

$$\ell^4\mathrm{k}^2 - 2\ell^2(\alpha + \delta_T)\mathrm{k} + (\alpha - \delta_T)^2 + 4\sigma\ell^2 = 0,$$

that represents the discriminant, we obtain the following roots for k

$$\begin{aligned} \mathrm{k}_c^{(A)} &= \beta\frac{(\alpha + \delta_T) \pm 2\sqrt{\alpha\delta_T - \sigma\ell^2}}{\ell^2} \\ &= \begin{cases} \beta\dfrac{(\alpha + \delta_T) + 2\sqrt{\alpha\delta_T - \sigma\ell^2}}{\ell^2} & =: \mathrm{k}_{c1}^{(A)} \\ \beta\dfrac{(\alpha + \delta_T) - 2\sqrt{\alpha\delta_T - \sigma\ell^2}}{\ell^2} & =: \mathrm{k}_{c2}^{(A)}. \end{cases} \end{aligned} \tag{9.48}$$

For these two roots to exist, the following inequality must hold

$$\alpha\delta_T - \sigma\ell^2 \geqslant 0. \tag{9.49}$$

If equation (9.49) is satisfied, then based on equation (9.45), the first root $\mathrm{k} = \mathrm{k}_{c1}^{(A)}$ is not physiologically relevant because

$$\mathcal{T}^*(k_{c1}^{(A)}) = -\frac{\delta_T + \sqrt{\alpha\delta_T - \sigma\ell^2}}{\ell^2} < 0.$$

Using the second root $k = k_{c2}^{(A)}$, on the other hand, we obtain

$$\mathcal{T}^*(k_{c2}) = \frac{-\delta_T + \sqrt{\alpha\delta_T - \sigma\ell^2}}{\ell^2}.$$

Thus, for $k = k_{c2}^{(A)}$ to be physiologically relevant, we must have $\ell > 0$ (an assumption we already made) and

$$-\delta_T + \sqrt{\alpha\delta_T - \sigma\ell^2} > 0. \tag{9.50}$$

Notice that if inequality (9.50) holds, then inequality (9.49) also holds. Inequalities (9.46), (9.47) and (9.50) thus imply that $\mathcal{T}^*(k_{c2}) > 0$ provided that

$$\sigma < \min\left\{\delta_T k/\beta,\ \delta_T\frac{\sqrt{\alpha} + \sqrt{\delta_T}}{\ell}\right\} = \delta_T k/\beta.$$

The above discussion reveals that the value of k, relative to $k_c^{(A)}$, determines the number of intersections between the $\mathcal{T}$- and $\mathcal{P}$-nullcines. Figure 9.11(A) shows that there could be one intersection at the healthy state $\boldsymbol{E}_0 = (0, 0)$ when $k > k_{c2}$ (inset), two intersections at $k = k_{c2}$ when $\mathcal{P}$-nullcline is tangential to the sigmoidal $\mathcal{T}$-nullcline, and three intersections at $\boldsymbol{E}_0$ along with two additional interior steady states $\boldsymbol{E}_1^{(A)}$ and $\boldsymbol{E}_2^{(A)}$ (where $\boldsymbol{E}_2^{(A)}$ represents the autoimmune state) for $k < k_{c2}$. Two of these three steady states $\boldsymbol{E}_0$ and $\boldsymbol{E}_2^{(A)}$, in the latter case, are stable nodes and the third one $\boldsymbol{E}_1^{(A)}$ is a saddle whose stable manifold is the separatrix that demarcates the basin of attraction of $\boldsymbol{E}_0$ from that of $\boldsymbol{E}_2^{(A)}$. In other words, $k = k_{c2}$ represents the critical value, or a threshold, for k at which the two steady states $\boldsymbol{E}_1^{(A)}$ and $\boldsymbol{E}_2^{(A)}$ merge at a saddle-node bifurcation point (see figure 9.12(A)).

When $\sigma > \delta_T k/\beta$, on the other hand, we have, based on equation (9.45), one negative and physiologically irrelevant steady state, and another positive and physiologically relevant steady state

$$\mathcal{T}^* = \frac{(\alpha - \delta_T - \ell^2 k/\beta) + \sqrt{(\alpha - \delta_T - \ell^2 k/\beta)^2 + 4(\sigma - \delta_T k/\beta)\,\ell^2}}{2\ell^2}.$$

In other words, according to this set up, there are two steady states that are always present, a healthy state $\boldsymbol{E}_0 = (0, 0)$ and an autoimmune state $\boldsymbol{E}^{(B)}$ (see figure 9.11). Interestingly, if we choose

$$\sigma > \delta_T\frac{\sqrt{\alpha} + \sqrt{\delta_T}}{\ell},$$

then increasing the value of k beyond the critical threshold $k_c^{(B)} = \sigma\beta/\delta_T$ causes the two steady states $\boldsymbol{E}_0$ and $\boldsymbol{E}^{(B)}$ to exchange stability at a transcritical bifurcation point (see figure 9.12(B)).

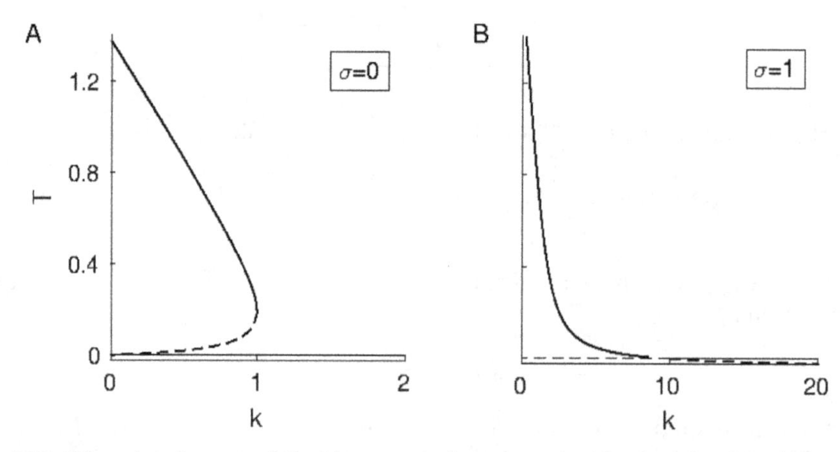

Figure 9.12. Bifurcation diagrams of $\mathcal{T}$ with respect to k, as determined by the reduced (scaled) one-clone model (9.42a)–(9.42b) for (A) $\sigma = 0$ and (B) $\sigma = 1$ ($\ell > 0$). Stable (unstable) steady states are plotted as solid (dashed) lines. Notice that in (A), the elevated (autoimmune) steady state $\boldsymbol{E}_2^{(A)}$ merges with the saddle $\boldsymbol{E}_1^{(A)}$ at a saddle node bifurcation point and eventually both disappear at higher values of k, whereas in (B), the autoimmune and healthy states, $\boldsymbol{E}_2^{(B)}$ and $\boldsymbol{E}_0$, respectively, collide at a transcritical bifurcation point and exchange stability.

Because the slope of the $\mathcal{P}$-nullcine ($\mathcal{T} = \mathcal{P}/\beta$), given by $1/\beta$, is slowly increasing during the autoimmune attack (due to beta-cell loss), solution trajectories of the full model (9.41a)–(9.41f), that initially converge to the elevated auotimmune state $\boldsymbol{E}_2^{(A)}$, will do so transiently (see dashed line in figure 9.11(A) and (B)) by following it along the $\mathcal{T}$-nullcline during the increase of the slope of the $\mathcal{P}$-nullcine until $\boldsymbol{E}_2^{(A)}$ disappears at a saddle node bifurcation point (as in panel (A)) or at a transcritical bifurcation point (as in panel (B)). In both cases, solution trajectories eventually converge to the only remaining attractor, $\boldsymbol{E}_0$, (that is no longer a 'healthy' state because it is preceded by a major autoimmune attack regardless of topology), but they do so abruptly in the former and gradually in the latter (see the bifurcation diagrams in figure 9.12). We observe such outcomes because in the saddle node case (panel (A)), the system is bistable for $k < k_{c2}^{(A)}$ and monostable for $k > k_{c2}^{(A)}$, whereas in the transcritical case (panel (B)), the system is monostable over the entire range of k.

These results illustrate the overall flow pattern of solution trajectories of the full model superimposed on the dynamics of the reduced model and show how one T-cell cycle (also called a wave in [53]) can be produced due to this transient stability of the autoimmune state. This type of behaviour is reminiscent of the remission-relapse phenomenon suggested by [60] about T1D. Other (auto)immune responses (e.g. multiple sclerosis and some infectious diseases), triggered by a transient rise in the expression level of (auto)antigen over time, would be also expected to have similar underlying dynamics. In the next section, we will verify some of these results analytically by applying linear stability analysis.

9.4.4 Stability analysis

It was suggested in [61] that the inclusion of thymus input to a T-cell model does not significantly alter its long term behaviour. In fact, we have shown in the previous section that the dynamics of the reduced (scaled) one-clone model (9.42a)–(9.42b) is minimally affected when varying σ within the range $[0, \delta_T \mathrm{k}/\beta)$. We therefore restrict hereafter the analysis of the stability properties of all possible steady states of this scaled model to $\sigma = 0$. This is done by first evaluating the Jacobian matrix of the model at the steady states with $\beta = 1$, as follows

$$J(\mathcal{T}, \mathcal{P}) = \begin{pmatrix} \frac{\partial F(\mathcal{T}, \mathcal{P})}{\partial \mathcal{T}} & \frac{\partial F(\mathcal{T}, \mathcal{P})}{\partial \mathcal{P}} \\ \frac{\partial G(\mathcal{T}, \mathcal{P})}{\partial \mathcal{T}} & \frac{\partial G(\mathcal{T}, \mathcal{P})}{\partial \mathcal{P}} \end{pmatrix} = \begin{pmatrix} \frac{\alpha \mathcal{P}}{\mathcal{P} + \mathrm{k}} - \delta_T - 2\left(\sqrt{\alpha} - \sqrt{\delta_T}\right)^2 \mathcal{T} & \frac{\alpha \mathcal{T}}{\mathcal{P} + \mathrm{k}} - \frac{\alpha \mathcal{T}\mathcal{P}}{(\mathcal{P} + \mathrm{k})^2} \\ \delta_P & -\delta_P \end{pmatrix}. \tag{9.51}$$

The eigenvalues of the Jacobian matrix at the healthy state $\boldsymbol{E}_0 = (0, 0)$ are both negative ($\lambda_1 = -\delta_T$ and $\lambda_2 = -\delta_P$), which means that $\boldsymbol{E}_0$ is always a stable node. The other two steady states $\boldsymbol{E}_1^{(A)}$ and the autoimmune state $\boldsymbol{E}_2^{(A)}$, on the other hand, will only exist if $\mathrm{k} < \mathrm{k}_{c2}^{(A)} = 1$ and $\ell > 0$. Assuming that these two conditions are satisfied and noting that the trace of the Jacobian matrix is always negative, it suffices to show that $\det(J(\boldsymbol{E}_1^{(A)})) < 0$ and $\det(J(\boldsymbol{E}_2^{(A)})) > 0$ to prove that $\boldsymbol{E}_1^{(A)}$ is a saddle and $\boldsymbol{E}_2^{(A)}$ is a stable node [62]. To achieve this goal, we will evaluate the determinant of the Jacobian matrix and determine when it changes sign. Since the non-trivial steady states of the reduced (scaled) one-clone model (9.42a)–(9.42b) must satisfy equations (9.43) and (9.44) (when $\sigma = 0$), then according to equation (9.51), we have

$$\begin{aligned} \det(J(\mathcal{T}, \mathcal{P})) &= \delta_P\left(\sqrt{\alpha} - \sqrt{\delta_T}\right)^2 \mathcal{T} - \delta_P\left(\frac{\alpha\,\mathcal{T}}{\mathcal{P} + \mathrm{k}} - \frac{\alpha\,\mathcal{T}\mathcal{P}}{(\mathcal{P} + \mathrm{k})^2}\right) \\ &= \delta_P\left(-\delta_T + \frac{\alpha\,\mathcal{T}^2}{(\mathcal{T} + \mathrm{k})^2}\right) \\ &= \delta_P\left(-\delta_T + \frac{\left(\delta_T + \left(\sqrt{\alpha} - \sqrt{\delta_T}\right)^2 \mathcal{T}\right)^2}{\alpha}\right). \end{aligned}$$

Thus $\det(J(\mathcal{T}, \mathcal{P})) = 0$, when $\mathcal{T} = \bar{\mathcal{T}} = \sqrt{\delta_T}/(\sqrt{\alpha} - \sqrt{\delta_T}) = \sqrt{\delta_T}/\ell$, which is identical to the value of $\mathcal{T}^*(\mathrm{k}_c = 1)$. This means that if $\mathcal{T} > \bar{\mathcal{T}}$, as is the case for the autoimmune state $\boldsymbol{E}_2^{(A)}$, then $\det(J(\mathcal{T}, \mathcal{P})) > 0$ and $\boldsymbol{E}_2^{(A)}$ is a stable node, whereas if $\mathcal{T} < \bar{\mathcal{T}}$, as is the case for $\boldsymbol{E}_1^{(A)}$, then $\det(J(\mathcal{T}, \mathcal{P})) < 0$ and $\boldsymbol{E}_1^{(A)}$ is a saddle.

At $\mathcal{T} = \bar{\mathcal{T}} = \mathcal{T}^*(k_c = 1)$, we have a saddle-node bifurcation as demonstrated in figure 9.12(A).

These latter stability results illustrate the local behavior of the reduced (scaled) one-clone model (9.42a)–(9.42b) near its steady states, when $\sigma = 0$. It remains to be checked whether these results are globally valid. In the following analysis, we show that they are indeed global. This is done by establishing first that the solutions of the reduced (scaled) one-clone model, whose initial conditions belong to the space $\mathbb{R}^+ \times \mathbb{R}^{\geqslant 0}$, are bounded and nonnegative, and by taking advantage of Dulac's criterion to rule out the existence of closed orbits in that space.

Result 4.1. Let $(\mathcal{T}(t), \mathcal{P}(t))$ be a solution to the (decoupled) reduced (scaled) one-clone model (9.42a)–(9.42b) satisfying $\sigma = 0$ and $\ell > 0$, with initial condition $(\mathcal{T}(t_0), \mathcal{P}(t_0)) \in \mathbb{R}^+ \times \mathbb{R}^{\geqslant 0}$. Then $(\mathcal{T}(t), \mathcal{P}(t))$ is bounded and nonnegative in the state-space $\mathbb{R}^+ \times \mathbb{R}^{\geqslant 0}$. [The result also holds for $\sigma \neq 0$.]

This result can be established by applying comparisons method [63]. First, notice that if $\mathcal{P}(t^*) = 0$ or $\mathcal{T}(t^*) = 0$, for some $t^* \geqslant t_0$, where $(\mathcal{T}(t_0), \mathcal{P}(t_0)) \in \mathbb{R}^+ \times \mathbb{R}^{\geqslant 0}$, then, according to equations (9.42a) and (9.42b), we have $\dot{\mathcal{P}}(t^*) \geqslant 0$ or $\dot{\mathcal{T}}(t^*) = 0$, respectively. This implies that solutions to system (9.42a)–(9.42b), with $\sigma = 0$ and $\ell > 0$, starting from the first quadrant will never leave the quadrant. In other words, $\mathcal{T}(t)$ and $\mathcal{P}(t)$ are both bounded below by zero.

Since $0 \leqslant \mathcal{P}/(\mathcal{P} + k) \leqslant 1$, we conclude that

$$\frac{d\mathcal{T}}{dt} < \alpha\, \mathcal{T} - \left(\sqrt{\alpha} - \sqrt{\delta_T}\right)^2 \mathcal{T}^2 - \delta_T\, \mathcal{T}.$$

This implies that $\mathcal{T}$ must be bounded by the solutions of the equation

$$\frac{d\tilde{\mathcal{T}}}{dt} = (\alpha - \delta_T)\tilde{\mathcal{T}} - \left(\sqrt{\alpha} - \sqrt{\delta_T}\right)^2 \tilde{\mathcal{T}}^2. \tag{9.52}$$

There are two steady states to equation (9.52): $\tilde{\mathcal{T}} = 0$ which is unstable, and $\tilde{\mathcal{T}} = (\sqrt{\alpha} + \sqrt{\beta}/\ell)$ which is stable. This implies that

$$\tilde{\mathcal{T}} \leqslant C_1 := \max\left(\tilde{\mathcal{T}}(0),\ \frac{\sqrt{\alpha} + \sqrt{\delta_T}}{\ell}\right),$$

where $\tilde{\mathcal{T}}(0) \geqslant 0$. It follows that $\mathcal{T} < C_1$. Substituting this upper bound into equation (9.42b), we get

$$\frac{d\mathcal{P}}{dt} < \delta_P\,(C_1 - \mathcal{P}).$$

Applying the same approach used above, we conclude that $\mathcal{P}$ is also bounded by the solutions of the equation

$$\frac{d\tilde{\mathcal{P}}}{dt} = \delta_P \left(C_1 - \tilde{\mathcal{P}}\right),$$

where

$$\tilde{\mathcal{P}} \leqslant C_2 := \max\left(\tilde{\mathcal{P}}(0),\ C_1 + C_2\right),$$

and $\tilde{\mathcal{P}}(0) \geqslant 0$. In other words, $\mathcal{P} < C_2$. This thus shows that $\mathcal{T}(t)$ and $\mathcal{P}(t)$ are both bounded above.

To show that there are no periodic orbits associated with this model, we will use Dulac's criterion [64].

Result 4.2. Let $\dot{\mathbf{x}} = \mathbf{f}(\mathbf{x})$. If $\mathbf{f}(\mathbf{x})$ is a continuously differentiable vector field defined on a simply connected subset S of the plane and if there exists a continuously differentiable real-valued function $g(\mathbf{x})$ such that $\nabla \cdot (g\mathbf{f})$ has one sign throughout S (where $g(\mathbf{x}) = g(\mathcal{T}, \mathcal{P})$ and $\nabla \cdot (g\mathbf{f}) = (\partial/\partial\mathcal{T}, \partial/\partial\mathcal{P}) \cdot (gf_1, gf_2)$), then there are no closed orbits lying entirely in S.

By choosing the continuously differentiable real-valued function $g(\mathcal{T}, \mathcal{P}) = 1/\mathcal{T}$ in $S = \mathbb{R}^+ \times \mathbb{R}^{\geqslant 0}$, and applying Dulac's criterion on $\dot{\mathbf{x}} = (\dot{\mathcal{T}}, \dot{\mathcal{P}})$ that satisfies the reduced (scaled) one-clone model (9.42a)–(9.42b) (with $\sigma = 0$ and $\ell > 0$), we obtain

$$\begin{aligned}
\nabla \cdot (g\dot{\mathbf{x}}) &= \left(\frac{\partial}{\partial\mathcal{T}}, \frac{\partial}{\partial\mathcal{P}}\right) \cdot \left(\frac{1}{\mathcal{T}}\left(\dot{\mathcal{T}}, \dot{\mathcal{P}}\right)\right) \\
&= \left(\frac{\partial}{\partial\mathcal{T}}, \frac{\partial}{\partial\mathcal{P}}\right) \cdot \left[\frac{1}{\mathcal{T}}\left(\alpha\,\mathcal{T}\frac{\mathcal{P}}{\mathcal{P} + \mathrm{k}} - \left(\sqrt{\alpha} - \sqrt{\delta_T}\right)^2 \mathcal{T}^2 - \delta_T\,\mathcal{T}\right), \frac{\delta_P}{\mathcal{T}}(\mathcal{T} - \mathcal{P})\right] \\
&= -\left(\sqrt{\alpha} - \sqrt{\delta_T}\right)^2 - \frac{\delta_P}{\mathcal{T}} < 0.
\end{aligned}$$

The control condition $\nabla \cdot (g\dot{\mathbf{x}})$ is always negative for $\mathcal{T} > 0$ and $\mathcal{P} \geqslant 0$. As a result, there are no closed orbits lying entirely in $\mathbb{R}^+ \times \mathbb{R}^{\geqslant 0}$. Furthermore, we already know from the previous result that solutions starting from $\mathbb{R}^+ \times \mathbb{R}^{\geqslant 0}$ will never leave this quadrant. Thus no periodic orbits would lie partially within $\mathbb{R}^+ \times \mathbb{R}^{\geqslant 0}$. These conclusions guarantee that the local stability properties of the steady states $\boldsymbol{E}_i^*$, $i = 0, 1, 2$, are also global (a similar result can be obtained for $\sigma \neq 0$). In this case, when $\mathrm{k} < \mathrm{k}_{c2}^{(A)} = 1$, we have a bistable system with the stable manifold of the saddle point $\boldsymbol{E}_1^{(A)}$ acting as a separatrix that divides the two basins of attraction of $\boldsymbol{E}_0$ and $\boldsymbol{E}_2^{(A)}$. This configuration remains almost the same when $\mathrm{k} = \mathrm{k}_{c2}^{(A)} = 1$. Here the separatrix becomes the boundary between the basin of attraction of $\boldsymbol{E}_0$ and the basin of attraction of $\boldsymbol{E}_1^{(A)} = \boldsymbol{E}_2^{(A)}$, a half-stable steady state.

The question that remains is how do these three steady states $\boldsymbol{E}_0$ and $\boldsymbol{E}_i^{(A)}$, $i = 1, 2$, manifest themselves in the full (scaled) model (9.41a)–(9.41f). Since $\dot{\beta} < 0$ for $\mathcal{T} > 0$ in equation (9.41e), we expect $\beta(t)$ to be an exponentially decaying

Table 9.2. Parameter values of the (scaled) one-clone model (9.41a)–(9.41f).

Parameter	Description	Value	Range	Ref.
σ	Influx rate of naïve T cells from thymus	≈ 0 day^{-1}	—	[32]
α	Expansion rate of T cells	4 day^{-1}	[2 – 20]	[19, 32, 61, 65, 66]
δ_T	Turnover rate of T cells	0.1 day^{-1}	[0.01–0.3]	[19, 32, 61, 65, 66]
k	pMHC-expression level for 50% maximum activation level of T cells	0.26	[0-1]	[19, 57, 67]
ϵ	Competition parameter for T cells	5×10^{-6} (day cell)$^{-1}$	—	[19, 32, 65]
δ_B	Turnover rate of B cells	0.02 day^{-1}	[0.017–0.02]	[19, 68]
η_1	Expansion rate of B cells	5.67×10^{-6} day^{-1}	—	[68]
η_2	Maturation rate of B cells	2.858 day^{-1}	—	[68]
δ_{P_c}	Turnover rate of plasma cells	0.2 day^{-1}	[0.116–0.23]	[19, 68]
δ_{I_g}	Degradation rate of autoantibodies	0.034 day^{-1}	[0.001–0.034]	[68]
a	Ratio of B-to-plasma autoantibody-release	0.1	—	[68]
κ	T-cell killing efficacy	7×10^{-10} (day cell)$^{-1}$	[10^{-11} – 10^{-7}]	[19, 32, 57]
$\hat{R}$	Ratio of T-cell net growth rate to competition factor	$5.670\,2 \times 10^5$ cells	—	[19, 32, 57]
δ_P	Degradation rate of autoantigen	0.1 day^{-1}	—	[19, 32, 59, 61, 65]

function of time, and the steady states of the full (scaled) model with elevated levels of T cells, labeled $\boldsymbol{S}_1$ and $\boldsymbol{S}_2$, to be transiently present when $\mathrm{k} < \mathrm{k}_c^{(A)} = 1$ and $\sigma = 0$. In other words, solution trajectories that start from the basin of attraction of $\boldsymbol{S}_2$ will initially approach this steady state, but eventually turn towards $\boldsymbol{S}_0 = (0, 1, 0, \mathrm{a}, \beta_\infty, 0)$, where $\beta_\infty \sim 0.3$ is the steady state level of β corresponding to a 30% beta-cell survival, when the slope of the hyperplane $\mathcal{T} = \mathcal{P}/\beta$ becomes large enough that the two steady states $\boldsymbol{S}_1^{(A)}$ and $\boldsymbol{S}_2^{(A)}$ merge and disappear (i.e. k becomes larger than $\mathrm{k}_c^{(A)} = 1$). This type of behaviour is illustrated in figure 9.11(A) showing a dashed line representing one solution trajectory superimposed on the phase-plane of the reduced (scaled) one clone model.

9.4.5 Model simulations

To illustrate the dynamics of the full (scaled) one-clone model in the presence of all of its components, we simulate system (9.41a)–(9.41f) in response to variations to two key parameters: the reciprocal of T-cell avidity k and T-cell killing efficacy κ (other parameter values are available in table 9.2). In these simulations, the scaled variables are used as representatives to the original variables of system (9.39a)–(9.39f). Furthermore, because we plan to ignore the effects of the thymus input, we set $\sigma = 0$ and assume that the initial scaled level of T cells, $\mathcal{T}$, is nonzero but $\ll 1$. One important aspect of the full (scaled) one-clone model is that the T- and B-cell avidity (represented by the parameter k^{-1}) are correlated through the activation term $\mathcal{F}_1$, defined by equation (9.34), appearing in both the T- and B-cell equations. We generalize this concept by assuming that this activation term is a Hill function

$$\mathcal{F}(\mathcal{P}) = \frac{\mathcal{P}^n}{\mathcal{P}^n + k^n}, \tag{9.53}$$

with a Hill coefficient $n = 1, 2, 3$. That will allow us to uncover not only how the correlation between T- and B-cell avidities may affect the dynamics, but also how the steepness of this curve alter outcomes.

Figure 9.13(A1) shows the heat-map of the (scaled) level of beta cells (β) at steady state (i.e. after 30 years of the autoimmune attack) when k and κ are varied within the ranges $[10^{-4}, 1]$ (unitless) and $[10^{-11}, 10^{-7}]$ (day cell)$^{-1}$, respectively. We see that the magnitude of beta-cell loss (shown as a gradual change in color from red to blue) increases steadily by increasing T-cell avidity and/or its killing efficacy in a manner identical to that seen in [19]. The presence of the red band on top of each panel demonstrates that if T-cell avidity is too small, then beta cells are safe from T-cell

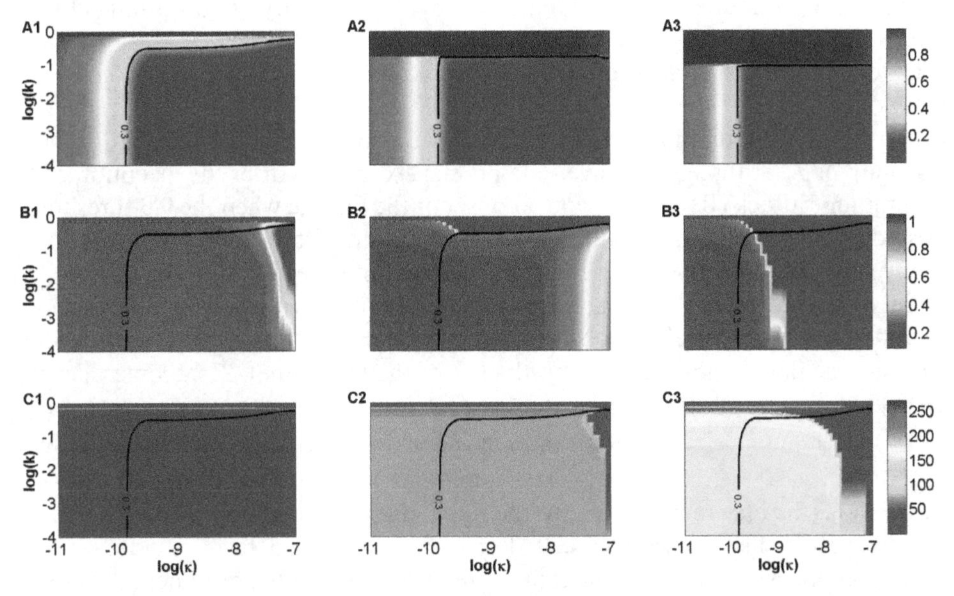

Figure 9.13. Heat-maps showing the response of the full (scaled) one clone model (9.41a)–(9.41f) to variations in the reciprocal of T-cell avidity (k in log-scale) and killing efficacy (κ in log scale) within the ranges $[10^{-4}, 1]$ and $[10^{-11}, 10^{-7}]$ (day cell)$^{-1}$, respectively. (A1–A3) The steady state level of beta cells (β) for $n = 1$ (A1), $n = 2$ (A2) and $n = 3$ (A3), where n is the Hill coefficient in equation (9.53), showing the magnitude of beta-cell loss in each case. (B1–B3) The autoantibody level $\mathcal{I}_g$ after six months of the autoimmune attack (B1), at the onset of the disease (for those that develop it) (B2) and at steady state (B3). (C1–C3) The time period between the start of the autoimmune attack and when $\mathcal{I}_g$ reaches the following detectability levels: $\mathcal{I}_g = 0.15$ (C1), $\mathcal{I}_g = 0.55$ (C2) and $\mathcal{I}_g = 0.95$ (C3). The black line in each panel represents the 30% threshold of surviving beta cells (0.3 critical threshold). The color-coding in each panel is quantified by the color-bars on the right.

destruction regardless of the level of T-cell killing efficacy κ. Increasing the Hill-coefficient to $n = 2$, in panel (A2), or $n = 3$, in panel (A3), does not alter the heat-map of β significantly. The only noticeable difference is the increase in the width of the red band as n increases. The increase in the steepness of the Hill function in equation (9.53), induced by larger n, means higher pMHC expression level on APCs is required for T-cell activation and thus wider red bands. The sudden change from red to blue when moving vertically downward across these red bands, however, is caused by the presence of two stable nodes whose basins of attraction are separated by the stable manifold of the saddle point (the separatrix discussed in the previous section) which generates the boundary of these red bands.

We also observe red bands on the left side of panels (A1–A3) that are less dark than those on top. These red bands demonstrate that most beta cells survive the autoimmune attack whenever T-cell killing efficacy is very small (i.e. when $\kappa \in [10^{-11}, 5 \times 10^{-10}]$ (day cell)$^{-1}$). For larger values of κ, on the other hand, a more significant decline in β is detected. In fact, increasing the value of κ beyond the critical threshold, highlighted by the thick black lines, labeled 0.3, makes beta-cell survival below 30%, indicating clinical manifestation of the disease (through the appearance of T1D-associated symptoms). In the extreme cases seen in the bottom right corner of panels (A1–A3), the survival of beta cells is very small due to the effective autoimmune assault dominated by high-avidity, high-killing efficacy T cells.

Given that B-cell activation also depends on the Hill function described by equation (9.53) with $n = 1$, we further characterize the effect of k and κ on the level and survival of circulating autoantibodies. To do so, we plot in panels (B1–B3) the heat-map of $\mathcal{I}_g$ at three successive time points: six months after the inception of the autoimmune attack (B1), at the clinical onset of the disease when the 0.3 threshold is crossed (which applies only to the points to the right of the black line) (B2) and at steady state (30 years after the inception of the autoimmune attack) (B3). Notice that model outcomes here are almost identical to those observed by the one clone model described in [19]. In brief, we observe four possible scenarios: $\mathcal{I}_g$ becomes elevated without reaching diagnostic T1D (to the left of the black line), $\mathcal{I}_g$ becomes elevated while reaching diagnostic T1D (to the right of the black line and with $\kappa \leqslant 10^{-9}$ (day cell)$^{-1}$), $\mathcal{I}_g$ remains elevated until disease onset (to the right of the black line with $\kappa \in [10^{-9}, 8 \times 10^{-8}]$ (day cell)$^{-1}$), or $\mathcal{I}_g$ remains undetectable throughout disease progression (the blue regimes close to the right edge of each panel). The high value of κ in the latter case causes beta-cell destruction to be too fast for autoantibody accumulation. Most of these outcomes are consistent with experimental observations in humans and animal models that screened positive for T1D-specific autoantibodies [41, 69].

To determine how fast the level of autoantibodies rise with respect to time while varying k and κ, we quantify in panels (C1–C3) the duration (in days) for the autoantibodies to reach the following detectability levels: 0.15 (C1), 0.55 (C2) and 0.95 (C3). As demonstrated by these panels and the color bar on the right, the rise in the level of autoantibodies in most cases is very fast and reaches its maximal level of

0.95 in less than 200 days after the engagement of T cells in the destruction of beta cells. It should be mentioned here that the blue regime in panels (C2) and (C3) for large κ, indicates that the maximal detectability levels chosen (0.55 and 0.95, respectively) are never attained in these two cases. This result could be used as a criterion to determine when high risk subjects could be tested for autoantibodies as a diagnostic tool.

The examination of the levels of beta cells (β) and autoantibodies ($\mathcal{I}_g$) in figure 9.13 can be better understood by tracking the time evolution of these two variables as well as T cells ($\mathcal{T}$) in response to variations to the same key parameters k and κ. As shown in figure 9.14, the 30 year time evolution of $\mathcal{T}$ (A1–A4), $\mathcal{I}_g$ (B1–B4) and β (C1–C4), after the start of the autoimmune attack, are plotted as heat-maps. These heat-maps are generated by taking k ∈ $[10^{-4}, 1]$ (unitless) and choosing the following values for the killing efficacy κ: 10^{-11} (A1–C1), 10^{-10} (A2–C2), 10^{-9} (A3–C3), 10^{-8} (A4–C4) $(\text{day cell})^{-1}$. The one consistent feature observed across all of these panels is the absence of any autoimmune attack at all times when k is close to 1 (i.e. T-cell avidity is too small to invoke an autoimmune response or even illicit T-cell expansion). However, when k is small enough, T-cell expansion becomes prominent as demonstrated by the fast increase (within few months) in $\mathcal{T}$ and the

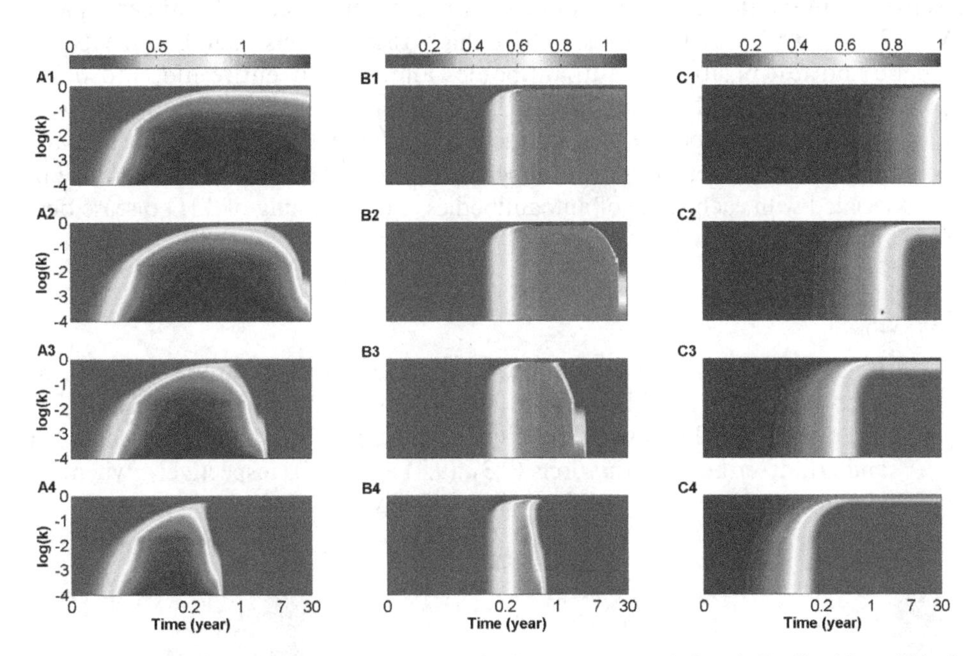

Figure 9.14. Time evolution of T1D disease progression in response to variations in T-cell avidity, within the range $[10^{-4}, 1]$ (unitless), at various values of T-cell killing efficacy: $\kappa = 10^{-11}$ (A1–C1), $= 10^{-10}$ (A2–C2), $= 10^{-9}$ (A3–C3) and $= 10^{-8}$ (A4–C4) $(\text{day cell})^{-1}$. The levels of autoreactive T cells $\mathcal{T}$ (A1–A4), autoantibodies $\mathcal{I}_g$ (B1–B4) and fraction of surviving beta cells (β) (C1–C4) are plotted as heat-maps with respect to both the reciprocal of T-cell avidity k (in log-scale) and time over 30 years. The color-coding in each panel is quantified by the color-bars on top of each column. Notice here that increasing κ gradually increases beta-cell loss and decreases the time duration of T-cell (survival) waves induced by the transient stability of the autoimmune state.

appearance of red regimes in panels (A1–A4). The increase in $\mathcal{T}$ does not always imply a significant loss in beta cells. In fact, panel (C1) shows that even after 30 years of follow up, we do not see much loss in beta cells, because κ is too small for T cells to cause any harm. By increasing κ, beta-cell loss becomes significant as demonstrated by the appearance of blue regimes in panels (C2–C4). The bigger the value of κ, the faster the loss in beta cells and the quicker the manifestation of the disease. T-cell survival, in these cases, is not maintained due to the decline in beta-cell specific peptides that are produced from apoptotic beta cells required for T-cell activation. The rise and decline in $\mathcal{T}$ create these (red) 'waves' that are due to the transient stability of the autoimmune state, and make T1D a remitting-relapsing disease as suggested in the previous section. This explains the appearance of blue regimes in panels (A3–A4) in later years which coincide with the onset of the blue regimes in panels (C3–C4). As for the level of circulating autoantibodies, panels (B1–B4) show that the time evolution of $\mathcal{I}_g$ is similar to that of $\mathcal{T}$, except for the delay in the rise of $\mathcal{I}_g$ to its peak when compared to that of $\mathcal{T}$. This delay suggests that major damage to beta cells could occur in susceptible individuals before they test positive for islet-specific autoantibodies.

One could interpret panels (A1–C1) to correspond to high risk subjects that test positive to autoantibodies their entire lives but never develop the disease, panels (A2–C2) and (A3–C3) to correspond to high risk subjects that become type 1 diabetic and test positive to autoantibodies almost their entire life, and panels (A4–C4) to correspond to individuals that develop the disease very quickly (due to the presence of very potent and destructive T cells), but test positive for autoantibodies only during a short window of time. Such criteria can be used to determine the risk associated with each clone of autoantibodies and the timing of T1D disease onset in individuals.

Exercises

1. It was assumed in the protein-processing model of section 9.2 that beta-cell killing is proportional to their surface display of autoantigen pMHC, P_1 (see equation (9.2)). Discuss the stability properties of this model for cases I, II and III, described by equations (9.3), (9.4) and (9.5), respectively, when $\mathcal{K}$ is assumed to be a saturated function of P_1 given by

$$\mathcal{K} = \kappa \frac{P_1}{P_1 + \mathcal{L}_1}.$$

2. Apply the substitutions described in equation (9.8) to reproduce the rescaled protein processing model given by equation (9.9).
3. Based on inequalities (9.13) and (9.17), we can let $z = (1 - f)(1 - \eta(1 - f))$ be the left-hand side of these inequalities.
   (a) Plot the level curves of this function with respect to f and η within the domain $[0, 1] \times [1, 2]$ and determine how increasing L affects the pathophysiology of T1D as determined by system (9.2). (These level

curves represent the two parameter bifurcations of model (9.2) with respect to f and η at various values of L.)

(b) Let η be a constant and plot the function $y(f) = (1 - f)(1 - \eta(1 - f))$ for various values of $\eta \in [0, 2]$. Determine analytically where the 'local' maximums of these curves occur with respect to f as a function η and establish a criterion (in terms of η) by which autoimmunity will be induced.

4. In modeling the phagocytic ability of macrophages, we developed three models, one of which (model 2) assumed that naïve macrophages, i.e. the class of macrophages with no apoptotic cells inside them (M_0), go through an activation step upon apoptotic-cell uptake (see equation (9.8)).

According to this new formalism, find estimates for the rate constants k_a and k_e by considering the time evolution of the model in the initial stages, then use QSS approximation to show that the percent phagocytosis ϕ and phagocytic index I_ϕ are approximately given by

$$\phi \approx 100\frac{\lambda_a}{1 - \lambda + \lambda_a},$$

and

$$I_\phi \approx 100\frac{\lambda_a}{(1 - \lambda + \lambda_a)(1 - \lambda)},$$

where $\lambda_a = k_a A / k_d$ and $\lambda = k_e A / k_d$.

Based on these two estimates, deduce that

$$k_e \approx \frac{k_a(I_\phi - \phi)(100 - \phi)}{\phi^2}.$$

5. We have shown in section 9.4 that the reduced (scaled) one-clone model (9.42a)–(9.42b) possesses at most three steady states when inequalities (9.46) and (9.47) are satisfied.

(a) The steady-state value of $\mathcal{T}$ is written in equation (9.45) as

$$\mathcal{T}^* = \frac{(\alpha - \delta_T - \ell^2 \mathrm{k}/\beta) \pm \sqrt{(\alpha - \delta_T - \ell^2 \mathrm{k}/\beta)^2 + 4(\sigma - \delta_T \mathrm{k}/\beta)\ell^2}}{2\ell^2}.$$

Show that, in order for these steady states to exist, the parameter k must satisfy the forms of $\mathrm{k}_{c1}^{(A)}$ and $\mathrm{k}_{c2}^{(A)}$ given in equation (9.48).

(b) Derive the condition showing that $\mathcal{T}^*(\mathrm{k}_{c1}^{(A)}) < 0$ and is therefore non-physiological, and that for $\mathrm{k}_{c2}^{(A)}$ to be physiologically relevant, the inequality

$$-\delta_T + \sqrt{\alpha\delta_T - \sigma\ell^2} > 0$$

must be satisfied.

(c) As was done in section 9.4.4, show that the local stability properties of these steady states remain the same when $\sigma \neq 0$ and that they are in fact global.

6. Use XPPAUTO to plot the two parameter bifurcation of the reduced (scaled) one-clone model (9.42a)–(9.42b) with respect to σ and k within the domain $[0, 1] \times [0, 20]$.

References

[1] Devendra D, Liu E and Eisenbarth G S 2004 Type 1 diabetes: recent developments *BMJ* **328** 750–4

[2] Beyan H, Buckley L R, Yousaf N, Londei M and Leslie R D G 2003 A role for innate immunity in type 1 diabetes? *Diabetes Metab. Res. Rev.* **19** 89–100

[3] Höglund P, Mintern J, Waltzinger C, Heath W, Benoist C and Mathis D 1999 Initiation of autoimmune diabetes by developmentally regulated presentation of islet cell antigens in the pancreatic lymph nodes *J. Exp. Med.* **189** 331–9

[4] Santamaria P 2001 Effector lymphocytes in autoimmunity *Curr. Opin. Immunol.* **13** 663–9

[5] Liblau R S, Wong F S, Mars L T and Santamaria P 2002 Autoreactive CD8 T cells in organ-specific autoimmunity: emerging targets for therapeutic intervention *Immunity* **17** 1–6

[6] Haskins K 2005 Pathogenic T-cell clones in autoimmune diabetes: More lessons from the NOD mouse *Adv. Immunol.* **87** 123–62

[7] Tsai S, Shameli A and Santamaria P 2008 CD8+ T cells in type 1 diabetes *Adv. Immunol.* **100** 79–124

[8] Zhang Y, O'Brien B, Trudeau J, Tan R, Santamaria P and Dutz J P 2002 *In situ* β cell death promotes priming of diabetogenic CD8 t lymphocytes *J. Immunol.* **168** 1466–72

[9] Georgiou H M, Constantinou D and Mandel T E 1995 Prevention of autoimmunity in nonobese diabetic (nod) mice by neonatal transfer of allogeneic thymic macrophages *Autoimmunity* **21** 89–97

[10] Tsai S and Santamaria P 2013 Mhc class ii polymorphisms, autoreactive T-cells, and autoimmunity *Front. Immunol.* **4** 321

[11] Turley S, Poirot L, Hattori M, Benoist C and Mathis D 2003 Physiological β cell death triggers priming of self-reactive T cells by dendritic cells in a type-1 diabetes model *J. Exp. Med.* **198** 1527–37

[12] Atkinson M A *et al* 2011 How does type 1 diabetes develop? the notion of homicide or β-cell suicide revisited *Diabetes* **60** 1370–9

[13] Lieberman S M and DiLorenzo T P 2003 A comprehensive guide to antibody and T-cell responses in type 1 diabetes *Tissue Antigens* **62** 359–77

[14] Noble J A and Erlich H A 2012 Genetics of type 1 diabetes *Cold Spring Harb. Perspect. Med.* **2** a007732

[15] Marée A F M, Komba M, Finegood D T and Edelstein-Keshet L 2008 A quantitative comparison of rates of phagocytosis and digestion of apoptotic cells by macrophages from normal (Balb/c) and diabetes-prone (NOD) mice *J. Appl. Physiol.* **104** 157–69

[16] Marée A F M, Kublik R, Finegood D T and Edelstein-Keshet L 2006 Modelling the onset of type 1 diabetes: can impaired macrophage phagocytosis make the difference between health and disease? *Philos. Trans. R. Soc. Lond. A: Math. Phys. Eng. Sci.* **364** 1267–82

[17] Roep B O 2003 The role of T-cells in the pathogenesis of type 1 diabetes: from cause to cure *Diabetologia* **46** 305–21

[18] Maclaren N *et al* 1999 Only multiple autoantibodies to islet cells (ica), insulin, gad65, IA-2 and IA-2β predict immune-mediated (type 1) diabetes in relatives *J. Autoimmun.* **12** 279–87
[19] Khadra A, Pietropaolo M, Nepom G T and Sherman A 2011 Investigating the role of T-cell avidity and killing efficacy in relation to type 1 diabetes prediction *PLoS One* **6** e14796
[20] Princiotta M F, Finzi D, Qian S-B, Gibbs J, Schuchmann S, Buttgereit F, Bennink J R and Yewdell J W 2003 Quantitating protein synthesis, degradation, and endogenous antigen processing *Immunity* **18** 343–54
[21] Nakayama M, Abiru N, Moriyama H, Babaya N, Liu E, Miao D, Yu L, Wegmann D R, Hutton J C and Elliott J F *et al* 2005 Prime role for an insulin epitope in the development of type 1 diabetes in NOD mice *Nature* **435** 220–3
[22] Lieberman S M, Evans A M, Han B, Takaki T, Vinnitskaya Y, Caldwell J A, Serreze D V, Shabanowitz J, Hunt D F and Nathenson S G *et al* 2003 Identification of the β cell antigen targeted by a prevalent population of pathogenic CD8+ T cells in autoimmune diabetes *Proc. Natl Acad. Sci.* **100** 8384–8
[23] Trudeau J D, Kelly-Smith C, Verchere C B, Elliott J F, Dutz J P, Finegood D T, Santamaria P and Tan R 2003 Prediction of spontaneous autoimmune diabetes in NOD mice by quantification of autoreactive t cells in peripheral blood *J. Clin. Investig.* **111** 217–23
[24] Mitre T M, Pietropaolo M and Khadra A 2016 The dual role of autoimmune regulator in maintaining normal expression level of tissue-restricted autoantigen in the thymus: A modeling investigation *Math. Biosci.* **287** 12–23
[25] Norbury C C, Basta S, Donohue K B, Tscharke D C, Princiotta M F, Berglund P, Gibbs J, Bennink J R and Yewdell J W 2004 CD8+ T cell cross-priming via transfer of proteasome substrates *Science* **304** 1318–21
[26] Monika M C, Brouwenstijn N, Bakker A H, Toebes M and Schumacher T N M 2004 Antigen bias in T cell cross-priming *Science* **304** 1314–7
[27] Shen L and Rock K L 2004 Cellular protein is the source of cross-priming antigen *in vivo* *Proc. Natl Acad. Sci. USA* **101** 3035–40
[28] Qian S-B, Reits E, Neefjes J, Deslich J M, Bennink J R and Yewdell J W 2006 Tight linkage between translation and mhc class i peptide ligand generation implies specialized antigen processing for defective ribosomal products *J. Immunol.* **177** 227–33
[29] Yewdell J W and Nicchitta C V 2006 The drip hypothesis decennial: support, controversy, refinement and extension *Trends Immunol.* **27** 368–73
[30] Yewdell J W 2007 Plumbing the sources of endogenous mhc class I peptide ligands *Curr. Opin. Immunol.* **19** 79–86
[31] Marée A F M, Santamaria P and Edelstein-Keshet L 2006 Modeling competition among autoreactive CD8+ T cells in autoimmune diabetes: implications for antigen-specific therapy *Int. Immunol.* **18** 1067–77
[32] Khadra A, Santamaria P and Edelstein-Keshet L 2009 The role of low avidity T cells in the protection against type 1 diabetes: a modeling investigation *J. Theor. Biol.* **256** 126–41
[33] Kurrer M O, Pakala S V, Hanson H L and Katz J D 1997 β cell apoptosis in T cell-mediated autoimmune diabetes *Proc. Natl Acad. Sci.* **94** 213–8
[34] Finegood D T, Scaglia L and Bonner-Weir S 1995 Dynamics of β-cell mass in the growing rat pancreas: estimation with a simple mathematical model *Diabetes* **44** 249–56
[35] O'Brien B A, Geng X, Orteu C H, Huang Y, Ghoreishi M, Zhang Y, Bush J A, Li G, Finegood D T and Dutz J P 2006 A deficiency in the *in vivo* clearance of apoptotic cells is a feature of the NOD mouse *J. Autoimmun.* **26** 104–15

[36] O'Brien B A, Huang Y, Geng X, Dutz J P and Finegood D T 2002 Phagocytosis of apoptotic cells by macrophages from NOD mice is reduced *Diabetes* **51** 2481–8

[37] Mitchell M 1998 *An Introduction to Genetic Algorithms* (Cambridge, MA: MIT Press)

[38] Marée A F M, Komba M, Dyck C, Labkecki M, Finegood D T and Edelstein-Keshet L 2005 Quantifying macrophage defects in type 1 diabetes *J. Theor. Biol.* **233** 533–51

[39] Akaike H 1998 Information theory and an extension of the maximum likelihood principle *Selected Papers of Hirotugu Akaike* (Berlin: Springer) pp 199–213

[40] Hurvich C M and Tsai C-L 1989 Regression and time series model selection in small samples *Biometrika* **76** 297–307

[41] Miao D, Yu L and Eisenbarth G S 2007 Role of autoantibodies in type 1 diabetes *Front. Biosci.* **12** 1889–98

[42] Janeway C A, Travers P, Walport M and Shlomchik M J 2005 *Immunobiology: The Immune System in Health and Disease* vol 1 (New York: Garland Publishing)

[43] Hawa M, Rowe R, Lan M S, Notkins A L, Pozzilli P, Christie M R and Leslie R D G 1997 Value of antibodies to islet protein tyrosine phosphatase-like molecule in predicting type 1 diabetes *Diabetes* **46** 1270–5

[44] Kawasaki E 2012 ZnT8 and type 1 diabetes [review] *Endocr. J.* **59** 531–7

[45] Peakman M, Tree T I, Endl J, van Endert P, Atkinson M A and Roep B O 2001 Characterization of preparations of GAD65, proinsulin, and the islet tyrosine phosphatase IA-2 for use in detection of autoreactive T-cells in type 1 diabetes report of phase II of the second international immunology of diabetes society workshop for standardization of T-cell assays in type 1 diabetes *Diabetes* **50** 1749–54

[46] Pietropaolo M, Barinas-Mitchell E and Kuller L H 2007 The heterogeneity of diabetes unraveling a dispute: is systemic inflammation related to islet autoimmunity? *Diabetes* **56** 1189–97

[47] Pietropaolo M, Becker D J, LaPorte R E, Dorman J S, Riboni S, Rudert W A, Mazumdar S and Trucco M 2002 Progression to insulin-requiring diabetes in seronegative prediabetic subjects: the role of two HLA-DQ high-risk haplotypes *Diabetologia* **45** 66–76

[48] Pietropaolo M, Surhigh J M, Nelson P W and Eisenbarth G S 2008 Primer: immunity and autoimmunity *Diabetes* **57** 2872–82

[49] Uibo R and Lernmark Å 2008 Gad65 autoimmunity–clinical studies *Adv. Immunol.* **100** 39–78

[50] Wenzlau J M, Moua O, Sarkar S A, Yu L, Rewers M, Eisenbarth G S, Davidson H W and Hutton J C 2008 Slc30Aa8 is a major target of humoral autoimmunity in type 1 diabetes and a predictive marker in prediabetes *Ann. N. Y. Acad. Sci.* **1150** 256–9

[51] DeNiro M and Al-Mohanna F A 2012 Zinc transporter 8 (ZnT8) expression is reduced by ischemic insults: a potential therapeutic target to prevent ischemic retinopathy *PLoS One* **7** e50360

[52] Pietropaolo M, Towns R and Eisenbarth G S 2012 Humoral autoimmunity in type 1 diabetes: prediction, significance, and detection of distinct disease subtypes *Cold Spring Harb. Perspect. Med.* **2** a012831

[53] Jaberi-Douraki M, Pietropaolo M and Khadra A 2014 Predictive models of type 1 diabetes progression: understanding T-cell cycles and their implications on autoantibody release *PLoS One* **9** e93326

[54] Bluestone J A, Herold K and Eisenbarth G 2010 Genetics, pathogenesis and clinical interventions in type 1 diabetes *Nature* **464** 1293–300

[55] Vendrame F *et al* 2010 Recurrence of type 1 diabetes after simultaneous pancreas-kidney transplantation, despite immunosuppression, is associated with autoantibodies and pathogenic autoreactive CD4 T-cells *Diabetes* **59** 947–57
[56] Pescovitz M D *et al* 2009 Rituximab, b-lymphocyte depletion, and preservation of beta-cell function *N. Engl. J. Med.* **361** 2143–52
[57] Skowera A *et al* 2008 Ctls are targeted to kill β cells in patients with type 1 diabetes through recognition of a glucose-regulated preproinsulin epitope *J. Clin. Invest.* **118** 3390–402
[58] Standifer N E, Ouyang Q, Panagiotopoulos C, Verchere C B, Tan R, Greenbaum C J, Pihoker C and Nepom G T 2006 Identification of novel HLA-A* 0201–restricted epitopes in recent-onset type 1 diabetic subjects and antibody-positive relatives *Diabetes* **55** 3061–7
[59] Mahaffy J M and Edelstein-Keshet L 2007 Modeling cyclic waves of circulating T cells in autoimmune diabetes *SIAM J. Appl. Math.* **67** 915–37
[60] Von Herrath M, Sanda S and Herold K 2007 Type 1 diabetes as a relapsing-remitting disease? *Nat. Rev. Immunol.* **7** 988–94
[61] Khadra A, Tsai S, Santamaria P and Edelstein-Keshet L 2010 On how monospecific memory-like autoregulatory CD8. T cells can blunt diabetogenic autoimmunity: a computational approach *J. Immunol.* **185** 5962–72
[62] Edelstein-Keshet L 2005 *Mathematical Models in Biology* (Philadelphia: SIAM)
[63] Miller R K and Michel A N 1982 *Ordinary Differential Equations* (New York: Academic)
[64] Strogatz S H 2014 *Nonlinear Dynamics and Chaos: With Applications to Physics, Biology, Chemistry, and Engineering* (Boulder, CO: Westview)
[65] Khadra A, Santamaria P and Edelstein-Keshet L 2010 The pathogenicity of self-antigen decreases at high levels of autoantigenicity: a computational approach *Int. Immunol.* **22** 571–82
[66] Kim P S, Lee P P and Levy D 2007 Modeling regulation mechanisms in the immune system *J. Theor. Biol.* **246** 33–69
[67] Standifer N E, Burwell E A, Gersuk V H, Greenbaum C J and Nepom G T 2009 Changes in autoreactive t cell avidity during type 1 diabetes development *Clin. Immunol.* **132** 312–20
[68] Vernino L, McAnally L M, Ramberg J and Lipsky P E 1992 Generation of nondividing high rate ig-secreting plasma cells in cultures of human β cells stimulated with anti-CD3-activated T cells *J. Immunol.* **148** 404–10
[69] Morran M P *et al* 2010 Humoral autoimmunity against the extracellular domain of the neuroendocrine autoantigen IA-2 heightens the risk of type 1 diabetes *Endocrinology* **151** 2528–37

Part V

Diabetes implications on kidneys

IOP Publishing

Diabetes Systems Biology
Quantitative methods for understanding beta-cell dynamics and function
Anmar Khadra

Chapter 10

Diabetes implications on kidneys

Anita T Layton and Rui Hu

In the United States, diabetes is one of the most common causes of end-stage renal diseases. But despite intense research, the underlying mechanisms remain incompletely understood. Diabetes impacts kidney function in a number of ways, including the differential regulation of glucose-related epithelial transporters. In this chapter, we will introduce mathematical modeling techniques that can be used to better understand the impacts of hyperglycemia on the kidney's reabsorption and secretion of water and solutes, mediated by specialized epithelial cells that form key nephron segments. To represent the dynamic exchange of water and solutes across tubular epithelia, we will derive the conservation and flux equations. These equations will then be used to build a model that describes transport across an epithelial cell.

10.1 Introduction

The kidneys are organs that serve a number of essential regulatory roles; the most important of which is when they function as filters, clearing metabolic wastes and toxins in the blood by excreting them in urine. The kidneys also serve other essential functions. Through a number of regulatory mechanisms, they help maintain the body's water balance, electrolyte balance, and acid–base balance. Additionally, the kidneys produce or activate hormones that are involved in erythrogenesis, calcium metabolism, and the regulation of blood flow [1, 2].

Diabetes is one of the most common causes of end-stage renal diseases in several countries. Despite intense research, the underlying mechanisms of diabetes-induced renal diseases remain incompletely understood [3]. Diabetes impacts kidney function in a number of ways, including the differential regulation of glucose-related epithelial transporters [4]. In this chapter, we will introduce mathematical modeling techniques that can be used to better understand the impacts of hyperglycemia on the kidney's reabsorption and secretion of water and solutes, mediated by specialized epithelial cells that form key nephron segments. To represent the dynamic exchange of water and solutes across tubular epithelia, we will derive the conservation and flux equations [5]. These equations will then be used to build a model that describes transport across an epithelial cell. Such models can be used to provide insights about renal diseases.

About a quarter of the blood that leaves the heart ends up in the kidneys. Part of that blood passes through tiny filters called the glomeruli. There are tens of thousands of glomeruli in our kidneys, connected to tiny blood vessels and tiny tubules (diameter of the order of 10 μm) called nephrons inside the kidney. Their job is to act as filters and accomplish the other tasks mentioned above. Despite the many levels of built-in redundancy, sometimes this filtering system breaks down. Indeed, diabetes can damage the kidneys and cause them to fail. Failing kidneys lose their ability to filter out waste products, resulting in kidney diseases, termed nephropathy. Gaining a better understanding of diabetes' effects on kidney function can thus provide insights into the development of diabetic nephropathy.

10.2 Basics of transepithelial transport

As blood flows through the nephron, the epithelial cells that form the tubular walls secrete or reabsorb water and solutes. Ultimately, what emerges at the end is urine, with composition adjusted just right so that urinary excretion matches daily intake. When blood volume and composition fluctuates, the kidney adapts by adjusting the amount of solute and water it transports; this is accomplished by modulating the expression and/or activity of its epithelial transporters. These changes are in turn mediated by a number of hormonal and neuronal signals, and involve rather complex signaling cascades. Transporter expressions and/or activities may also change under pathophysiological conditions, such as diabetes.

10.2.1 Paracellular versus transcellular pathways

Transepithelial transport across a nephron segment can take place via transcellular and/or paracellular pathways, as illustrated in figure 10.1. Solutes that move across the paracellular route traverse through tight junctions, also known as zonula occludens. These are junctional structures that mediate adhesion between cells. The main components of tight junctions are two types of proteins: claudins and

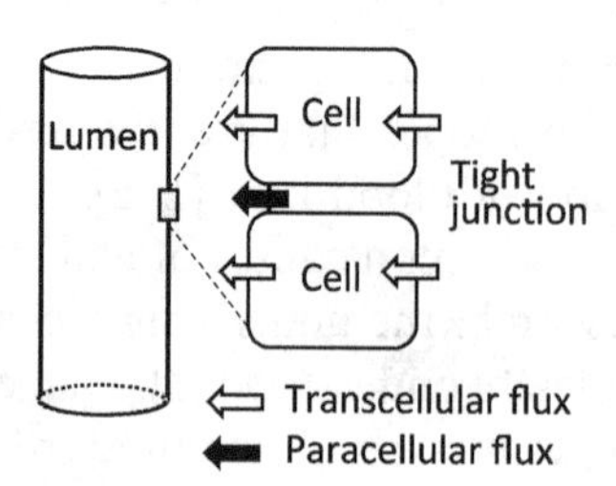

Figure 10.1. Schematic diagram of transport along a nephron segment. The nephron segment is represented as as a straight cylinder, with fluid flow along the tubular lumen. Solutes and water are reabsorbed or secreted through the transcellular pathway, which consists of the apical and basolateral cellular membranes as barriers, or the paracellular pathway, via the tight junctions.

occludins. Together these proteins determine the permeability of the paracellular barrier.

Solutes moving across the transcellular pathway must traverse through the apical and basolateral membranes. Thus, this pathway consists of two resistances in series. Even in the presence of specialized membrane transporters, the transcellular pathway is typically much less conductive than the paracellular pathway. As such, the transepithelial resistance predominantly reflects the resistance of the tight junction. Leaky epithelia such as the proximal tubule (the first nephron segment) have low tight junctional resistance, on the order of 10 Ω cm^2. In contrast, the resistance of tight epithelia, such as that of the collecting tubule (the terminal nephron segment), is about 100 times higher, on the order of 1000 Ω cm^2. In this chapter, we focus on transcellular transport.

10.2.2 Passive versus active transport

Active transport requires energy, whereas passive transport by definition does not. Energetically favorable processes decrease the Gibbs free energy (G) of the system. Thus, the spontaneous (or passive) direction of a transport process corresponds to that which reduces G. Given an open system with constant temperature and pressure, G is proportional to the change in the electrochemical gradients of the system components. The electrochemical potential of a solute S in compartment j (μ_S^j) is given by

$$\mu_S^j \equiv \left[\frac{\partial(nG)}{\partial n_S}\right], \tag{10.1}$$

where n is the total number of moles and n_S denotes specifically the number of moles of S. Different compartments at the same temperature and pressure are in equilibrium when the chemical potential of each solute is the same in all compartments, i.e. when $\Delta G = 0$. Based on thermodynamic principles, for dilute solutions where solute–solute interactions are negligible, μ_S^j is given by [6]:

$$\mu_S^j = \mu_S^0 + RT\ln(C_S^j) + z_S F\psi^j. \tag{10.2}$$

In the above equation, the chemical potential of the pure solute in a reference state is denoted by μ_S^0, solute valence is given by z_S, the concentration of solute S in compartment j is denoted C_S^j, and the electric potential in compartment j is denoted ψ^j. The usual physical constants are the Faraday constant F, the gas constant R, and absolute temperature T. To transport a solute spontaneously from compartment j to compartment k, this process should lower the Gibbs free energy of the system. In other words, passive transport is the movement of solute down its electrochemical gradient. This happens when the following condition is met:

$$\Delta\mu_S = \mu_S^k - \mu_S^j < 0, \quad \text{i.e.} \quad \mu_S^j > \mu_S^k. \tag{10.3}$$

This is the major difference from active transport, where solutes may move against an electrochemical gradient, in which case energy is required. There are three forms of passive transport:

- Simple diffusion: here a solute diffuses across the cellular membrane without involving an integral membrane transport protein. However, the solute must be substantially hydrophobic to cross the hydrophobic core of the lipid bilayer unassisted. Solutes that can be transported via simple diffusion include small gas molecules, e.g. oxygen and carbon dioxide.
- Channel diffusion: here a solute diffuses through ion channels, which are membrane proteins with a hydrophilic pore. They are selectively permeable to certain solutes, and their opening is gated (i.e., regulated) by different factors, such as the transmembrane voltage or specific ligands. Detailed description of the kinetics of some of these ion channels is presented in chapter 2 [5].
- Facilitated diffusion: here diffusion is mediated by passive carriers, membrane proteins that are solute specific and that does not have a hydrophobic pore. Facilitated diffusion happens in three steps: (i) binding of the solute to the carrier, (ii) the carrier then undergoes a conformational change that moves the solute across the membrane, and finally (iii) dissociation of the solute from the carrier. Note that not only is facilitated diffusion much slower than channel diffusion, it is also more rapidly saturated.

As for active transport, we distinguish between primary and secondary active transport. Primary active transport, or direct active transport, utilizes the energy stored in chemical bonds to transport solutes across membranes. Most of the enzymes that drive primary active transport are transmembrane ATPases. These are enzymes that catalyze the hydrolysis of adenosine triphosphate (ATP) into adenosine diphosphate (ADP) and phosphate. The energy that is released by this reaction allows solute transport to move in an energetically unfavorable direction.

Like primary transport, secondary transport too uses energy to move solutes across membranes. However, the source of energy is not ATP; instead, the electrochemical gradient generated by primary transport is used to transport the solute. Specifically, in secondary transport the energy of the transmembrane potential difference of one solute is coupled to that of another to move the latter uphill. Secondary active transporters are either co-transporters (a.k.a. symporters) or exchangers (a.k.a. antiporters). In co-transport, two solute species are carried in the same direction; in contrast, in antiport, the two species move in opposite directions. A well-known example of antiporter is the sodium–calcium exchanger, where the (energetically unfavorable) efflux of one Ca^{2+} ion is coupled to the (energetically favorable) entry of three Na^+ ions into the cell [7].

The secondary transport that is particularly relevant to diabetes is the sodium–glucose co-transport (SGLT). The two most well known members of SGLT family are SGLT1 and SGLT2, which are expressed along the proximal tubule (the initial segment of the nephron) of the kidney [8]. SGLT is also found in the intestinal mucosa of the small intestine. In 1960, Robert K Crane (1919–2010, an American biochemist) presented for the first time his discovery of the SGLT as the mechanism for intestinal glucose absorption [9]. By harnessing the Na^+ electrochemical gradient (or simply the Na^+ gradient), the glucose symporter SGLT1 co-transports glucose and sodium into the cell. Further analysis of the kinetics of this transport system is provided in the next section.

10.3 Principal classes of transepithelial transporters

In this section, we briefly describe two major categories of transepithelial transporters, with a focus on flux calculations.

10.3.1 Na/K-ATPase pumps

The Na/K-ATPase, referred to in chapter 2 as Na pump, is a ubiquitous membrane enzyme. It was discovered by the Danish scientist Jens Christian Skou, the 1997 Nobel Prize in Chemistry winner [10]. The Na^+ pump actively transport three Na^+ ions out of the cell for every two K^+ ions pumped into the cell. Without this pump, it would not be possible for the cell to maintain a high intracellular level of K^+ and a low level of Na^+. Thus, this pump is essential for maintaining the resting potential and regulating cellular volume. Equally importantly, it provides most of the energy needed to induce secondary active transport.

We will derive a formula for computing ionic fluxes across the Na^+ pump. Consider the binding of each ion to the enzyme as an independent process. If we assume that the binding of one intracellular Na^+ ion to the free enzyme (denoted E) to form the complex NaE is a first-order, reversible reaction, then we get

$$\mathrm{Na}^+ + E \underset{k_{\mathrm{Na}E}^{\mathrm{off}}}{\overset{k_{\mathrm{Na}E}^{\mathrm{on}}}{\rightleftharpoons}} \mathrm{Na}E, \tag{10.4}$$

where $k_{\mathrm{Na}E}^{\mathrm{on}}$ and $k_{\mathrm{Na}E}^{\mathrm{off}}$ are the on and off rates of this reaction, respectively. The generation rate of the NaE complex is given by

$$\frac{dC_{\mathrm{Na}E}}{dt} = k_{\mathrm{Na}E}^{\mathrm{on}} C_{\mathrm{Na}}^{i} C_E - k_{\mathrm{Na}E}^{\mathrm{off}} C_{\mathrm{Na}E}, \tag{10.5}$$

where $C_{\mathrm{Na}E}$, C_E and C_{Na}^{i} are the concentrations of NaE, E and intracellular Na^+, respectively. Using quasi-steady state approximation, we obtain

$$C_{\mathrm{Na}E} = \frac{C_E^{\mathrm{tot}} C_{\mathrm{Na}}^{i}}{C_{\mathrm{Na}}^{i} + K_{\mathrm{Na}}^{i}}, \tag{10.6}$$

where total concentration of enzyme is denoted by $C_E^{\mathrm{tot}} = C_E + C_{\mathrm{Na}E}$ and is assumed to remain constant, and $K_{\mathrm{Na}}^{i} = k_{\mathrm{Na}E}^{\mathrm{off}} / k_{\mathrm{Na}E}^{\mathrm{on}}$ is the dissociation constant of the NaE complex. Equation (10.6) means that the probability (p_{Na}) of having one pump unit bound to one Na^+ ion is proportional to

$$p_{\mathrm{Na}} \equiv \frac{C_{\mathrm{Na}E}}{C_{\mathrm{tot}}} = \frac{C_{\mathrm{Na}}^{i}}{C_{\mathrm{Na}}^{i} + K_{\mathrm{Na}}^{i}}. \tag{10.7}$$

If all the Na^+ binding sites have the same affinity and there is no cooperativity between them, the probability of having three intracellular Na^+ ions bound to the pump is simply p_{Na}^3. Similarly, the probability of having two extracellular K^+ ions bound to the pump is p_K^2, where p_K is given by $C_K^e/(C_K^e + K_K^e)$, with C_K^e denoting the external K^+ concentration and K_K^e the dissociation constant of the KE complex. Taken together, we can express the flux of Na^+ ions across the pump as

$$J_{Na}^{NaK} = J_{Na}^{NaK,max}\left[\frac{C_{Na}^i}{(C_{Na}^i + K_{Na}^i)}\right]^3\left[\frac{C_K^e}{(C_K^e + K_K^e)}\right]^2, \tag{10.8}$$

and that of K^+ ions as:

$$J_K^{NaK} = -(2/3)J_{Na}^{NaK}, \tag{10.9}$$

10.3.2 Non-equilibrium thermodynamic formalism

Many transporters have not been sufficiently well characterized experimentally. As a result, no kinetic model is available for these transporters. In this case, the non-equilibrium thermodynamic formalism can be used to determine the fluxes. This approach was first adapted to renal tubular transport models by Alan Weinstein (Weill Cornell Medicine) [11]. Using this formalism, the diffusive component of the flux of solute S can be written as:

$$J_S^{diff} = \sum_{k=1}^{n} L_{Sk}\Delta\mu_k, \tag{10.10}$$

where n is the number of solutes and $\Delta\mu_k$ is the electrochemical gradient of each solute k across the membrane, given by

$$\Delta\mu_k = RT\Delta\ln(C_k) + z_k F\Delta\psi, \tag{10.11}$$

derived from equation (10.2). The coefficient L_{Sk} couples the flux of solute S to the driving force exerted on solute k. When S and k do not interact, $L_{Sk} = 0$. For the special case $S = k$, L_{SS} is related to the membrane permeability to solute S by the equation

$$L_{SS} = \frac{RTP_S}{\bar{C}_S}, \tag{10.12}$$

where $\bar{C}_S$ is the mean transmembrane concentration, given by

$$\bar{C}_S = \frac{\Delta C_S}{\Delta\ln C_S} = \frac{C_S^e - C_S^i}{\ln C_S^e - \ln C_S^i}. \tag{10.13}$$

As an example, consider the electroneutral Cl^-/HCO_3^- exchanger, with a 1:1 stoichiometry. Using the non-equilibrium thermodynamic formalism, the Cl^- and HCO_3^- fluxes across the exchanger are given by:

$$\begin{bmatrix} J_{Cl} \\ J_{HCO_3} \end{bmatrix} = L_{Cl/HCO_3} \begin{bmatrix} +1 & -1 \\ -1 & +1 \end{bmatrix} \begin{bmatrix} \Delta\mu_{Cl} \\ \Delta\mu_{HCO_3} \end{bmatrix}. \tag{10.14}$$

Note the opposite signs (± 1) are due to this transporter being an exchanger, with Cl^- and HCO_3^- going in opposite directions. The fluxes are therefore calculated as:

$$J_{Cl} = L_{Cl/HCO_3}(\Delta\mu_{Cl} - \Delta\mu_{HCO_3}) \tag{10.15}$$

$$J_{HCO_3} = L_{Cl/HCO_3}(\Delta\mu_{HCO_3} - \Delta\mu_{Cl}). \tag{10.16}$$

Another example that is particularly relevant to diabetes is the sodium–glucose cotransporter, SGLT. The isoform SGLT2 is expressed along the proximal convoluted tubule and has a 1:1 Na^+:glucose stoichiometry [12]. Under the non-equilibrium thermodynamics formulation, SGLT2-mediated fluxes are given by

$$\begin{bmatrix} J_{NaI} \\ J_G \end{bmatrix} = L_{SGLT2} \begin{bmatrix} 1 & 1 \\ 1 & 1 \end{bmatrix} \begin{bmatrix} \Delta\mu_{Na} \\ \Delta\mu_G \end{bmatrix}. \tag{10.17}$$

Downstream along the proximal straight tubule, the other isoform SGLT1 is expressed which has a 2:1 Na^+:glucose stoichiometry [12]. The fluxes associated with it are thus given by

$$\begin{bmatrix} J_{NaI} \\ J_G \end{bmatrix} = L_{SGLT1} \begin{bmatrix} 4 & 2 \\ 2 & 1 \end{bmatrix} \begin{bmatrix} \Delta\mu_{Na} \\ \Delta\mu_G \end{bmatrix}. \tag{10.18}$$

The non-equilibrium thermodynamic formalism is particularly helpful then kinetic measurements are not available. This formulation yields equations that satisfy thermodynamic conditions while requiring specification of only one parameter, namely, the coupling coefficient L_{Sk}. Its simplicity notwithstanding, some important limitations of this approach must be noted. For instance, it cannot account for ion-specific binding affinities and translocation rates. Also, it does not consider internal and external asymmetries.

10.4 Model equations

To predict transport rates across the tubular epithelium, we must compute solute concentrations and volume of the cell, and electrical potential of the epithelial membranes. These variables can be determined by solving conservation equations for mass, volume, and electrical charge. Below we derive the corresponding conservation equations for epithelial cells.

10.4.1 Conservation for water and solute

Intracellular solute concentrations and volume change as the cell exchanges solutes and fluid with the tubular lumen on the apical side, and with the interstitium (or, peritubular fluid) as well as the lateral space between cells (i.e. the paracellular pathway) on the basolateral side. To formulate the model equations, we denote the epithelial cell cytosolic compartment by M, and its surrounding compartments by N. The volume of the cell cytosol (V^M) can expand and contract, and the amount of fluid accumulated during the time interval $[t, t + dt]$ is equal to the net amount of volume flowing into the cytosol from all adjacent compartments during dt, i.e.

$$V^M(t + dt) - V^M(t) = \int_t^{t+dt} \sum_N A^{NM}(\tau)J_V^{NM}(\tau)d\tau, \tag{10.19}$$

where A^{NM} is the surface area at the interface between compartments M and N, and J_V^{NM} is the flux of volume from N to M (see below). The differential form of equation (10.19) yields the intracellular volume conservation equation

$$\frac{dV^M}{dt} = \sum_N A^{NM} J_V^{NM}. \tag{10.20}$$

Similarly, the amount of a given solute S that accumulates within the cell cytosol during the time interval $[t, t + dt]$ is given by the net amount of S entering the cell, plus the net amount of S generated by chemical reaction within the cell during dt. It follows that

$$V^M(t + dt)C_S^M(t + dt) - V^M(t)C_S^M(t) = \int_t^{t+dt} \sum_N A^{NM}(\tau)J_s^{NM}(\tau)d\tau, \tag{10.21}$$

where J_S^{NM} is the molar flux of solute S from N to M (see below). The differential form of equation (10.21) yields the intracellular solute conservation equation:

$$\frac{dV^M C_S^M}{dt} = \sum_N A^{NM} J_S^{NM}. \tag{10.22}$$

Besides the cytosol, solute concentrations and volume in the lateral space between cells must be updated as well. The corresponding conservation equations are similar to those for the cytosol.

10.4.2 Conservation of electric charge

The next step is to derive model equations that capture the conservation of electric charge across cellar membrane. To do so, we represent the cell membrane as an electrical circuit with a capacitor in parallel with a resistor, and assume that variations in the membrane potential are inversely proportional to the membrane capacitance (C_m). The electrical potential of the peritubular space (superscript 'P') is always taken to be the ground potential (i.e. $V^P = 0$). Assuming that the capacitance of the tight junction is zero, the apical (V^{ML}) and basolateral (V^{MP}) transmembrane electric potentials as be described as follows

$$C_m^{ML}\frac{dV^{ML}}{dt} = -\sum_S z_S F\left(A^{ML} J_S^{ML} + A^{PL} J_S^{PL}\right) \tag{10.23}$$

$$C_m^{MP} \frac{dV^{MP}}{dt} = -\sum_S z_S F\left(A^{MP} J_S^{MP} - A^{PL} J_S^{PL}\right), \tag{10.24}$$

The capacitance C_m is assumed to be proportional to the surface area, with a proportionality constant of 1 μF cm^{-2}. Because capacitance values are typically very small, these electrical potential equations give rise to a stiff system. That is, a small change in the right hand side can result in rapid changes in membrane potentials. As such, the time scale of the different conservation equations varies widely. To circumvent this issue, most epithelial transport models avoid solving these stiff differential equations. Instead, they approximate and determine membrane potentials V^{ML} and V^{MP} based upon two conditions: (i) electroneutrality within each epithelial compartment M, and (ii) open-circuit conditions (i.e. no net current into the lumen). In other words, the model assumes

$$\sum_S z_S C_S^M = 0 \tag{10.25}$$

$$\sum_S z_S F\left(A^{ML} J_S^{ML} + A^{PL} J_S^{PL}\right) = 0. \tag{10.26}$$

10.5 A simple cell model

We apply the above concepts and formulate a simple epithelial cell model based on the conservation of mass and electric charge. The target cell model represents the transport of water and four solutes (Na^+, K^+, Cl^-, glucose); see figure 10.2. The apical membrane expresses Na^+, K^+, and Cl^- channels, and SGLT2 co-transporters, whereas the basolateral membrane expresses Na/K-ATPase pumps, K^+ channels, KCl co-transporters and GLUT2 transporters. For simplicity, we do not consider the paracellular pathway in this example.

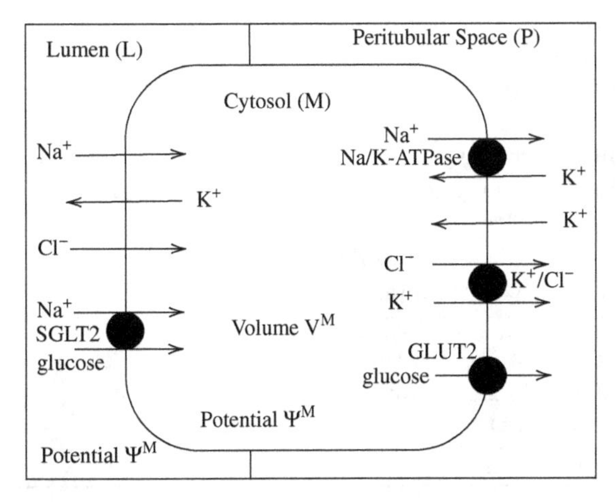

Figure 10.2. Schematic diagram of model epithelial cell.

The composition of the lumen and peritubular space is assumed to be the same: $[Na^+]$ = 144 mM, $[K^+]$ = 5 mM, $[Cl^-]$ = 149 mM, and [glucose] = 5 mM. (This situation is approximately true at the entrance of the proximal tubule.) The hydraulic and oncotic pressures are taken to be zero in these two compartments. The concentration of impermeant proteins (of valence −1) in the cytosol is set to 60 mM, and the reference cell volume is V_o^M = 0.005 11 cm^3 cm^{-2} epithelium. Note that the epithelial surface area is usually based upon the inner diameter of the tubule. Other cell properties can be found in table 10.1.

We will consider the steady-state model solution, which consists of steady-state intracellular concentrations of Na^+, K^+, Cl^- and glucose, the intracellular volume, and the electric potentials in the lumen and cytosol. Model solution can be computed by solving seven conservation equations at steady state: volume conservation equation (10.20), solute conservation (equation (10.22)), and mass electroneutrality (equations (10.25) and (10.26)).

Using Newton's method, one can compute an approximate steady-state model solution: intracellular $[Na^+]$ = 166 mM, $[K^+]$ = 10 mM, $[Cl^-]$ = 116 mM, [glucose] = 4 mM; cytosolic and luminal electrical potentials $\Psi^M = -5$ V and $\Psi^L = -5$ V, with Ψ^P taken to be 0; and cell volume is 0.04 cm^3 cm^{-2} epithelium.

10.6 Perspectives

The simple cell model can be extended to include additional solutes and membrane transporters to yield more comprehensive models of different types of renal epithelial cells [13]. The resulting cell models can then be connected in series to simulate solute and water transport along a nephron [14–18]. Such models have been used to predict renal and cardiovascular effects of SGLT2 inhibition [4, 19–21].

Table 10.1. Model cell parameters.

Parameter	Apical	Basolateral
Membrane surface area (cm^2 cm^{-2} epith)	36	20
Water permeability (cm s^{-1})	0.000 1	0.000 1
Na^+ channel permeability (cm s^{-1})	0.4×10^{-5}	0
K^+ channel permeability (cm s^{-1})	1.0×10^{-5}	1.0×10^{-5}
Cl^- channel permeability (cm s^{-1})	1.0×10^{-5}	0
Maximum Na^+, K^+-ATPase flux (mmol s^{-1} cm^{-2})	0	1.0×10^{-5}
Binding affinity of Na/K-ATPase to Na^+ (mM)	—	4
Binding affinity of Na/K-ATPase to K^+ (mM)	—	1
KCl cotransporter coupling coefficient (mmol^2 J^{-1} s^{-1} cm^{-2})	0	1.0×10^{-5}
SGLT2 activity level (mmol^2 J^{-1} s^{-1} cm^{-2})	1.3×10^{-5}	0
Binding affinity of SGLT2 to Na^+ (mM)	25	—
Binding affinity of SGLT2 to glucose (mM)	4.9	—
GLUT2 activity level (mmol^2 J^{-1} s^{-1} cm^{-2})	0	$0.162\,5 \times 10^{-5}$
Binding affinity of GLUT2 to glucose (mM)	—	17

Exercises

1. Derive equation (10.6), the steady-state expression for the concentration of bound Na^+, C_{NaE}.
2. The sodium–hydrogen antiporter or sodium–hydrogen exchanger (NHE) is a membrane protein found in the kidney. The isoform NHE3 is expressed along the brush border of the proximal tubule and transport Na^+ and H^+ with a stoichiometry of 1:1. Express the Na^+ and H^+ flux in terms of the electrochemical potential difference of Na^+ and H^+ ($\Delta\mu_{Na}$ and $\Delta\mu_H$) using the non-equilibrium thermodynamic formulation.

References

[1] Eaton D C 2009 *Vander's Renal Physiology* (New York: McGraw-Hill)

[2] Alpern R J and Hebert S C 2007 *Seldin and Giebischas The Kidney: Physiology & Pathophysiology 1-2* (Amsterdam: Elsevier)

[3] Reidy K *et al* 2014 Molecular mechanisms of diabetic kidney disease *J. Clin. Invest.* **124** 2333–40

[4] Layton A T, Vallon V and Edwards A 2016 Predicted consequences of diabetes and sglt inhibition on transport and oxygen consumption along a rat nephron *Am. J. Physiol.-Renal Physiol.* **310** F1269–83

[5] Layton A T and Edwards A 2014 *Mathematical Modeling in Renal Physiology* (Berlin: Springer)

[6] Job G and Herrmann F 2006 Chemical potential—a quantity in search of recognition *Eur. J. Phys.* **27** 353

[7] Quednau B D, Nicoll D A and Philipson K D 2004 The sodium/calcium exchanger family-slc8 *Pflügers Arch.* **447** 543–8

[8] Ghezzi C, Loo D D F and Wright E M 2018 Physiology of renal glucose handling via SGLT1, SGLT2 and GLUT2 *Diabetologia* **61** 2087–97

[9] Crane R K 1960 Intestinal absorption of sugars *Physiol. Rev.* **40** 789–825

[10] Skou J C and Esmann M 1992 The |Na, k-atpase *J. Bioenerg. Biomembr.* **24** 249–61

[11] Weinstein A M 1983 Nonequilibrium thermodynamic model of the rat proximal tubule epithelium *Biophys. J.* **44** 153–70

[12] Mackenzie B, Loo D D F, Panayotova-Heiermann M and Wright E M 1996 Biophysical characteristics of the pig kidney Na^+/glucose cotransporter SGLT2 reveal a common mechanism for SGLT1 and SGLT2 *J. Biol. Chem.* **271** 32678–83

[13] Edwards A and Layton A T 2017 Cell volume regulation in the proximal tubule of rat kidney *Bull. Math. Biol.* **79** 2512–33

[14] Layton A T, Vallon V and Edwards A 2016 A computational model for simulating solute transport and oxygen consumption along the nephrons *Am. J. Physiol.-Renal Physiol.* **311** F1378–90

[15] Layton A T, Laghmani K, Vallon V and Edwards A 2016 Solute transport and oxygen consumption along the nephrons: effects of Na^+ transport inhibitors *Am. J. Physiol.-Renal Physiol.* **311** F1217–29

[16] Edwards A, Castrop H, Laghmani K, Vallon V and Layton A T 2014 Effects of NKCC2 isoform regulation on NaCl transport in thick ascending limb and macula densa: a modeling study *Am. J. Physiol.-Renal Physiol.* **307** F137–46

[17] Layton A T, Edwards A and Vallon V 2017 Adaptive changes in GFR, tubular morphology, and transport in subtotal nephrectomized kidneys: modeling and analysis *Am. J. Physiol.-Renal Physiol.* **313** F199–209

[18] Layton A T, Edwards A and Vallon V 2018 Renal potassium handling in rats with subtotal nephrectomy: modeling and analysis *Am. J. Physiol.-Renal Physiol.* **314** F643–57

[19] Layton A T, Vallon V and Edwards A 2015 Modeling oxygen consumption in the proximal tubule: effects of NHE and SGLT2 inhibition *Am. J. Physiol.-Renal Physiol.* **308** F1343–57

[20] Layton A T and Vallon V 2018 SGLT2 inhibition in a kidney with reduced nephron number: modeling and analysis of solute transport and metabolism *Am. J. Physiol.-Renal Physiol.* **314** F969–84

[21] Layton A T and Vallon V 2018 Cardiovascular benefits of SGLT2 inhibition in diabetes and chronic kidney diseases *Acta Physiol.* **222** e13050

CPSIA information can be obtained
at www.ICGtesting.com
Printed in the USA
BVHW010753240621
609943BV00025BA/65